Fundamentals of
FRACTURE MECHANICS

'Antipov had been pestering the repair shops about the quality of the replacements for the tracks. The steel was not sufficiently tensile, the rails failed the test for strains and Antipov thought that they would crack in frosty weather. The management paid no attention to his complaints. Evidently someone was making money on the contracts.'

Doctor Zhivago – Boris Pasternak
(Reproduced by courtesy of W. Collins and Co. Ltd.)

Fundamentals of FRACTURE MECHANICS

J. F. Knott, B.Met., Ph.D., F.I.M., C.Eng.
Department of Metallurgy and Materials
Science, Cambridge University

BUTTERWORTHS
London Boston Sydney Wellington Durban Toronto

First published 1973
Reprinted 1976
Reprinted 1979
Reprinted 1981

ISBN 0 408 70789 5

Printed in Great Britain by Page Bros., Norwich

PREFACE

The aim of macroscopic fracture mechanics is to provide for engineers a means of designing against fracture in engineering structures in a more quantitative manner than is possible using traditional toughness testing techniques. At the same time, an understanding of the micro-mechanics of fracture provides for metallurgists a scientific basis for the development of materials of high toughness. Since both engineers and metallurgists are striving to produce structures and materials which are more resistant to fracture, it is important for them to gain a mutual understanding of the others' concepts, techniques and problems. A metallurgist has to learn that the very way in which a piece of metal breaks is inherently affected by the stress state in service: an engineer must appreciate that different materials have different properties, not all of which can be neatly represented by algebraic symbols in equations. This book is designed to deal mainly with the macroscopic aspects of fracture mechanics, so that the idiosyncrasies of particular materials are not discussed in detail. The micro-mechanics of fracture processes in simple systems are, however, described in some depth to show that the models, devised to explain why cracks propagate or voids coalesce at the particular stresses or strains that they do, are often of a similar nature, whether the events are on a scale measured in microns or metres.

The selection and presentation of material has turned out to be highly personalised; in effect, defining a view of what the field of fracture mechanics should properly include. Selection is very necessary, because the literature connected with all possible aspects of the subject is overwhelming in its sheer mass and volume. There is more in the book relating to notched bars and micromechanisms than might be expected from a 'conventional' interpretation of its title, because an understanding of crack tip phenomena on the microscopic scale can be more easily gained by analogy with fracture events near blunt stress concentrators.

It is inevitable that the book has a fairly large mathematical content. Fracture mechanics is concerned with the quantitative treatment of fracture, and this must involve precise and accurate stress analysis. There is no easy way round such analysis, but attempts have been made to explain the physical principles from which the equations are developed and the formal mathematics has been kept to a minimum. On a first reading, it may be wise to move on to Chapter 4 as soon as the going becomes difficult, but even Chapter 3 should not present too many problems to a final year science or engineering undergraduate. An impression of the course of the subject can be gained fairly rapidly from Chapters 1, 4, 5 and parts of 6 and 9, but the reader then has to take for granted many pieces of information which are discussed in detail elsewhere.

I am indebted to many friends and colleagues who have helped, knowingly or unknowingly, in the production of this work. My thanks are due primarily to members and ex-members of my research group, particularly Richard Smith and Rob Ritchie. The book includes many of their experimental results and micrographs. I was extremely fortunate in having the opportunity to discuss some of my ideas with Professor James Rice, during the year that he came as an Overseas Fellow to Churchill College. Within the Department of Metallurgy and Materials Science, fracture mechanics research has been continually encouraged and supported by Professor Robert Honeycombe. Thanks for helpful discussion are due also to Professor Edwin Smith, Dr Graham Oates, Dr Michael May, Dr David Elliott, Mr Frank Walker, Dr Richard Dolby and Mr Len Wortley. I have had many opportunities to sample the most recent advances in fracture mechanics and I hope that I have been able, in this book, to present a balanced and clear account of all the work that has been done and is being done.

March 1973 J.F.K.

AUTHOR'S NOTE TO THE REVISED IMPRESSION

I am grateful to Fred Nichols and colleagues, of the Bettis Atomic Power Laboratory, Westinghouse Electric Corporation, for their detailed comments on the text. As a result, several minor changes have been made and there is now, I hope, consistency between *Figure 2.16a, Table 2.2* and *Table 7.2*. The penultimate section of Chapter 4 has been re-worded, but, to meet in full their comments on elastic energy release in mixed elastic/plastic stress fields, it would require more extensive re-writing than could be contemplated at this time. Comments from Dr. J. P. Benthem and Dr. Keith Nuttall are also gratefully acknowledged.

J.F.K.

CONTENTS

One

MODES OF FAILURE

1.1 Introduction

One of the fundamental requirements of any engineering structure is that it should not fail in service and much of the skill of the structural engineer lies in recognising that there are several possible modes of failure and in guarding against them in his design. In addition to chemical failure by environmental corrosion, the main modes of mechanical failure in which the designer is usually interested may be listed as:

failure by elastic instability (buckling);
failure by excessively large elastic deformation (jamming);
failure by gross plastic deformation (yielding);
failure by tensile instability (necking);
failure by fast fracture (cracking).

We shall consider each of these in turn.

1.2 Buckling and Jamming

Failure by elastic buckling occurs classically in circumstances where a long, slender strut is subjected to large compressive forces (see *Figure 1.1*). In practice, the line of action of the forces does not lie exactly along the central longitudinal axis of the strut and a bending moment is produced about its centre. For small end loads, the elastic restoring forces in the strut are sufficient to allow it to straighten back into its original position, if it is subjected to a slight sideways deflection, i.e. the system is in stable equilibrium. As the end loads are increased, a point of neutral equilibrium is reached at a critical load, where the buckling and restoring forces are exactly matched, so that any value of sideways deflection is maintained. Loads higher than this critical value induce elastic failure: the slightest deviation of the strut from its initial linearity produces buckling which increases catastrophically until the bent strut either yields or fractures, or until its ends touch. In this last case, the strut is said to be an elastica. The critical buckling load is that needed to produce neutral equilibrium and depends on the slenderness of the strut, on the elastic modulus of the material from which it is made and on the means by which the loads are transmitted to its ends.

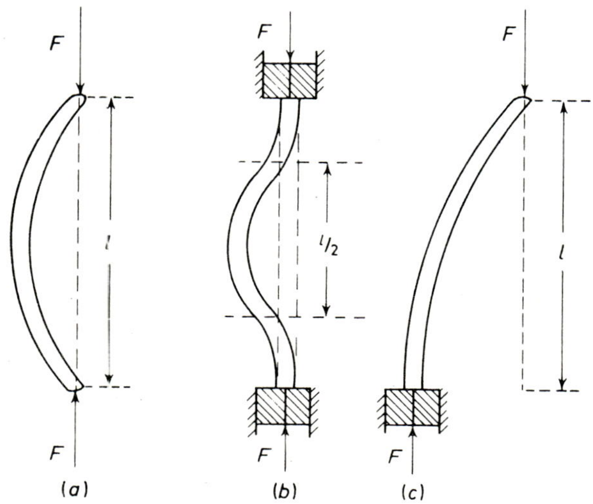

Figure 1.1 Buckling of a strut. (a) Both ends pin-jointed; (b) both ends fixed; (c) one end fixed; one end pin-jointed

For pin-jointed ends (*Figure 1.1a*), the theoretical buckling load, F, is given by Euler's formula:

$$F = \frac{\pi^2 EI}{l^2} \qquad (1.2.1)$$

where E is Young's modulus, and l and I are respectively the length and the moment of inertia of the strut. If the ends are fixed, rather than pin-jointed (*Figure 1.1b*), points of inflection are produced at positions one- and three-quarters the way along the length and the central portion behaves like a pin-jointed strut of length $l/2$. The buckling load is therefore increased by a factor of four. Conversely, if one end is fixed and the other end is pin-jointed and free to move sideways (*Figure 1.1c*), the strut behaves like half a double pin-jointed strut of length $2l$ and the buckling load is decreased by a factor of four. It is this form of loading which is demonstrated in pole-vaulting and the employment of the elastic restoring forces in a pole to give the vaulter lift represents one of the few deliberate uses of large elastic deflections in a slender strut. The use of relatively low-modulus fibre-glass poles has, in recent years, led to substantial improvements in pole-vaulting records.

More commonly, elastic strains in structures are kept to the very low values associated with traditional design in materials such as mild steel. Excessively large elastic deflections can produce failure by jamming: for example, in engines, where the clearances between rotors and stators are small, too large a deflection when load is applied would cause the engine to seize. A major aim of modern fibre technology, outside pole-vaulting, is to produce high modulus fibres so that the engineer can utilise high stresses without being embarrassed by high elastic strains. In conventional structural design, the ratio of modulus to specific gravity is fairly constant for most materials and low deflections can be obtained only by the use of thick struts or suitable framework configurations. A recent development has been the introduction of 'box-girder' sections which increase the moment of inertia for a given weight of material. In thick struts, ultimate failure is produced by simple crushing.

1.3 Yielding

If the critical load is exceeded, a slender strut begins to buckle catastrophically and may eventually suffer complete collapse by yielding. Regarding the section of a strut undergoing sideways deflections as if it were a simple beam subjected to pure bending (*Figure 1.2*), yielding starts on the tension and compression faces when the bending moment, M, has attained a value such that:

$$M = \frac{\sigma_Y W^2 B}{6} \tag{1.3.1}$$

where W is the width and B is the thickness of the strut (assumed to be rectangular in cross-section) and σ_Y is the material's yield stress. Because there is a gradient of stress, from tension to compression, across the width, the complete

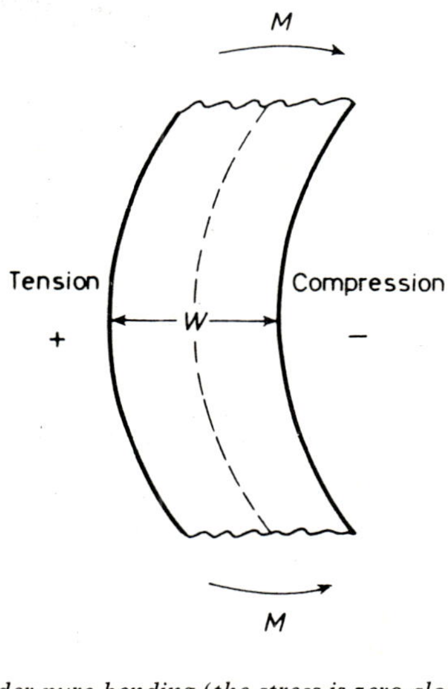

Figure 1.2 Beam under pure bending (the stress is zero along the dashed line in the centre of the beam)

cross-section does not yield at this value of bending moment. *Plastic collapse* can be said to have occurred only at a higher value, given by:

$$M = \frac{\sigma_Y W^2 B}{4} \tag{1.3.2}$$

We shall call this general collapse *general yield* and define it simply as follows: 'A body is said to have undergone general yielding when it is no longer possible to trace a path, across the load-bearing cross-section, from loading point to loading point, through elastically deformed material only'.

This is a most important definition, because 'yielding', as a phenomenon, is commonly identified with general yielding and general yielding only. As will be seen later, such an identification can prove to be confusing when talking about fracture behaviour.

It is clear that, for the case of a simple rectangular beam in pure bending, the load required to produce yielding initially in the outermost fibres is only two-thirds that needed to spread the yield completely across the cross-section. For a British Standard 'I'-beam, similarily subjected to pure bending, the load required for plastic collapse is 1.15 times greater than that required for initial yielding.

Plastic collapse is familiar to us as the failure mechanism observed when a tin can is crushed. It may occasionally be used deliberately as a feature of engineering design, for example in cars, where the reaction on the driver of the large forces experienced in a crash is reduced by allowing the energy of collision to be absorbed in plastic crumpling of the bodywork.

In an ideally elastic/plastic solid which exhibited no work-hardening, complete plastic instability would occur in a simple tension test when the applied tensile stress was equal to the material's yield stress. In general, this behaviour is not observed in annealed metals, because their capacity for work-hardening enables them to deform in a stable manner at all stresses up to the ultimate tensile stress in the tension test. Only if a metal has been heavily cold-worked before testing will the onset of yielding immediately produce plastic *instability* (i.e. continuing deformation under constant load) and then only if the test is controlled by loading-rate (a 'soft' testing system) rather than by displacement-rate (a 'hard' system). Similarly, the sharp drop in load that may occur when a tensile bar of mild steel first yields is observable only in a hard system. In the ideally soft loading system, annealed mild steel shows no decrease in load-bearing capacity until the ultimate tensile stress is reached.

If an engineer wishes to design against failure by plastic collapse, his variables are no longer restricted to those associated with elastic constants and with the geometry of the structure, because he has a choice of materials which possess different yield strengths. The difference between an annealed mild steel and a quenched and lightly tempered alloy steel represents about a six-fold increase in load-bearing capacity if a piece is designed to be resistant to failure by plastic instability on the basis of the initial yield stress, and about a three-fold increase if the design is based on the ultimate tensile stress.

Early principles of structural design ensured that the material's yield stress was nowhere exceeded in a structure, but, in recent years, it has become widely recognised that localised yielding can be permitted, provided that large deformations of the total structure are prevented. This condition, that the structure shall not deform generally as a 'mechanism', is known as the 'Plastic Design' concept and is equivalent to saying that the beam in bending (*Figure 1.2*) may be stressed at levels above that required for initial yielding (equation 1.3.1), provided that the stress level is below that needed to produce general yielding (equation 1.3.2). Using plastic design analysis, tolerable applied stresses are calculated by considering the effects of stress distribution on the spread of yield. Although the work-hardening capacity of a material may permit the applied loads to be increased still further, this factor is not generally taken into account in calculations. The development of higher strength structural materials has therefore been based on the achievement of higher yield strength, often at the expense of high work-hardening capacity.

For simple framework structures, it is, in principle, relatively easy, although, in practice, rather tedious, to calculate the load required to produce general plastic collapse. The calculations become more complicated for large, thick pieces because allowance has to be made for effects of triaxial stresses on the spread of yielding.

1.4 Necking

The likelihood of failure occurring by plastic instability (necking) should be small, because it requires the presence of tension members which are subjected to a soft loading system. Such configurations may arise when wire ropes are used as stays or to carry crane hooks, but the designer would automatically ensure that the stress in such a wire rope was substantially less than the material's ultimate tensile stress, so that failure would occur only as a result of a gross overload. Heavily cold-drawn wires will of course possess very little uniform elongation, and if they do fail on overloading they may well give the impression that the failure has been produced by a crack propagation mechanism, rather than by plastic instability as such. A typical example is the failure of a metal violin string which has been over-tautened. The general feeling which once persisted, that a material should have a high work-hardening capacity to obviate failure by plastic instability, seems now, as described above, to have been replaced by a more rational appreciation of the yield stress as the material property required to prevent failure by general yielding.

The overall behaviour depends on the interaction of material properties with the structure's geometry and the applied stress system. The decrease in uniform elongation produced by cold-working, which could give rise to catastrophic failure in a violin string, does not imply that cold-worked conditions are always to be avoided in service. Indeed, in a process such as the autofrettage of a pressure tube, large plastic deformations are deliberately introduced near the internal diameter so that a favourable, compressive residual stress system is present in service. Much the same philosophy underlies the 'proof testing' of large pressure vessels. The theme of interaction between several variables is one which will be pursued later with regard also to failure by fast fracture.

1.5 Principles of Engineering Design

Assuming that failure by necking is prevented simply by ensuring that tensile stresses in the structure are less than the material's ultimate tensile stress, and that buckling is prevented by design of the structural configuration, practical design trends will be made most clear if we concentrate on prevention of failure by general yielding. The applied design stresses are related to the material's yield stress through an appropriate design code. On the basis of pure plastic design reasoning, the design code ought simply to be the relationship between the uniaxial yield stress and the general yield stress of the structure, but the practical engineer is more conservative and applies a further 'safety factor' to his design, to reduce the allowable levels of applied stress. The main reason for the use of a safety factor is that, even when detailed stress analyses have been carried out, there still exists a degree of uncertainty concerning the situation in a real structure. One problem which arises is that the operative values of local stress are often not known at all well. Residual stresses, remaining in metal plates or sections after forming operations, give rise to one major uncertainty and other unknowns can arise from the stresses induced by the fabrication procedure or by uneven loading and bracing during erection. The engineer therefore usually follows a design code which ensures that the calculable stresses in his structure shall nowhere exceed the material's yield or ultimate tensile strength, divided by an appropriate figure for the safety, or 'ignorance' factor. The figure usually lies in the range 1.5–2 for designs based on the yield stress, depending on the particular application, and, hopefully, compensates for any local enhancement of calculated stresses.

The safety factor applied to the material's yield stress is therefore used as part of the rationale of plastic design in that it assists in the prevention of failure by general yielding. The use of a safety factor applied to the ultimate tensile stress seems to be of less value, particularly since the factor applied is often in the range 2.5–4, and therefore produces rather pessimistic results for the magnitude of tolerable applied stress. Failure by plastic instability (necking) will normally not be the critical failure mode, but, presumably, could become so if there were a situation where the local enhancement of stress were not uniform throughout the piece. If plastic design had allowed a tensile member to be strained sustantially whilst still preventing the deformation of a framework as a mechanism, non-uniform stress enhancement could imply that necking would precede general yielding. There does not seem to be any reason, however, why the safety factor applied to the ultimate tensile stress should be any larger than that applied to the yield stress, if the contribution from work-hardening in the structure is similar to that in the tensile test. The design code ought then to revert automatically to one based on uniaxial yield stress.

Whether based on yield stress or ultimate tensile stress, an engineer's design code implies that, if higher applied stresses are to be put on a structure of given geometrical configuration, material of higher uniaxial strength must be used. It has been the availability of higher strength structural materials, together with the bold use of plastic design concepts, that has enabled engineers in recent years to produce slender, graceful structures which are subjected to stresses much higher than those of previous generations. On the other hand, counterbalancing this elegance and efficiency, it has become distressingly apparent that such structures are particularly prone to failure by fast fracture.

1.6 Cracking

The essence of fast fracture is that it is a failure mechanism which involves the *unstable* propagation of a crack in a structure. In other words, once the crack has started to move, the loading system is such that it produces accelerating growth. In the history of failure by fast fracture in service structures, the fractures have almost always been produced by applied stresses less than the design stress calculated from the appropriate code and safety factor. This has naturally enhanced the catastrophic nature of the fractures and has led to the general description of them as 'brittle'. As we shall see later, the microscopic mechanisms by which the cracks propagate may be anything from low strain cleavage or intergranular fracture to fully ductile shear separation (in thin sheet material). However, in practical terms, the engineering definition of 'brittle' must be retained and it must refer to the onset of *instability* when the applied stress is less than the general yield stress. In this book, we shall interpret a brittle fracture as '*one in which the onset of unstable crack propagation is produced by an applied stress less than the general yield stress of the uncracked ligament remaining when instability first occurs*'.

Numerous brittle fractures have occurred in service and examples may be drawn from most fields of structural and mechanical engineering. Details of catastrophic service failures in ships, bridges, turbogenerators, structural steelwork, pressure vessels and gas pipe-lines may be found in references 1–8 listed at the end of this chapter. The sum total of failures and the amount of damage done is awesome to contemplate. As far as the present book is concerned, we consider only the major features which these failures hold in common. These are, primarily, the presence of gross stress-concentrators in fairly large pieces and a loading system which does not relax the applied stress as any crack formed begins to grow. Brittle fractures in steels occur particularly at low temperatures and in thick sections but, both in steel and in aluminium alloys, it is equally possible to suffer disastrous fractures, 'brittle' in the engineering sense of the word, in very thin sheets. One example of this occurred in the final fracture of 'Comet' aircraft fuselage skins which contained large fatigue cracks. In all cases, the introduction of defects or metallurgical embrittlement during manufacture worsens the situation.

If a piece is to fracture in a brittle manner (i.e. before general yield) it must contain a stress concentrator, because it is necessary to confine the mechanism by which cracking is produced to a small localised region. We shall therefore be interested predominantly in the ways in which fracture can be produced ahead of a pre-existing crack or other stress concentrator and in how these may be related to the applied stress system. Before we study these matters in more detail in the following chapters and show how they lead to the general formulation of Fracture Mechanics, it is, first, worth describing the more traditional methods by which resistance to failure by fast fracture has been assessed in the past, to understand why they are not entirely satisfactory.

1.7 Notched Impact Testing

It has long been accepted that measurements of ductility in the tensile test, such as elongation or reduction in area, whilst providing convenient figures for specifying material quality in general terms, are not really suitable for predicting the material's resistance to fast crack propagation in service. For steels particularly, measurements of 'brittleness' have usually been obtained by observing the amount of energy absorbed when a notched specimen is broken, generally under conditions of impact loading. Specimens differ in their detailed design, but the principle is employed in many well-known tests, including those of Charpy, Izod, Schnadt and Van der Veen. Typical results for a low-strength structural steel are indicated schematically in *Figure 1.3* where the energy absorbed and the percentage of 'crystalline' fracture on the broken surfaces are plotted as functions of test temperature. It can be seen that there is a fairly sharp transition from low- to

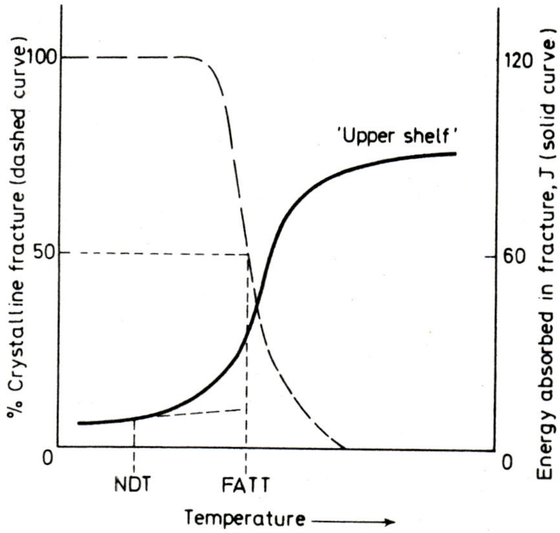

Figure 1.3 Schematic impact transition curves (behaviour typical of mild steel)

high-energy fracture over a very narrow temperature range and that this is closely associated with the change from bright crystalline to dull fibrous fracture. As will be described in detail in Chapters 7 and 8, the bright crystalline appearance is produced by reflection from $\{100\}$ cleavage facets and the dull fibrous appearance is due to a fracture mechanism which involves the very localised 'internal' plastic necking of material between non-metallic inclusions.

Material quality is defined in terms of such an impact energy transition curve either by specifying that a certain minimum amount of energy (say 30 J) must be absorbed at a given test temperature or by determining a 'transition temperature' at which the sharp change in behaviour occurs. Various 'transition temperatures' have been employed, including the temperature at which the energy curve first begins to rise [the nil ductility temperature (NDT)], that at which the fracture appearance is 50% crystalline, 50% fibrous [the fracture appearance transition temperature (FATT)], and that at which the fracture appearance first becomes entirely crystalline.

The information obtained from notched impact tests cannot be applied directly to assess the resistance to fast crack propagation of a piece in service, because neither the fracture appearance nor the amount of energy absorbed can be related in a quantitative manner to the applied design stress, even if the geometry and strain rate associated with the impact test could be said to produce effects identical to those produced by service conditions. Impact-test information should really be used only to correlate with known performance in service. Provided that such correlations are made, the impact test provides comparative measurements of the toughness of different batches of steels of the same nominal composition, i.e. it may be used to give figures for quality control. Extensive correlations of this type have, however, been made for only a few specific applications.

One example is given by the tests carried out on steel plates from 'Liberty Ship' hulls, which fractured by brittle cleavage fracture during or after the Second World War[5]. Here, it was found that notched-impact testpieces cut from plates which had failed in a brittle manner absorbed less than 20 J (15 ft lb) at 10°C, whereas those cut from 'tough' plates absorbed more than 20 J. The figure of 20 J energy absorption at 10°C was therefore suggested as a critical value for ensuring that ship steel is not susceptible to brittle fracture. Much more extensive analysis of failures suggested that a criterion of greater than 47 J (35 ft lb) energy absorption and less than 70% crystallinity at 0°C was necessary to prevent brittle fractures[9]. Present specifications relate the required impact properties both to the strength level of the steel and to the thicknesses of plate used in service.

An alternative treatment applied to impact results is the measurement and specification of the transition temperature. This is primarily a feature of the impact test itself, but can again be correlated, for a given material and application, with a transition temperature which is characteristic of the service conditions. Often, an intermediary large-scale test is used to simulate the anticipated extremes of service environment in the testing laboratory. The transition curve for the large test then represents (hopefully) the worst case that could occur and the 'worst case' transition temperature is compared with the transition temperature in notched impact. Typically, a specification for material quality would then be that its notched impact transition temperature should be greater than, say, 30°C below the operating temperature in service. We shall consider this type of specification in more detail in Chapter 8.

To summarise, it is clear that, by operating within closely defined limits, the notched impact test may be used to give comparative values of the toughness of nominally identical steels. It is, therefore, generally suitable for routine quality control testing. The information obtained cannot, however, be used to calculate the magnitude of applied stress to produce rapid crack propagation in a structure containing defects of varying size and geometry and so the designer is forced to seek an alternative, quantitative measurement of the resistance of a material to such crack propagation. This resistance is said to be a material's 'fracture toughness' and may be said to control failure by fast fracture in much the same way as the general yield stress controls failure by yielding. Both parameters depend very much on test temperature, strain-rate, geometrical configuration, and on the material's microstructure. The following chapters provide a basis for the characterisation of fracture toughness in terms of both the macroscopic mechanics and microscopic mechanisms of crack propagation, beginning with an analysis of the stresses and deformation around the stress concentrators at which fracture initiates.

1.8 Conclusions

The present chapter has emphasised the several possible modes of mechanical failure and has drawn attention to the ways in which they may be guarded against in engineering design. If we regard the geometrical considerations as part of the calculative processes we are left with the following properties of a material which control resistance to failure in practice.

Mode of failure	*Control*
Buckling or jamming	Elastic modulus
Yielding	Yield stress
Crack propagation	Fracture toughness

The 'necking' failure mode has been omitted, for reasons given earlier.

To optimise resistance to failure in conventional structural materials, where the ratio of modulus to specific gravity is approximately constant, it is necessary to develop both high yield stress and high toughness. The ideal balance can be struck only with regard to a particular structure, and represents the point where failures by gross yielding or crack propagation are equally unlikely to occur.

REFERENCES

1. Boyd, G. M., *Brittle Fracture in Steel Structures*, Butterworths, London (1970)
2. Biggs, W. D., *The Brittle Fracture of Steel*, Macdonald and Evans, London (1960)
3. Parker, E. R., *Brittle Fracture of Engineering Structures*, Wiley, New York (1957)
4. Tipper, C. F., *The Brittle Fracture Story*, Cambridge University Press (1962)
5. ASTM, *Special Technical Publication*, No. 158 (1954)
6. *Special Report on Failure of a Boiler during Static Test at Sizewell Nuclear Power Station*, West of Scotland Iron and Steel Institute (1964)
7. BWRA, *Bulletin*, 7, No. 6 (June, 1966)
8. *Report of Royal Commission into the Failure of Kings Bridge*, Presented to both Houses of (Australia's) Parliament, A. C. Brooks, Government Printer, Melbourne (1963)
9. Hodgson, J. and Boyd, G. M., *Trans. R. Instn. nav. Archit.*, **103** (1961)

Two

STRESS CONCENTRATIONS

2.1 Introduction

Since the occurrence of failure by fast fracture is necessarily associated with the presence of high local stresses and strains near stress concentrators, any realistic measurement of a material's fracture toughness must be fully appreciative of the precise magnitudes and distributions of these stresses and strains. The full mathematical stress analyses, for anything but the most simple geometries, are extremely complex and it is the aim of this chapter and the succeeding one to try to explain the principles of the calculations in a manner such that the important results can be made acceptable to the non-specialist. Inevitably, this leads to loss of rigour and those who wish to follow the subject in depth must refer to more advanced texts.

2.2 Pictorial Representation – Stress Trajectories

The local enhancement of stress and strain near a sharp change in geometrical cross-section, such as a fillet or notch, or ahead of the tip of a crack, is by no means intuitively obvious. The phenomenon was, indeed, not generally recognised by engineers until nearly the end of the nineteenth century. The most plausible non-mathematical description of the stress–concentration effect is given in terms of 'lines of force' or 'stress trajectories'. We suppose, as indicated in *Figure 2.1*, that we have a plate, containing a central elliptical hole, which is subjected to a uniform tensile stress. Regard the stress as being transmitted from one end of the plate to the other by means of lines of force (rather like magnetic lines of force). At the uniformly stressed ends of the plate, the lines are spaced uniformly (stress = force/unit area = force/spacing x plate thickness) and remain at this spacing, provided that they traverse the plate well away from the ends of the hole. The more central lines are severely distorted by the presence of the hole (the stress field is said to be 'perturbed') and, because they are supposed to behave like elastic strings and therefore try to minimise their lengths, they cluster together near the ends of the hole to give a decrease in the local spacing and therefore an increase in the local stress (more lines of force in the same area).

The appeal of this model is increased if we regard the central hole as a sharp crack and sketch the hypothetical local arrangement of atoms around one of the tips, as in *Figure 2.2.* Here, we may give some physical reality to the lines of force by regarding them as the sequences of atomic bonds running from top to bottom of the plate. It can then be seen that the transmission of force around

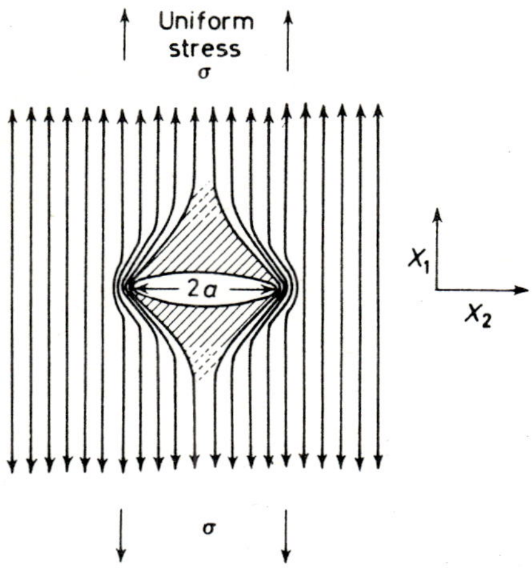

Figure 2.1 Perturbation of lines of force by an elliptical hole in a plate

the crack tip entails heavy loading and hence large straining of the bond AB. Smaller loads and strains will be carried by CD, but only at a position fairly remote from the crack tip will a bond (e.g. PQ) be loaded by virtually the same stress as that applied to the ends of the plate. It should be noted that bond AB can extend to a length greater than CD only by creating some tensile strain (and hence stress) along AC and BD. We are therefore led to suspect that the uniaxial stressing of a plate which contains a crack produces, not only a high tensile stress in the X_1 direction, but also a tensile stress in the X_2 direction. We shall return to this point later.

The situation depicted in *Figure 2.2* is intended only to act as a very approximate model. It shows unbalanced forces acting normal to the crack faces and these must disappear in practice to give unloaded regions immediately above and below the crack, as shown by the shaded regions in *Figure 2.1.* General tensions from regions V and W will, however, load bonds AB, CD, etc., as described above and we might expect that, in some way, much of the strain energy which has been released by unloading the shaded regions is now stored in the highly stressed region at the crack tip. Again, we shall examine the mathematical features of this argument later.

The most beautiful visual method of demonstrating the presence of stress concentrations is to make use of the stress birefringence properties of certain resins or plastics, such as perspex. Here, the passage of polarised light through a stressed, notched piece produces coloured interference fringes. The detailed analysis of the fringes to determine the magnitudes of the stresses relies, however, on even more complicated mathematics than that used for direct analytical calculations, because the birefringence theory has also to be included.[1] Further analogies of stress distributions have been produced, utilising magnetic or electrostatic field distributions, and arise from the similarity of the basic mathematical equations for material behaviour.

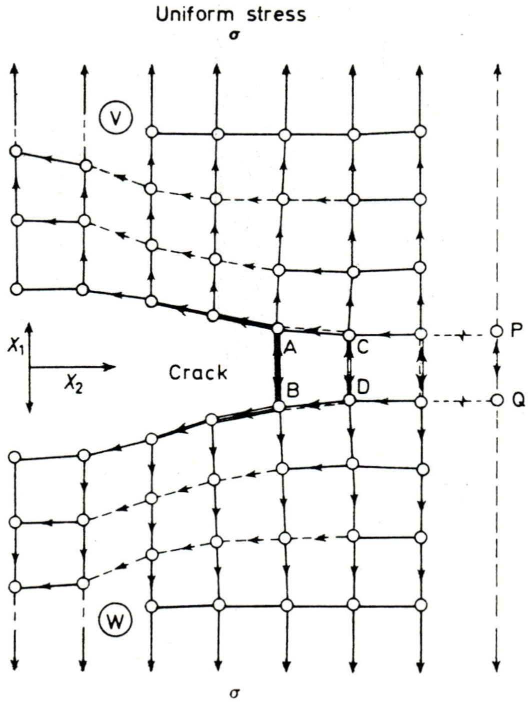

Figure 2.2 Schematic loading of atomic bonds near a crack tip

We now begin to examine the mathematical analysis of stress concentrations and the first step is to define stress and strain and the relationships between them.

2.3 Stress

The stress at a point O is formally defined in terms of the (positive) body force acting across an infinitesimally small planar area passing through O. Since the orientation of the area can be described by its unit normal vector, and the force is also a vector, it is convenient to describe stress in terms of the components of these two vectors parallel to suitable co-ordinate axes. In three dimensions, each vector will possess three components and so we expect stress to be defined by nine terms.

In right-handed Cartesian co-ordinates, stress is defined in terms of the components of the force acting across the faces of an infinitesimally small cube situated at O, with these faces normal to the three axes, X_1, X_2 and X_3 respectively. The resultant components of stress are then shown as in *Figure 2.3*. Each

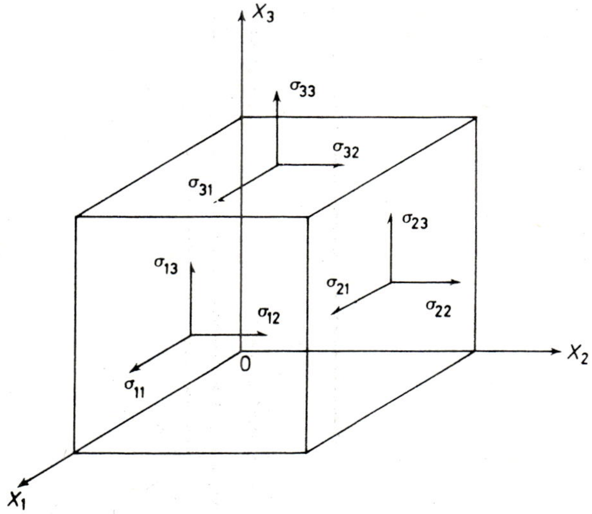

Figure 2.3 Stresses referred to Cartesian axes

stress component is defined by two subscripts. The first indicates the direction of the outward-facing normal to the appropriate face of the cube. The second indicates the direction of the particular component of force. For example, the stress component σ_{11}, defined as $\lim_{A_1 \to 0} \frac{F_1}{A_1}$, is a tensile stress acting in the X_1 direction across the face whose outward facing normal also lies in the X_1 direction. The components $\sigma_{11}, \sigma_{22}, \sigma_{33}$ are all tensile stresses; $\sigma_{21}, \sigma_{23}, \sigma_{12}, \sigma_{13}, \sigma_{31}, \sigma_{32}$ are all shear stresses. The full state of stress (or stress *tensor*) at a point O is written for convenience as σ_{ij}; here σ_{ij} represents nine components, by allowing both i and j to take any of the values 1, 2 or 3 independently. If the stress components are in equilibrium, it is apparent that certain pairs of shear stresses must have the same magnitude ($\sigma_{12} = \sigma_{21}, \sigma_{23} = \sigma_{32}, \sigma_{31} = \sigma_{13}$) to prevent any unbalanced couple from acting on the cube.

It is often convenient to refer stress components to co-ordinate axes other than Cartesian. In cylindrical polar co-ordinates, (z, r and θ), for example, the components are as shown in *Figure 2.4*. Here, the positive sense of θ is given by a clockwise rotation when sighting along the positive z direction. The components $\sigma_{rr}, \sigma_{\theta\theta}, \sigma_{zz}$ are tensile stresses; $\sigma_{r\theta}, \sigma_{rz}, \sigma_{\theta z}, \sigma_{\theta r}, \sigma_{z\theta}, \sigma_{zr}$ are shear stresses. In equilibrium, $\sigma_{\theta z} = \sigma_{z\theta}, \sigma_{zr} = \sigma_{rz}, \sigma_{\theta r} = \sigma_{r\theta}$. Again, the shorthand form σ_{ij}

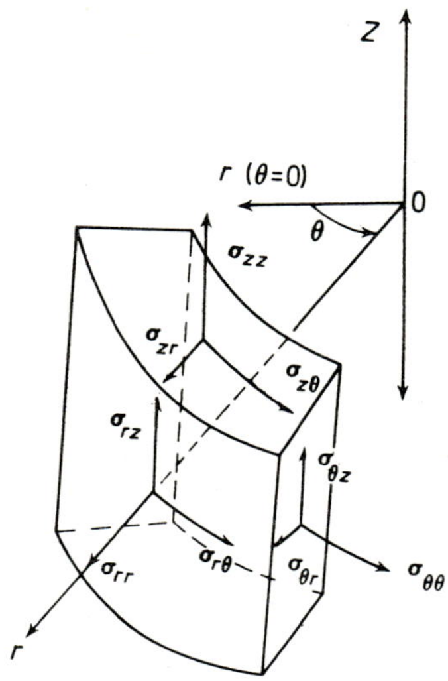

Figure 2.4 Stresses referred to cylindrical co-ordinates

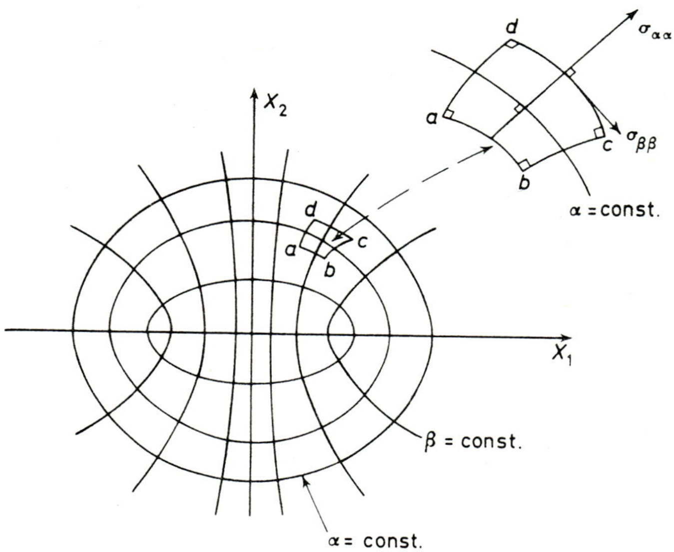

Figure 2.5 Definition of stresses in elliptical co-ordinates

may be used to denote the stress at a point, where i and j now take independently the values r, θ or z.

A third co-ordinate system sometimes used is that of curvilinear co-ordinates in which, in the plane $x_3 = 0$, the point x_1, x_2 is given by:

$$x_1 = c \cosh \alpha \cos \beta$$

$$x_2 = c \sinh \alpha \sin \beta$$

where c is a constant.

Lines of constant α, with β varying from 0 to 2π, are confocal ellipses; lines of constant β are confocal hyperbolae. The two sets of curves intersect orthogonally. The co-ordinate system is sketched in *Figure 2.5* and has the advantage that, by suitable choices of constant, either an ellipse may be made long and narrow to represent an internal crack or a pair of hyperbolae may be adjusted to take the geometrical form of external notches. A stress such as $\sigma_{\alpha\beta}$ acts on a face whose normal is orthogonal to the tangent to the curve α = constant, in the tangential direction.

2.4 Strain

Strain is produced by the displacement of points in a body *relative to one another*, i.e. the total displacement of points in response to a stress system, once rigid body translations or rotations have been eliminated. Suppose that a point P, defined by Cartesian co-ordinates x_i (x_1, x_2, x_3), is displaced to a new point P$'$, given by x_i', where $x_i' = x_i + u_i$. Then $u_i(u_1 u_2 u_3)$ is the displacement of P and to be a *strain* must differ from one point to another. Hence u_i is a function of x_i. For small strains, linear behaviour is assumed and u_i may be written simply as u_i = const. x_i. If all displacements are linear functions of the initial co-ordinates, we have $x_1, x_2, x_3 \rightarrow x_1 + u_1, x_2 + u_2, x_3 + u_3$, where $u_1 u_2$ and u_3 are given by:

$$\begin{aligned} u_1 &= e_{11}x_1 + e_{12}x_2 + e_{13}x_3 \\ u_2 &= e_{21}x_1 + e_{22}x_2 + e_{23}x_3 \\ u_3 &= e_{31}x_1 + e_{32}x_2 + e_{33}x_3 \end{aligned} \tag{2.4.1}$$

or, in shorthand form,

$$u_i = e_{ij}x_j \tag{2.4.2}$$

where each term e_{ij} is a constant of proportionality corresponding to a displacement in the direction of axis X_i of points on the axis X_j. We can relate the values of e_{ij} to engineering tensile and shear strains by considering a two-dimensional deformation system.

Let point $P(x_1, x_2)$ move to $P'(x_1', x_2')$, where:

$$\left.\begin{array}{ll} x_1' = x_1 + u_1 & x_2' = x_2 + u_2 \\ \text{and} \quad u_1 = e_{11}x_1 + e_{12}x_2 & u_2 = e_{21}x_1 + e_{22}x_2 \end{array}\right\} \quad (2.4.3)$$

Now consider the movement of a point on the X_1 axis: $P(x_1, 0)$. The initial distance of the point from the fixed origin O is $OP = x_1$. After deformation, the co-ordinates of P′ are x_1', x_2', where:

$$\left.\begin{array}{l} x_1' = x_1 + u_1 = x_1 + e_{11}x_1 \\ x_2' = x_2 + u_2 = e_{21}x_1 \end{array}\right\} \quad (2.4.4)$$

It is clear from the first equation that e_{11} is a simple tensile strain:

$$e_{11} = \frac{u_1}{x_1} = \frac{\text{increase in length}}{\text{original length}}$$

The term e_{21} is related to the engineering shear strain γ, but is not automatically equated to γ, because it contains a rotational component.

The working definition of engineering shear strain, γ, is that it is the change in angle between two lines which are initially orthogonal (see *Figure 2.6a*). The

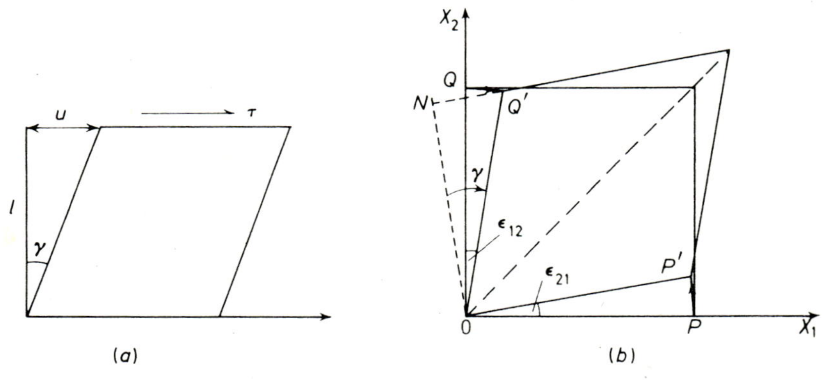

Figure 2.6 (a) Definition of shear strain (shear strain = $u/l = \gamma$). (b) Definition of pure shear ($\epsilon_{21} = \epsilon_{12} = \gamma/2$ for infinitesimal strains)

total change in angle between two points initially situated on the X_1 and X_2 axes is given by $(e_{12} + e_{21})$ and a state of *pure shear* is defined when $e_{12} = e_{21}$, the positive sign being taken for a rotation of a point on a positive axis towards the other positive axis. The situation in *Figure 2.6b* is then found. The total shear in the material is γ (imagine the rigid body rotation of OP round to OP′ as a starting point, so that ON is then sheared to OQ′) and the pure shears, written now as $\epsilon_{12} = \epsilon_{21}$, are therefore found to be $\gamma/2$. The formal definition of a pure shear strain is then

$$\epsilon_{ij} = \frac{1}{2}\left(\frac{\partial u_i}{\partial x_j} + \frac{\partial u_j}{\partial x_i}\right) \quad (2.4.5)$$

and this also gives the correct form for tensile strains, when $i = j$. Equation 2.4.5 is then the definition of the strain *tensor*, ϵ_{ij}, which, by the way it is defined is a symmetrical tensor ($\epsilon_{ji} = \epsilon_{ij}$) like the stress tensor.

2.5 Principal Stresses and Strains

In three dimensions, the stress acting across a planar area of arbitrary orientation passing through the point O may be written in terms of its components as σ_{ij}. It is possible to determine three specific orientations for this plane such that no shearing stresses act on it. These three planes are referred to as *principal planes* and are found to be orthogonal to one another. The three vectors normal to the planes form the *principal axes* and the three tensile stresses which act across the planes are the *principal stresses*, written as σ_1, σ_2 and σ_3 (single subscripts) where it is conventional to regard σ_1 as the algebraically largest and σ_3 as the algebraically smallest principal stress. The stress σ_1 is the largest tensile stress in the body, produced by the applied stress system. The values of σ_1, σ_2 and σ_3 may be found, by calculating the values of σ for which the determinant

$$\begin{vmatrix} (\sigma_{11} - \sigma), \sigma_{21}, \sigma_{31} \\ \sigma_{12}, (\sigma_{22} - \sigma), \sigma_{32} \\ \sigma_{13}, \sigma_{23}, (\sigma_{33} - \sigma) \end{vmatrix} \tag{2.5.1}$$

is equal to zero, where $\sigma_{11}, \sigma_{21}, \ldots$, etc., are the nine components of stress referred to Cartesian co-ordinates. The direction cosines $l_1 l_2 l_3$ of the principal axes may be determined by substituting the known values of σ ($\sigma_1 \sigma_2 \sigma_3$) into the equations

$$\begin{aligned} l_1(\sigma_{11} - \sigma) + l_2\sigma_{21} + l_3\sigma_{31} &= 0 \\ l_1\sigma_{12} + l_2(\sigma_{22} - \sigma) + l_3\sigma_{32} &= 0 \\ l_1\sigma_{13} + l_2\sigma_{23} + l_3(\sigma_{33} - \sigma) &= 0 \end{aligned} \tag{2.5.2}$$

with the additional information that

$$l_1^2 + l_2^2 + l_3^2 = 1 \tag{2.5.3}$$

The visualisation of the stress at a point is obviously simplified considerably if stresses can be referred to principal axes. Various *states of stress* may be said to exist, depending on the values of σ_1, σ_2 and σ_3. In general, the three principal stresses are unequal, and the state of stress is said to be *triaxial*. If all three are equal, it is *hydrostatic*; if two are equal, and the third is non-zero, it is *cylindrical*; if two are zero, it is *uniaxial*; if one is zero, it is *biaxial*, or, more generally, a state of *plane stress*. This last state is a situation commonly encountered in the loading of very thin sheets, which do not develop any tensile stress through their thickness.

The maximum, or *principal*, shearing stresses act on planes whose normals bisect the (90°) angles between pairs of principal axes. Since σ_1 is the largest and σ_3 is the smallest principal stress, the maximum shear stress, τ_{max}, acts on the two planes whose normals make angles of 45° to the σ_1 and σ_3 directions. The magnitude of τ_{max} is given by:

$$\tau_{max} = \frac{\sigma_1 - \sigma_3}{2} \tag{2.5.4}$$

The maximum principal shearing stress in a system is of great importance in deciding at what applied stress yielding will be produced.

By analogy with stress, it is also possible to define a set of orthogonal co-ordinate axes along which, or on planes normal to which, there are no shear strains. For an isotropic body it may be shown that the principal axes of stress and strain are identical, i.e. an element aligned along a principal stress axis undergoes only simple extension or contraction in response to a tensile or compressive principal stress. The formulation for the determination of principal strains and the maximum shear strains follows that for stress, with the replacement of stress components by the appropriate strain components.

It should be noted that, if the determinant in equation 2.5.1, is expanded as a cubic equation in σ, it will take the form:

$$\sigma^3 - I_1\sigma^2 - I_2\sigma - I_3 = 0 \tag{2.5.5}$$

where

$$I_1 = \sigma_{11} + \sigma_{22} + \sigma_{33}$$

$$I_2 = -(\sigma_{11}\sigma_{22} + \sigma_{33}\sigma_{11} + \sigma_{22}\sigma_{33}) + \sigma_{12}^2 + \sigma_{31}^2 + \sigma_{23}^2$$

$$I_3 = \sigma_{11}\sigma_{22}\sigma_{33} + \sigma_{12}\sigma_{31}\sigma_{23} - \sigma_{11}\sigma_{23}^2 - \sigma_{22}\sigma_{31}^2 - \sigma_{33}\sigma_{12}^2 \tag{2.5.6}$$

The quantities I_1, I_2 and I_3 remain unchanged, whatever the choice of co-ordinate axes, and so are referred to as *invariants* of the stress system (or *tensor*).

2.6 Stress–Strain Relationships in Elastic Solids

It is an empirically observed fact that the strain produced in a thin wire by the application of a small uniaxial stress is directly proportional to the magnitude of the applied stress. This observation, made initially by Hooke, describes *linear elastic* behaviour (stress and strain are related by a linear equation) and provides a definition for Young's Modulus, E, as the elastic coefficient which relates stress to strain:

$$\sigma = E\epsilon \tag{2.6.1}$$

Alternatively, the coefficient (equal to $1/E$) which relates strain to stress is termed the *elastic compliance* of the system. Poisson's ratio, ν, for such an unrestricted uniaxially loaded wire, is defined as the ratio of lateral contraction to longitudinal extension, and has a value lying in the range 0.28–0.33 for most metals.

The three-dimensional formulation of the relationships between stress and strain for a given stress system assumes that similar linear relationships hold for the individual components of stress and strain. In the most general form, the six independent stress components must be related to the six independent strain components by six linear equations, involving 36 coefficients:

$$
\begin{aligned}
\sigma_{11} &= C_{11}\epsilon_{11} + C_{12}\epsilon_{22} + C_{13}\epsilon_{33} + C_{14}\epsilon_{12} + C_{15}\epsilon_{13} + C_{16}\epsilon_{23}\\
\sigma_{22} &= C_{21}\epsilon_{11} + C_{22}\epsilon_{22} + C_{23}\epsilon_{33} + C_{24}\epsilon_{12} + C_{25}\epsilon_{13} + C_{26}\epsilon_{23}\\
\sigma_{33} &= C_{31}\epsilon_{11} + C_{32}\epsilon_{22} + C_{33}\epsilon_{33} + C_{34}\epsilon_{12} + C_{35}\epsilon_{13} + C_{36}\epsilon_{23}\\
\sigma_{12} &= C_{41}\epsilon_{11} + C_{42}\epsilon_{22} + C_{43}\epsilon_{33} + C_{44}\epsilon_{12} + C_{45}\epsilon_{13} + C_{46}\epsilon_{23}\\
\sigma_{23} &= C_{51}\epsilon_{11} + C_{52}\epsilon_{22} + C_{53}\epsilon_{33} + C_{54}\epsilon_{12} + C_{55}\epsilon_{13} + C_{56}\epsilon_{23}\\
\sigma_{31} &= C_{61}\epsilon_{11} + C_{62}\epsilon_{22} + C_{63}\epsilon_{33} + C_{64}\epsilon_{12} + C_{65}\epsilon_{13} + C_{66}\epsilon_{23}
\end{aligned}
\qquad (2.6.2)
$$

or

$$
\begin{aligned}
\epsilon_{11} &= S_{11}\sigma_{11} + S_{12}\sigma_{22} + S_{13}\sigma_{33} + S_{14}\sigma_{12} + S_{15}\sigma_{13} + S_{16}\sigma_{23}\\
\epsilon_{22} &= S_{21}\sigma_{11} + S_{22}\sigma_{22} + S_{23}\sigma_{33} + S_{24}\sigma_{12} + S_{25}\sigma_{13} + S_{26}\sigma_{23}\\
\epsilon_{33} &= S_{31}\sigma_{11} + S_{32}\sigma_{22} + S_{33}\sigma_{33} + S_{34}\sigma_{12} + S_{35}\sigma_{13} + S_{36}\sigma_{23}\\
\epsilon_{12} &= S_{41}\sigma_{11} + S_{42}\sigma_{22} + S_{43}\sigma_{33} + S_{44}\sigma_{12} + S_{45}\sigma_{13} + S_{46}\sigma_{23}\\
\epsilon_{23} &= S_{51}\sigma_{11} + S_{52}\sigma_{22} + S_{53}\sigma_{33} + S_{54}\sigma_{12} + S_{55}\sigma_{13} + S_{56}\sigma_{23}\\
\epsilon_{31} &= S_{61}\sigma_{11} + S_{62}\sigma_{22} + S_{63}\sigma_{33} + S_{64}\sigma_{12} + S_{65}\sigma_{13} + S_{66}\sigma_{23}
\end{aligned}
\qquad (2.6.3)
$$

where C_{11}, etc., are elastic coefficients and S_{11}, etc., are elastic compliances. An important feature of any elastically deforming body is that it possesses a uniquely defined strain-energy density $W = W(\epsilon_{mn})$

$$W = W(\epsilon_{mn}) = \int_0^{\epsilon_{mn}} \sigma_{ij}\, d\epsilon_{ij} \qquad (2.6.4)$$

For the simple case of an elastically deforming uniaxial bar, the expression reduces to the familiar form:

$$W = \tfrac{1}{2} \text{ stress x strain per unit volume}$$

The implication of the uniqueness of the strain energy density is that specific relationships must exist between the coefficients: $S_{ij} = S_{ji}$; $C_{ij} = C_{ji}$, to maintain a constant value for the integral of the product of stress and strain components. For isotropic solids, it is possible to simplify the relationships further by making use of the fact that the principal axes of stress and strain coincide, and that the displacements must be symmetrical with respect to these axes. In terms of principal axes, we may write, for example:

$$
\begin{aligned}
\sigma_1 &= (\lambda + 2\mu)\epsilon_1 + \lambda\epsilon_2 + \lambda\epsilon_3\\
\sigma_2 &= \lambda\epsilon_1 + (\lambda + 2\mu)\epsilon_2 + \lambda\epsilon_3\\
\sigma_3 &= \lambda\epsilon_1 + \lambda\epsilon_2 + (\lambda + 2\mu)\epsilon_3
\end{aligned}
\qquad (2.6.5)
$$

where λ and μ are known as Lamé's constants. If the dilatation $\Delta = \epsilon_1 + \epsilon_2 + \epsilon_3$ is introduced, equations 2.6.5 become simply

$$\begin{aligned} \sigma_1 &= \lambda\Delta + 2\mu\epsilon_1 \\ \sigma_2 &= \lambda\Delta + 2\mu\epsilon_2 \\ \sigma_3 &= \lambda\Delta + 2\mu\epsilon_3 \end{aligned} \tag{2.6.6}$$

The stress/strain relations referred to axes $X_1X_2X_3$ become simply

$$\begin{aligned} \sigma_{11} &= \lambda\Delta + 2\mu\epsilon_{11} & \sigma_{12} &= \mu\gamma_{12} = 2\mu\epsilon_{12} \\ \sigma_{22} &= \lambda\Delta + 2\mu\epsilon_{22} & \sigma_{23} &= \mu\gamma_{23} = 2\mu\epsilon_{23} \\ \sigma_{33} &= \lambda\Delta + 2\mu\epsilon_{33} & \sigma_{31} &= \mu\gamma_{31} = 2\mu\epsilon_{31} \end{aligned} \tag{2.6.7}$$

It can be seen immediately that μ is the shear modulus, which relates shear stress to shear strain. The second of Lamés constants, λ, is related to Young's modulus, E, by the expression:

$$E = \frac{\mu(3\lambda + 2\mu)}{(\lambda + \mu)} \tag{2.6.8}$$

and to Poisson's ratio, ν, by:

$$\nu = \frac{\lambda}{2(\lambda + \mu)} \tag{2.6.9}$$

by considering equations 2.6.7 applied to the case of simple uniaxial loading of a thin wire specimen which is allowed to contract in an unrestricted manner as it extends ($\sigma_2 = \sigma_3 = 0$). By eliminating λ from equations 2.6.8 and 2.6.9 above, it is possible to derive a further commonly used relationship between the elastic constants:

$$\mu = \frac{E}{2(1 + \nu)} \tag{2.6.10}$$

Hence, ν must be greater than -1. The relationship between λ and μ is given by:

$$\frac{\lambda}{\mu} = \frac{2\nu}{1 - 2\nu} \tag{2.6.11}$$

Hence, ν is less than 0.5.

For many solids undergoing elastic deformation, $\lambda \simeq \mu$, hence $\nu = 0.25$. For metals which are deforming plastically, the volume remains essentially constant because the deformation occurs by slip processes. The deformation is then said to be *incompressible*, when $\nu = 0.5$.

It is equally possible to rearrange the equations to give strain in terms of stress. For example, the principal strains in a body subjected to principal stresses σ_1, σ_2 and σ_3 are

$$\epsilon_1 = \frac{1}{E}\left[\sigma_1 - \nu(\sigma_2 + \sigma_3)\right]$$

$$\epsilon_2 = \frac{1}{E}\left[\sigma_2 - \nu(\sigma_3 + \sigma_1)\right] \qquad (2.6.12)$$

$$\epsilon_3 = \frac{1}{E}\left[\sigma_3 - \nu(\sigma_1 + \sigma_2)\right]$$

It is now possible to define a further state of stress, termed *plane strain*, in which one of the principal strains is zero. A typical example of plane strain deformation occurs in the central region of a wide strip which is being rolled: here the strip extends as its thickness is reduced, but its width remains virtually constant. If we set ϵ_2 equal to zero, we obtain, from equation 2.6.12 above

$$\sigma_2 = \nu(\sigma_1 + \sigma_3) \qquad (2.6.13)$$

where ν lies in the range 0.25–0.33 for elastic deformation and is equal to 0.5 for incompressible (plastic) deformation. Setting ϵ_2 equal to zero merely enables us to persist with the definition of σ_1 as the largest and σ_3 as the smallest principal stress.

2.7 Elastic Triaxiality

Given the general relationships between stress and strain, we are now in a position to anticipate further the likely distribution of stress near a concentrator. We examine the situation depicted in *Figure 2.7a*, which shows a schematic distribution as deduced from the model in Section 2.2 of the σ_{11} (principal) tensile stress ahead of a notch as a function of x_2 along the axis $x_1 = 0$. Unfortunately, this choice of axes, which agrees with those generally used in papers on fracture mechanics, demands that $\sigma_{11} > \sigma_{33} > \sigma_{22}$ in plane strain. It is hoped that this change in nomenclature will not prove too confusing.

Imagine a set of elements: $a, b, c, d, \ldots, p, q, \ldots$, etc., arranged along the X_2 axis. The average stress, $\bar{\sigma}_{11}(a)$, acting on element a is substantially greater than that acting on element b, and so the longitudinal strain $\epsilon_{11}(a)$ would be greater than $\epsilon_{11}(b)$ if the elements were deforming freely. However, in free deformation, there would also be Poisson's ratio contractions, $\epsilon_{22}(a)$, $\epsilon_{22}(b)$, . . , etc. Since X_1, X_2, X_3 are the directions of the axes of principal stress, and assuming plane strain deformation; $\epsilon_{33} = 0$, $\sigma_{33} = \nu(\sigma_{11} + \sigma_{22})$; we may use equations 2.6.12 to show that ϵ_{22} would be equal to $-\frac{\nu}{(1-\nu)} \cdot \epsilon_{11}$ if the elements were free from restraint in the X_2 direction. However, because $\epsilon_{11}(a) > \epsilon_{11}(b)$, ..., etc. it would then follow that $\epsilon_{22}(a) > \epsilon_{22}(b) > \epsilon_{22}(c)$, . . . , etc., and the a/b, b/c, c/d, . . . , etc., interfaces would separate. To prevent this happening and to allow the material to deform as an elastic continuum, a tensile stress, σ_{22}, must be present, to hold the elements together. The form of this stress (see *Figure 2.7b*) must be such that it is zero at the free surface (element '*a*' feels no restraint from the empty notch), that it increases steeply in response to the steep elastic stress

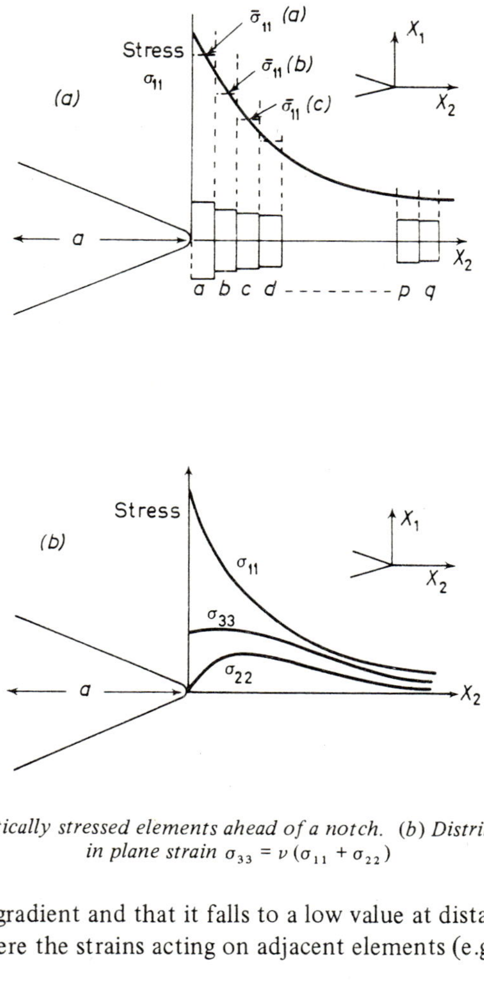

Figure 2.7 (a) Elastically stressed elements ahead of a notch. (b) Distribution of stresses in plane strain $\sigma_{33} = \nu(\sigma_{11} + \sigma_{22})$

(and hence strain) gradient and that it falls to a low value at distances remote from the notch where the strains acting on adjacent elements (e.g. '*p*' and '*q*') are virtually identical.

The production of the σ_{22} stress agrees with the conjectures based on a consideration of the stretching of bonds near a crack tip (*Figure 2.2*) and we therefore expect this stress to be a feature of mathematical solutions.

The stress σ_{33} is equal to $\nu(\sigma_{11} + \sigma_{22})$ in plane strain and its distribution is also shown schematically in *Figure 2.7b*. In plane stress, σ_{33} is zero, and the distribution of σ_{22} is similar to that already discussed, although its magnitude is smaller, because contraction can occur in the X_3 direction. The plane strain deformation therefore gives rise to a state of tensile triaxiality near a notch. At one time this was thought to be an important feature in the production of brittle fracture of metals. Nowadays, however, pure elastic triaxiality effects would be expected to affect only very brittle materials such as concrete, plasters and glasses. It is important to realise that, in plane strain, $\sigma_{11} > \sigma_{33} > \sigma_{22}$, and so σ_{22} is the smallest principal stress; in plane stress, $\sigma_{11} > \sigma_{22} > \sigma_{33}$ (= 0) and σ_{33} is the smallest stress. It will be shown later that this has a controlling influence with regard to the form of plastic deformation in thin and thick pieces. In thin pieces, σ_{33}, which is zero at the free surfaces, is virtually equal to zero throughout the thickness; the maximum shear stress consequently lies on planes at 45° to X_1 and X_3 (through-the-thickness). Throughout most of the thickness of a thick piece, ϵ_{33} is virtually zero, and the yielding in the central regions is confined to the X_1X_2 plane (see Section 2.11).

The continuity of material during elastic deformation, which we have shown to give rise to a stress σ_{22}, is one of the major conditions imposed in the mathematical theory of elasticity which is described below.

2.8 The Basis of the Theory of Elasticity

The two main principles on which the theory of elasticity is founded are first that the stresses shall be in equilibrium and secondly that the deformations produced by these stresses must be such that elastic continuity is preserved throughout the body. The second requirement is known as the compatibility condition. The equilibrium condition may be derived by consideration of *Figure 2.8* which shows three of the nine pairs of force components acting on an

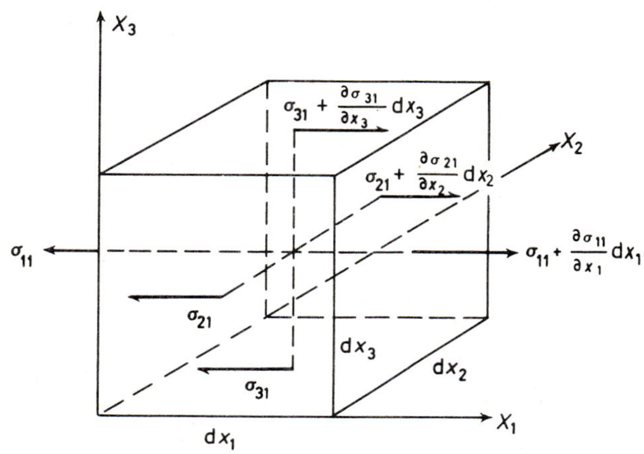

Figure 2.8 Equilibrium of stresses. If forces *are resolved in the X_1 direction:*

$$\left(\sigma_{11} + \frac{\partial\sigma_{11}}{\partial x_1} dx_1\right) dx_2 dx_3 - \sigma_{11} dx_2 dx_3 + \left(\sigma_{21} + \frac{\partial\sigma_{21}}{\partial x_2} dx_2\right) dx_3 dx_1 - \sigma_{21} dx_3 dx_1$$

$$+ \left(\sigma_{31} + \frac{\partial\sigma_{31}}{\partial x_3} dx_3\right) dx_1 dx_2 - \sigma_{31} dx_1 dx_2 = 0, \qquad \text{i.e.}\ \frac{\partial\sigma_{11}}{\partial x_1} + \frac{\partial\sigma_{21}}{\partial x_2} + \frac{\partial\sigma_{31}}{\partial x_3} = 0$$

element of dimensions $dx_1 dx_2 dx_3$. By considering all nine pairs of force components, it may be seen that, for equilibrium, the following relationships must be satisfied:

$$\frac{\partial \sigma_{11}}{\partial x_1} + \frac{\partial \sigma_{21}}{\partial x_2} + \frac{\partial \sigma_{31}}{\partial x_3} = 0$$

$$\frac{\partial \sigma_{12}}{\partial x_1} + \frac{\partial \sigma_{22}}{\partial x_2} + \frac{\partial \sigma_{32}}{\partial x_3} = 0 \qquad (2.8.1)$$

$$\frac{\partial \sigma_{13}}{\partial x_1} + \frac{\partial \sigma_{23}}{\partial x_2} + \frac{\partial \sigma_{33}}{\partial x_3} = 0$$

ignoring any body forces which may act on the element.

The operation of the compatibility condition is most clearly illustrated by reference to a two-dimensional system. For infinitesimal strains, we may proceed from the results of Section 2.4 to write:

$$\epsilon_{11} = \frac{\partial u_1}{\partial x_1} \qquad \epsilon_{22} = \frac{\partial u_2}{\partial x_2}$$

$$2\epsilon_{12} = 2\epsilon_{21} = \gamma_{12} = \gamma_{21}$$

$$= \frac{\partial u_2}{\partial x_1} + \frac{\partial u_1}{\partial x_2} \qquad (2.8.2)$$

Differentiating ϵ_{11} twice by dx_2; ϵ_{22} twice by dx_1; and $2\epsilon_{12} = \gamma_{12}$ by dx_1 and dx_2, we obtain:

$$\frac{\partial^2 \epsilon_{11}}{\partial x_2^2} + \frac{\partial^2 \epsilon_{22}}{\partial x_1^2} = \frac{\partial^2 \gamma_{12}}{\partial x_1 \partial x_2} \qquad (2.8.3)$$

If the strains satisfy this equation, the continuity of the body is preserved. The equation may be expressed in terms of stress by differentiating equations 2.6.7, with all stresses of the type σ_{3i} put equal to zero, since we are considering only two dimensions, and making the appropriate substitutions, to obtain:

$$\frac{\partial^2 \sigma_{11}}{\partial x_2^2} - \nu \frac{\partial^2 \sigma_{22}}{\partial x_2^2} + \frac{\partial^2 \sigma_{22}}{\partial x_1^2} - \nu \frac{\partial^2 \sigma_{11}}{\partial x_1^2} = 2(1 + \nu) \frac{\partial^2 \sigma_{21}}{\partial x_1 \partial x_2} \qquad (2.8.4)$$

The solution of a problem in elasticity requires that expressions be found for the stress components, which satisfy both equilibrium (equations 2.8.1) and compatibility (equation 2.8.4) conditions and which give the boundary conditions appropriate to the situation being considered. As in simple integration with respect to one variable, the answer when differentiated must give the original form, so the answer to an elasticity problem must satisfy the appropriate original equations. As for many standard integration solutions, the practising mathematician has a knowledge of the types of functions which, when differentiated, are likely to satisfy these equations. Any analytical expressions are likely to be extremely complicated unless the geometrical form of the body can be described by simple mathematical functions. Even so, general solutions for three dimensions are difficult to obtain, unless simplifications such as rotational symmetry are admissible, and the main body of calculations has been carried out for idealised plane states of stress: either plane stress ($\sigma_3 = 0$) or plane strain ($\epsilon_3 = 0$).

2.9 Airy's Stress Function

The problem is to find a suitable function ϕ in x_1 and x_2 which satisfies the equilibrium and compatibility equations and which enables the stresses to be related to the applied loads. Airy first demonstrated that such a function could exist for two-dimensional deformation[2] and showed that the stresses could be derived from it as follows:

$$\sigma_{11} = \frac{\partial^2 \phi}{\partial x_2^2}$$

$$\sigma_{22} = \frac{\partial^2 \phi}{\partial x_1^2} \qquad (2.9.1)$$

$$\sigma_{12} = -\frac{\partial^2 \phi}{\partial x_1 \partial x_2}$$

Direct substitution in equation 2.8.1, with stresses of the type σ_{3i} set equal to zero shows that equilibrium is satisfied.

To satisfy compatibility (equation 2.8.4) further requires that

$$\frac{\partial^4 \phi}{\partial x_1^4} + 2\frac{\partial^4 \phi}{\partial x_1^2 \partial x_2^2} + \frac{\partial^4 \phi}{\partial x_2^4} \equiv \nabla^2(\nabla^2 \phi) = 0 \qquad (2.9.2)$$

If the function satisfies the compatibility equation, the stresses are automatically determined by equations 2.9.1 above, provided that the boundary conditions are also satisfied. It is this further restriction that has rather limited the number of analytical solutions for stresses in bodies of complicated shape. However, it is also possible to solve such problems in two dimensions fairly readily by means of finite element methods or finite difference equations, using numerical methods such as relaxation. These are particularly suited to the use of computers, which are now increasingly applied to the exact solution of elasticity problems in bodies of the geometries actually used in service (see Sections 3.16, and 3.17).

To demonstrate the use of the analytical method, it may be profitable first to consider the development of the solution for the stress distribution in a plate of infinite dimensions, which contains a circular hole of radius a, when it is subjected to a uniaxial stress σ. The deformation is assumed to occur in a plane state of stress.

It is convenient to work in cylindrical co-ordinates, where the equations for equilibrium (cf. equations 2.8.1) in two dimensions (polar co-ordinates r and θ) become:

$$\frac{\partial \sigma_{rr}}{\partial r} + \frac{1}{r}\frac{\partial \sigma_{r\theta}}{\partial \theta} + \frac{\sigma_{rr} - \sigma_{\theta\theta}}{r} = 0$$

$$\frac{1}{r}\frac{\partial \sigma_{\theta\theta}}{\partial \theta} + \frac{\partial \sigma_{r\theta}}{\partial r} + 2\frac{\sigma_{r\theta}}{r} = 0 \qquad (2.9.3)$$

assuming that there are no body forces. The compatibility equation becomes:

$$\nabla^2(\nabla^2\phi) = \left(\frac{\partial^2}{\partial r^2} + \frac{1}{r}\frac{\partial}{\partial r} + \frac{1}{r^2}\frac{\partial^2}{\partial \theta^2}\right)\left(\frac{\partial^2 \phi}{\partial r^2} + \frac{1}{r}\frac{\partial \phi}{\partial r} + \frac{1}{r^2}\frac{\partial^2 \phi}{\partial \theta^2}\right) = 0 \qquad (2.9.4)$$

The stresses σ_{rr}, $\sigma_{\theta\theta}$ and $\sigma_{r\theta}$ may be obtained from a suitable stress function ϕ by means of the relationships:

$$\sigma_{rr} = \frac{1}{r}\frac{\partial \phi}{\partial r} + \frac{1}{r^2}\frac{\partial^2 \phi}{\partial \theta^2}$$

$$\sigma_{\theta\theta} = \frac{\partial^2 \phi}{\partial r^2} \qquad (2.9.5)$$

$$\sigma_{r\theta} = -\frac{\partial}{\partial r}\left(\frac{1}{r}\frac{\partial \phi}{\partial \theta}\right) = \frac{1}{r^2}\frac{\partial \phi}{\partial \theta} - \frac{1}{r}\frac{\partial^2 \phi}{\partial r \partial \theta}$$

The problem is solved by finding a stress function which satisfies the compatibility condition and which also satisfies the boundary conditions for the outer boundary and for the surface of the hole, which must be free from external forces. For an infinite plate, the stress function which fits these conditions is:

$$\phi = \left(-\frac{\sigma}{4}r^2 - \frac{a^4}{4}\frac{\sigma}{r^2} + \frac{a^2}{2}\sigma\right)\cos 2\theta + \frac{\sigma r^2}{4} - \frac{\sigma a^2}{2}\, ln r \qquad (2.9.6)$$

from which σ_{rr}, $\sigma_{\theta\theta}$ and $\sigma_{r\theta}$ may be determined by use of equations 2.9.5 above. The maximum value of the tangential stress, $\sigma_{\theta\theta}$, is obtained when $\theta = \pi/2$ or $3\pi/2$ and $r = a$, the radius of the hole. The variation of $\sigma_{\theta\theta}$ with r ($\equiv x_2$) when $\theta = \pi/2$ or $3\pi/2$ is given by

$$\sigma_{\theta\theta} = \sigma\left(1 + \frac{a^2}{2r^2} + \frac{3a^4}{2r^4}\right) \qquad (2.9.7)$$

and this distribution is drawn schematically in *Figure 2.9*. We see that when $r = \infty$, $\sigma_{\theta\theta}$ becomes equal to the uniform applied tensile stress, σ; when $r = a$, $\sigma_{\theta\theta} = 3\sigma$. The redistribution of the stress in the region of the hole has thus occasioned the development of a tangential tensile stress of magnitude three times that of the applied stress. The hole acts as a stress concentrator with an *elastic stress concentration factor* (ESCF) of 3.

It is important also to note that a radial tensile stress, σ_{rr}, is developed in the region near the hole, as was anticipated in Section 2.7. When $\theta = \pi/2$ or $3\pi/2$, σ_{rr} is given by:

$$\sigma_{rr} = \frac{3\sigma}{2}\left(\frac{a^2}{r^2} - \frac{a^4}{r^4}\right) \tag{2.9.8}$$

which is zero when $r = a$ or ∞ (normal to the free surface of the hole at $r = a$; normal to the free boundary surface at $r = \infty$) and which has a maximum value of $(\frac{3}{8})\sigma = (\frac{1}{8})\sigma_{\theta\theta(\max)}$ when $r = \sqrt{2}\,a$. The variation of σ_{rr} with r is shown also in *Figure 2.9.*

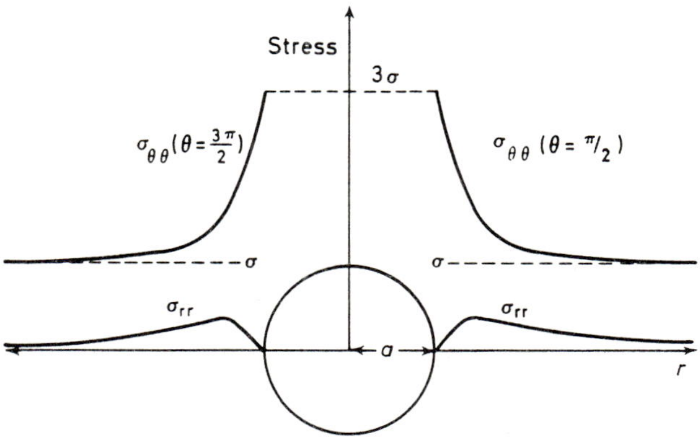

Figure 2.9 Distribution of stress around a circular hole in an infinite plate, subjected to a uniform stress, σ (plane stress)

In plane stress, σ_{zz} is zero: in plane strain, ϵ_{zz} is zero and so σ_{zz} is given by:

$$\sigma_{zz} = \nu(\sigma_{rr} + \sigma_{\theta\theta}) \tag{2.9.9}$$

In this last case, a stress state of triaxial tension is produced some little distance below the surface of the hole. The general consequences of this on yielding have been discussed earlier and will be taken up further in Section 2.11.

First, we will try to speculate on what the form of stress concentration around holes of different geometry might be, in anticipation of the more detailed analyses presented in Chapter 3.

2.10 Rough Estimate of Stress Concentration

The stress concentration at the edge of a circular hole may be derived using a real stress function, whereas the functions used to derive results for elliptical holes or sharp cracks usually contain both real and imaginary parts (see Chapter 3). Since lack of physical reality may be felt by the introduction of imaginary numbers, the aim of this section is to produce a rough, but plausible, comparison between results deducible from real and complex functions. To do this, we compare the maximum tensile stress calculated for an elliptical hole of semi-major axis a and semi-minor axis b (see Section 3.5) with an estimate of the maximum tensile stress developed ahead of the series of contiguous holes shown in *Figure 2.10a.*

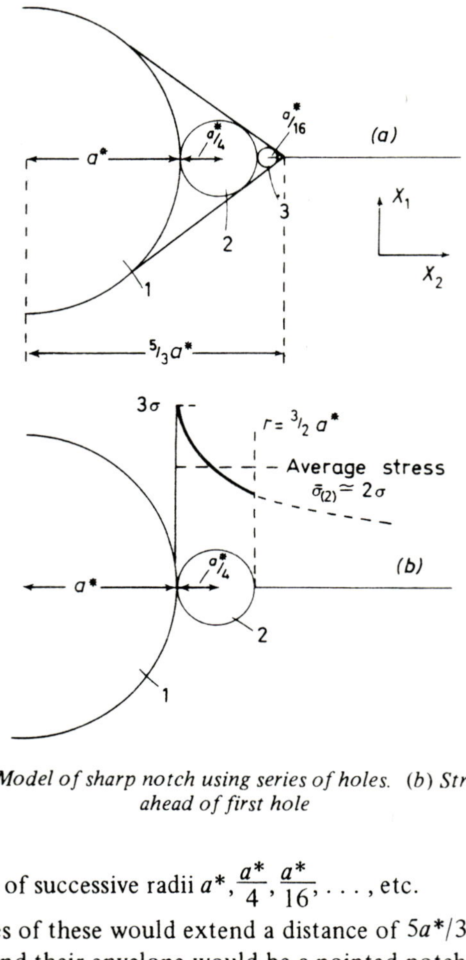

Figure 2.10 (a) Model of sharp notch using series of holes. (b) Stress distribution ahead of first hole

Imagine a series of successive radii $a^*, \dfrac{a^*}{4}, \dfrac{a^*}{16}, \ldots$, etc.

An infinite series of these would extend a distance of $5a^*/3$ from the origin along the X_2 axis and their envelope would be a pointed notch of total included angle approximately 75°. We may regard the second hole (radius $\dfrac{a^*}{4}$) as being subjected to the non-uniform stress field of the first hole as shown in *Figure 2.10b.* The average stress, $\bar{\sigma}_{(2)}$ acting on the second hole is then given by

$$\bar{\sigma}_{(2)} = \frac{2\sigma}{a^*} \int_{a^*}^{\frac{3}{2}a^*} \left(1 + \frac{a^{*2}}{2r^2} + 3\frac{a^{*4}}{2r^4}\right) \mathrm{d}r \qquad (2.10.1)$$

(see equation 2.9.7) which gives a result very close to $\bar{\sigma}_{(2)} = 2\sigma$. The peak tensile stress, $\sigma_{11(3)}$, ahead of the second hole is then, approximately $\sigma_{11(3)} = 3 \times 2\sigma$. The average stress acting on the third hole, $\bar{\sigma}_{(3)}$, is $2^2\sigma$, and the peak stress ahead of it is $\sigma_{11(4)} = 3 \times 2^2 \sigma$. In general, if we imagine n holes in the series, the peak stress ahead of the nth hole is:

$$\sigma_{11(n)} = 3.2^{n-1}.\sigma = 1.5 \times 2^n \sigma \qquad (2.10.2)$$

We now examine the results deduced for an ellipse: $\sigma_{\beta\beta} = \sigma\left(1 + 2\sqrt{\frac{a}{\rho}}\right)$ where ρ is the radius of the tip, and a is the major semi-axis, and compare them with the figures given above, in *Table 2.1* (see equation 3.5.9). Agreement is obviously not completely perfect, but is sufficient to give general credibility to future results.

Table 2.1

No. of holes	*Peak stress*	$\left(1 + 2\sqrt{\frac{a}{\rho}}\right)\sigma$	a	ρ	a/ρ
1	3σ	3σ	a^*	a	1
2	6σ	5.9σ	$\left(a^* + \frac{a^*}{2}\right)$	$\frac{a^*}{4}$	6
3	12σ	11.2σ	$\left(a^* + \frac{a^*}{2} + \frac{a^*}{8}\right)$	$\frac{a^*}{16}$	26

If n, the number of holes, is large, we approach the solution for a pointed notch, where $a \gg \rho$, $2\sqrt{\frac{a}{\rho}} \gg 1$. Then we have, on the one hand,

$$\sigma_{11(\max)} = 1.5 \times 2^n \sigma \qquad (2.10.2)$$

and, from the formula for an elliptical hole:

$$\sigma_{\beta\beta} = 2\sqrt{\left[\frac{\sum_n\left(a^* + \frac{a^*}{2} + \frac{a^*}{8} + \ldots\right)}{a^*/2^{2n-2}}\right]}.\sigma$$

$$= 2\sqrt{\left(\frac{5}{3}.2^{2n-2}\right)}.\sigma \qquad (2.10.3)$$

$$= 1.3 \times 2^n \sigma$$

Despite the approximations, we can perhaps feel some confidence in the general form of the results obtained using complex stress functions. The most important result is that for the pointed notch, where:

$$\sigma_{11(\max)} = 2\sqrt{\frac{a}{\rho}}.\sigma \qquad (2.10.4)$$

We shall examine this in detail in Chapter 3, but first we will consider the development of plasticity in a piece which contains a stress concentrator.

2.11 Plastic Deformation at Stress Concentrators

When a notched piece is stressed elastically, it is possible to produce high stresses near the notch and these may locally exceed the material's yield stress to produce a small plastic zone or 'enclave'. The stress distribution within this enclave depends very much on whether the deformation is occurring in plane stress or in plane strain.

In plane stress, the smallest principal stress is that through the thickness, σ_{33} (see Section 2.7) and yielding occurs on planes at 45° to the X_1 and X_3 axes. The Tresca criterion[3] for yielding tells us that yield occurs when the maximum shear stress is equal to a critical constant value, τ_Y, and this gives:

$$\sigma_{11} - \sigma_{33}(= 0) = 2\tau_Y = \sigma_Y \qquad (2.11.1)$$

throughout the plastic zone, where τ_Y is the yield stress in shear and σ_Y is the uniaxial yield stress. Thus, the maximum stress in the plastic zone is equal to the material's uniaxial yield stress, see *Figure 2.11a*.

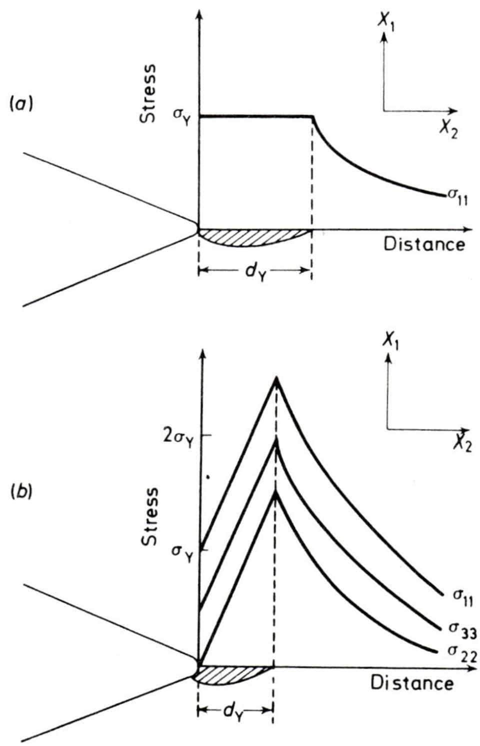

Figure 2.11 Schematic elastic/plastic stress distribution near a notch. (a) in plane stress (σ_{33} = 0); (b) in plane strain (ϵ_{33} = 0)

In plane strain the stress distribution is altered markedly. The smallest principal stress is now σ_{22} (see Section 2.7) and the yield must consequently spread in the X_1X_2 plane, with:

$$\sigma_{11} - \sigma_{22} = 2\tau_Y = \sigma_Y$$

i.e.
$$\sigma_{11} = \sigma_Y + \sigma_{22} \tag{2.11.2}$$

For elastic loading, σ_{22} increases from zero at the free surface of the notch and it would therefore be expected that σ_{11} (and σ_{33}) would also increase. In fact, the rate of rise is much greater than would be expected from the elastic stress distribution shown in *Figure 2.7b*. This may be understood by further considering the deformation of elements lying along the X_2 axis, as in *Figure 2.7a*.

In normal elastic deformation, the stress σ_{22} is generated in order to maintain the continuity of elements a and b, b and c, etc., which would otherwise be disrupted by the incompatible lateral contraction strains: $\epsilon_{22} = -\dfrac{\nu}{1-\nu}\epsilon_{11} \simeq -0.4\epsilon_{11}$, for $\nu = 0.3$. However, if we suppose that elements a and b, for example, yield, even greater strains ϵ_{22} are produced, because plastic deformation must occur at constant volume (plastic deformation proceeds by slip processes, with no volume dilatation). If $\epsilon_{33} = 0$ (plane strain) the lateral strain ϵ_{22} must now be equal to $-\epsilon_{11}$ and so a much larger stress σ_{22} is generated. This in turn entails a large σ_{11} stress (equation 2.11.2). It is important to notice that the peak value of σ_{11} now occurs some distance below the notch, where σ_{22} has its maximum value (*Figure 2.11b*).

The solving of problems involving both elastic and plastic deformation around notches in plane strain becomes complicated, because both elastic and plastic compatibility must be satisfied; stress must be related to elastic strain via equations 2.6.7 and to the plastic strain increments (see Section 3.16).

It is usual to see analytical solutions rather for plastic/rigid deformation, where the elastic components are reduced to zero. A solution takes the form of a slip-line field which comprises two sets of orthogonal slip-lines (lines of constant shear stress, τ_Y, or, identically in isotropic material, lines of velocity discontinuities) which satisfy equilibrium, compatibility and boundary conditions.

For a circular hole, the slip-lines are logarithmic spirals and it is possible to write σ_{11} as a function of distance x measured from the edge of the hole, as:

$$\sigma_{11} = \sigma_Y\left[1 + \ln\left(1 + \frac{x}{a}\right)\right] \tag{2.11.3}$$

If we examine the expression for the radial stress, $\sigma_{rr} = \sigma_{22}$ near a circular hole, whilst the deformation is elastic, we have (equation 2.9.8)

$$\sigma_{rr} = \sigma_{22} = \frac{3}{2}\sigma\left(\frac{a^2}{r^2} - \frac{a^4}{r^4}\right) \tag{2.9.8}$$

The initial gradient, at $r = a$, is then:

$$\frac{d(\sigma_{rr})}{dr} = \frac{3\sigma}{a} \tag{2.11.4}$$

As the surface of the hole first yields, we may equate 3σ to the yield stress σ_Y. Hence, for a very small yielded zone of length δa we have:

$$\sigma_{22} = \sigma_Y \frac{\delta a}{a} \tag{2.11.5}$$

or, from equation 2.11.2,

$$\sigma_{11} = \sigma_Y + \sigma_{22} = \sigma_Y\left(1 + \frac{\delta a}{a}\right) \tag{2.11.6}$$

Similarly, if $x(= \delta a)$ is very small, equation 2.11.3 reduces to

$$\sigma_{11} = \sigma_Y\left[1 + \ln\left(1 + \frac{\delta a}{a}\right)\right] = \sigma_Y\left(1 + \frac{\delta a}{a}\right) \tag{2.11.7}$$

We therefore deduce that the increase of σ_{11} for a very small plastic zone is governed by the increase of the elastic value of σ_{22}.

As the zone grows bigger, the larger values of ϵ_{22} for plastic/elastic deformation may be expected to increase the value of σ_{22}, and hence σ_{11}, substantially above that for purely elastic deformation. In the plastic enclave, $\epsilon_{22} = -\epsilon_{11}$; in the elastic case, $\epsilon_{22} = -0.4\epsilon_{11}$. As a rough estimate, we may therefore suppose that the value of σ_{22} at $r = \sqrt{2}\,a$ (equation 2.9.8) is increased from $3\sigma/8$ to $3\sigma/8 \times 1/0.4 = \dfrac{7.5\sigma}{8}$. Equating 3σ to σ_Y as above, we have, approximately,

$$\sigma_{11} = \sigma_Y + \sigma_Y \frac{2.5}{8} = 1.31\sigma_Y \tag{2.11.8}$$

The accurate slip-line field equation (2.11.3) with $x = (\sqrt{2}-1)a$ gives $\sigma_{11} = \sigma_Y(1 + \ln 1.414) = 1.35\sigma_Y$. Again, these figures are used simply to demonstrate that the physical picture is reasonably compatible with the mathematical answer.

2.12 General Yield Loads

The introduction of slip-line fields to describe the form of the plastic zone around a circular notch brings us to an appropriate point for describing how the general yield (or limit) loads of notched or cracked specimens are calculated. For a cross-section bounded by two semi-circular notches of radius a, as in *Figure 2.12*, the tensile yield load in plane strain is given simply by integrating equation 2.11.3 across the notched width of the piece, to obtain the expression:

$$P_{GY} = 2\sigma_Y(W-a)B\left[1 + \frac{a}{(W-a)}\right]\ln\left[1 + \frac{(W-a)}{a}\right] \tag{2.12.1}$$

where B is the thickness, and $2W$ is the total width of the piece. The distributions of σ_{11} and σ_{22} are indicated (*Figure 2.12*). It can be seen that the maximum tensile stress and maximum hydrostatic stress are found in the centre of the specimen. Bridgman[4] has used this plane-strain distribution to predict the stress

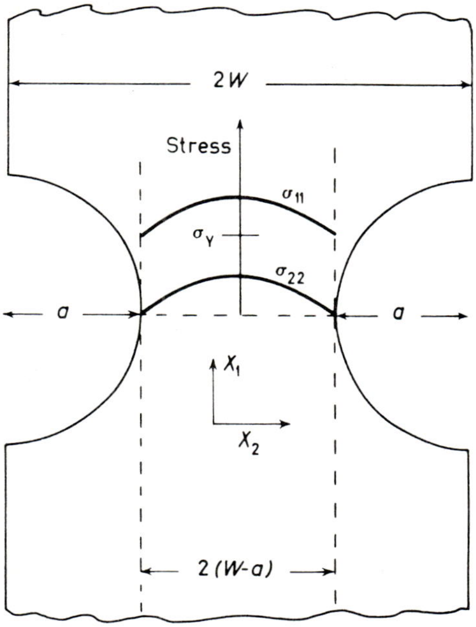

Figure 2.12 Distribution of longitudinal stress, σ_{11}, and transverse stress, σ_{22}, in an externally notched bar (plane strain deformation)

distribution in a necked tensile specimen, using a work-hardened flow stress, $\bar{\sigma}$, rather than the yield stress in equation 2.12.1. The ratio of general yield load, P_{GY}, to the load which would be required to yield a smooth specimen of the dimensions of the notched cross-section is called the *constraint factor*, L:

$$L = P_{GY}/\sigma_Y . 2(W-a)B \tag{2.12.2}$$

For the geometry indicated in *Figure 2.12*, L has a value of about 1.3. High values of L will be observed in practical test specimens only if the notches are sufficiently deep. With shallow notches, it may become possible to yield the gross cross-section (at a load $\sigma_Y\ 2W.B$) before the constrained yielding can cross the notched width (at a load P_{GY}). The critical ratio of notch depth to width $W/(W-a)$ to obtain fully constrained yielding in specimens containing exterior semi-circular notches is calculated to be two in plane strain[5].

The general yield load of a plane strain tensile specimen bounded by sharp cracks rather than semi-circular notches is best approached by a model due to Orowan[6], which treats the configuration as the reverse of a 'plastic punching' process: see *Figures 2.13a* and *2.13b*. In *Figure 2.13a*, the flow of material from under the punch round to the free surfaces can be seen to be accomplished by sliding round the slip-lines. The theory of slip-lines tells us that the pressure is related to the angle turned through by a line, e.g.

along an α line
$$p - 2\tau_Y\theta = \text{const.} \qquad (2.12.3)$$

At the free surface, $\theta = 0$ and the pressure is $2\tau_Y$. Thus, in the centre, where $\theta = \pi/2$, we obtain:

$$p = 2\tau_Y\left(1 + \frac{\pi}{2}\right) \qquad (2.12.4)$$

For the specimen in tension, we find that the general yield load, P_{GY}, is, similarly:

$$P_{GY} = 2\tau_Y\left(1 + \frac{\pi}{2}\right).2(W - a).B \qquad (2.12.5)$$

On the Tresca yield criterion, $\sigma_Y = 2\tau_Y$, on the Mises criterion[7], $\sigma_Y = \sqrt{3}\,.\,\tau_Y$

Hence
$$P_{GY} = 2.57\sigma_Y\,.\,2(W - a)\,.\,B \quad \text{(Tresca)}$$
$$P_{GY} = 2.96\sigma_Y\,.\,2(W - a)\,.\,B \quad \text{(Mises)} \qquad (2.12.6)$$

The constraint factor is often taken as 3. Again, the result holds only if the cracks are sufficiently deep and, for this geometry in plane strain, the critical ratio of $W/(W - a)$ which must be exceeded has been calculated as 9:1. This very high ratio is seldom achieved in practical testpieces.

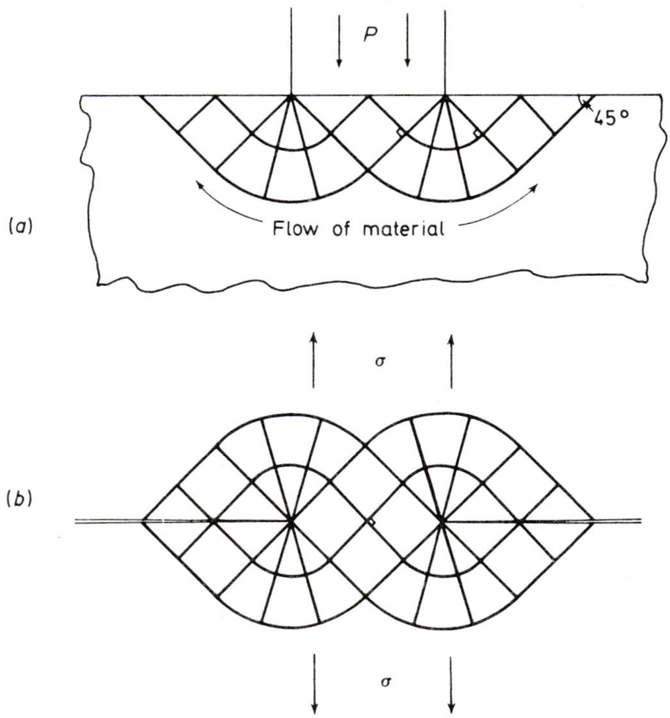

Figure 2.13 Orowan's model for general yield in a cracked specimen. (a) The 'plastic punch'; (b) the Prandtl field

It is important also to notice that, for this one configuration, the maximum tensile stress, σ_{11}, in the centre of the specimen is uniformly equal to $\sim 3\sigma_Y$ so that the elevation, or 'intensification', of tensile stress is equal to the constraint factor all the way across the cross-section. This is an exceptional situation: in all other constrained situations (e.g. *Figure 2.12*), σ_{11} varies across the section and stress intensification and constraint are clearly separate phenomena. In loose terms, constraint is the integrated effect of stress intensification across the cross-section.

The differences are clearly illustrated in pure bending. Here, constraint is referred to bending moments. The bending moment for general yield on the net section (*Figure 2.14a*), is given simply by:

$$M = 2\int_0^{(W-a)/2} 2\tau_Y By\,dy \qquad \text{(where } y \text{ is the height above the neutral axis)}$$

$$= \tau_Y . B . (W-a)^2/2$$

$$= \sigma_Y . B . (W-a)^2/4 \qquad (2.12.7)$$

assuming the Tresca criterion. Equation (2.12.7) has a form equivalent to that quoted in equation 1.3.2, and provides a lower bound solution for the notched case. An upper bound[8] may be obtained by calculating the rate of dissipating

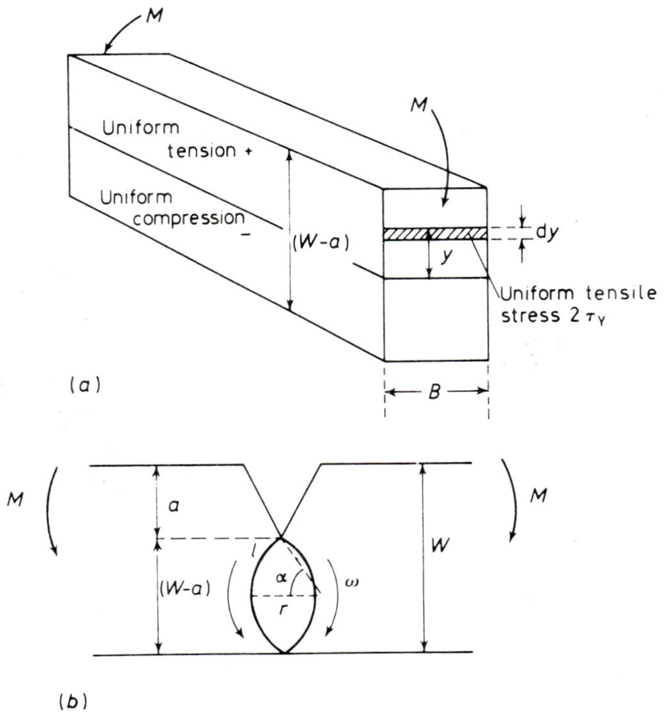

Figure 2.14 Lower and upper bounds for general yield load. (a) Lower bound stress field; (b) upper bound velocity field

energy for the situation in *Figure 2.14b* where the ends of the notched bar are supposed to pivot around circular arcs of length l and radius r at an angular velocity ω. Then, the rate of internal energy dissipation is given by $2\tau_Y l r \omega B$ and the rate of doing work by the applied couples is $2M\omega$. As an upper bound, we therefore obtain: $M = \tau_Y l r B$. In terms of the central angle 2α of the arcs, we set:

$$M_{(\text{upper bound})} = \tau_Y B \frac{(W-a)^2}{2} \operatorname{cosec}^2\alpha \, . \, \alpha \qquad (2.12.8)$$

To find the minimum value, we set $(dM/d\alpha) = 0$, i.e. $\tan\alpha = 2\alpha$. This is satisfied when $\alpha = 66°50'$, so that the upper bound to M is given by:

$$M = 0.69\tau_Y . B . (W-a)^2 \qquad (2.12.9)$$

M

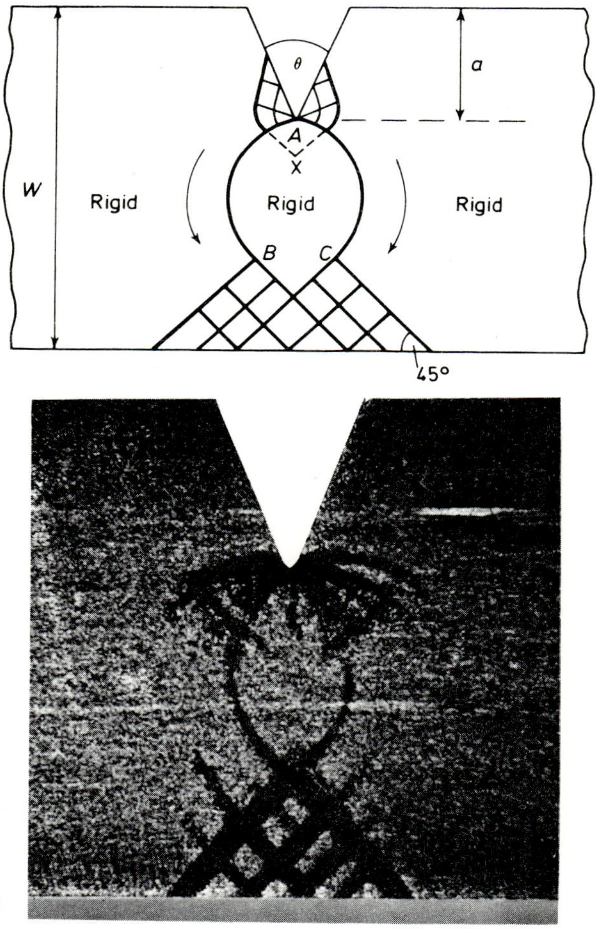

M

(a)

(b)

Figure 2.15 (a) Plane strain slip-line field for the general yield of a notched bar ($6.4° \leqslant \theta \leqslant 114.6°$). AB and AC are the 'plastic hinges'. (b) General yield deformation pattern revealed by etching in Fry's reagent (x6) (after Knott and Cottrell[11])

We therefore have limits on the constraint factor:

$$1 \leqslant L \leqslant 1.38 \tag{2.12.10}$$

The stress and velocity fields are made compatible by the choice of the appropriate slip-line field. Green's upper-bound slip-line field[9,10] shown in *Figure 2.15a* gives a value for L of 1.26 from the associated velocity field. Fry's etching of the yield zones shows good agreement with a field of this type[11] (*Figure 2.15b*). Lianis and Ford[12] have shown that a similar result is obtained from a lower bound stress field. The slip-line field shows that the bar rotates around the central rigid region on two 'plastic hinges'. The action is rather like that of a double ball-and-socket joint or of the bending of an elbow. For included notch angles greater than two radians, the central hinges vanish and the constraint factor then varies in what is virtually a linear manner with angle to the limiting

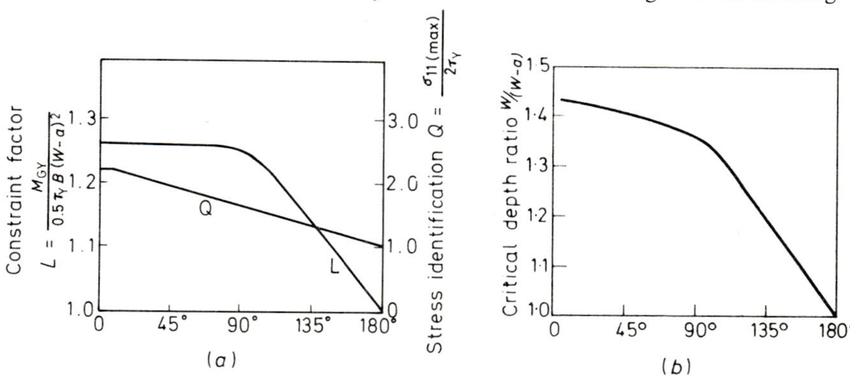

Figure 2.16 (a) Variation of constraint and stress intensification with included notch angle in notch-bending. (b) Variation of critical depth ratio with notch angle (after Knott[14])

value of unity for an unnotched bar (an angle, $\theta = \pi$), as shown in *Figure 2.16a*. For comparison, the critical notch/specimen depth ratio is shown in *Figure 2.16b* It can be seen that deeper notches are required to maintain constrained yielding when the constraint is high[13,14].

The maximum value of σ_{11} is found by extending the slip-lines around the notch into the rigid region of Green's field (*Figure 2.15a*, where the dotted lines represent this extension). Then at general yield, we have at point X:

$$\sigma_{11(\mathrm{max})} = 2\tau_{\mathrm{Y}}\left(1 + \frac{\pi}{2} - \frac{\theta}{2}\right) \tag{2.12.11}$$

cf. equations 2.12.3 and 2.12.4. This solution holds for all values of θ greater than 6.4° and predicts a linear variation of σ_{11} with θ, even in the range where L is virtually independent of θ (see *Figure 2.16a*). Constraint depends primarily on the far-reaching yield (the 'hinges'): stress intensification depends on the local situation near the notch.

The general situation in three-point bending is similar, but, here, shear forces are produced across the notched section. Also, the width of the indenter situated below the notch has an effect on constraint, because it affects the distribution of compression and shear stresses. The constraint factor for a Charpy specimen (45° 'V'-notch with 0.25 mm root radius; $W = 10$ mm, $B = 10$ mm, $(W - a) = 8$ mm, moment arm = 20 mm), has been calculated as 1.24 if the width of the indenter is taken as 0.5 mm[13]. An average constraint factor for acute notches in three-point bending is about 1.22; the value of $\sigma_{11(\max)}$ at general yield in a Charpy specimen is given by:

$$\sigma_{11(\max)} = 1.94 \,.\, 2\tau_Y \tag{2.12.12}$$

rather than

$$\sigma_{11(\max)} = 2.18 \,.\, 2\tau_Y \tag{2.12.13}$$

as calculated for the deep 45° V-notch bar in pure bending (*Figure 2.16a*).

The critical depth ratio, $\dfrac{W}{W - a}$, required to sustain constrained yielding in three-point bending has been calculated as 1.22 for a Charpy (45° V-notch) specimen. This contrasts with the value of 1.41 for pure bending[10,15]. The lower value is associated with a lower constraint value and with a lower value of $\sigma_{11(\max)}$.

It is possible to devise plane strain slip-line fields for cracked specimens in which the constraint factor is virtually unity. For example, the thick centre-cracked specimen, shown in *Figure 2.17*, with the associated slip-line field, has a general yield load given by:

$$P_{GY} = 2\tau_Y \,.\, 2(W - a)B \tag{2.12.14}$$

On our definition of constraint, L is unity. Only if one refers to the uniaxial yield stress as the reference and uses Mises yield criterion does the specimen appear to give any constraint. Then, L is $2/\sqrt{3}$.

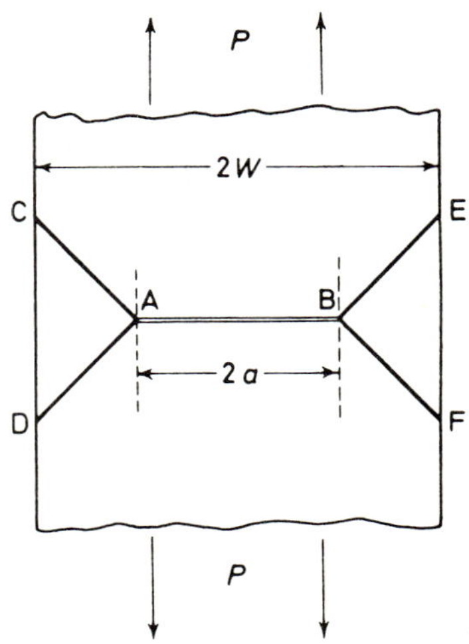

Figure 2.17 General yield slip-line field for a centrally-cracked specimen in plane strain. (Slip-lines are AC, AD, BE, BF)

In plane stress, again, apparent constraint exists only if the uniaxial yield stress is the reference. The plane stress slip-line field for acute notches in bending is shown in *Figure 2.18.* On the Mises criterion, L is 1.072 for all acute notches;

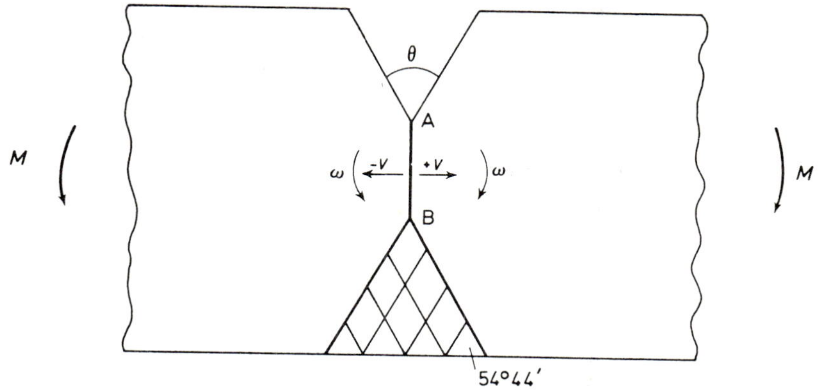

Figure 2.18 Plane stress slip-line field for the general yield of a notched bar ($0° \leqslant \theta \leqslant 141°\ 4'$) *AB is a local neck, across which there is a discontinuity of the normal component of the velocity along the neck (after Lianis and Ford*[16]*)*

the maximum value of σ_{11} is: $\sigma_{11} = 2\sigma_Y/\sqrt{3} = 2\tau_Y$. Using Tresca's criterion, L is unity and $\sigma_{11(max)}$ is equal to σ_Y. Similar values are obtained for L and $\sigma_{11(max)}$ in plane stress tension. Because the constraint factor is virtually unity, identical slip-line fields can be found for deep and shallow notches in plane stress.

Slip-line fields have also been devised for 'shallow' (less than the critical $\frac{W}{W-a}$ ratio) notches in plane strain. *Figure 2.19* shows a field for a shallow acute

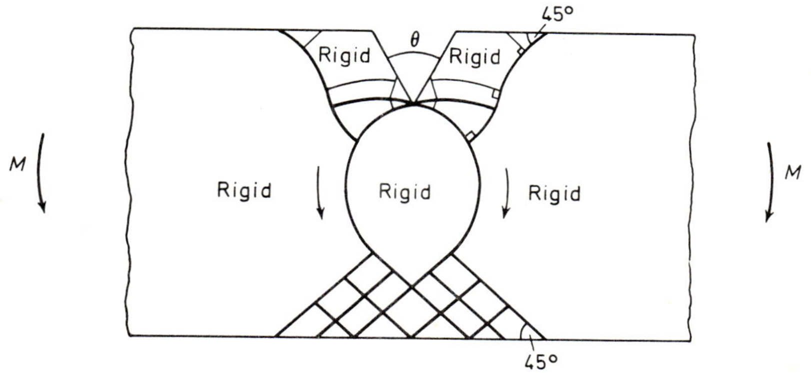

Figure 2.19 Plane strain general yield slip-line field for a shallow-notched bar ($6.4° \leqslant \theta \leqslant 114.6°$) *(after Ewing*[13]*)*

notch in bending[13]; *Table 2.2* gives the values of constraint factor and stress intensification ($Q = \sigma_{11(max)}/2\tau_Y$; see following section) for shallow notches of 45° included angle. Agreement with the theory is shown by the experimental values[14] quoted in *Table 2.2.* The field for shallow-notched tension specimens[17] is shown in *Figure 2.20* and the appropriate values of constraint factor and associated notch depth ratio are given in *Table 2.3*.

Limit loads are therefore well documented for most common testpiece geometries. In deep, double-notched tension testpieces, it is found that

Table 2.2. CONSTRAINT AND STRESS INTENSIFICATION IN SHALLOW-NOTCHED BARS IN BENDING

Notch depth ratio W/(W–a)	1.43	1.23	1.11
Constraint factor $L_{(theor)}$	1.26	1.21	1.14
Constraint factor $L_{(exp)}$*	1.26	1.17	1.08
Stress intei sification $Q_{(theor)}$	2.18	2.08	1.70
Stress intensification $Q_{(exp)}$*	2.18	2.04	1.79

* The experimental values for shallow notches assume the theoretical solution for deep notches.

$\sigma_{11(max)} = L.\sigma_Y$. This is, however, a consequence of the slip-line field for this type of testpiece. It is of importance to distinguish clearly between the general elevation of applied load by the triaxial stresses (L) and the local effect near the tip of the stress concentrator, which can raise the tensile stress, σ_{11}, to a value substantially greater than the yield stress in uniaxial tension. We shall define a

Table 2.3a CONSTRAINT FACTORS FOR SLIT NOTCHES ($\theta = 0$) AS A FUNCTION OF NOTCH DEPTH RATIO

$W/(W-a)$	8.62	7.06	5.77	4.69	4.34
L	2.57	2.39	2.22	2.05	1.98

(b) CRITICAL NOTCH DEPTH RATIOS $W/(W-a)$ FOR DOUBLE-NOTCHED TENSION SPECIMENS

Total notch angle (θ)	*Deep notch constraint factor (L)*	*Critical ratio [W/(W–a)]*
0	2.57	8.62
20	2.40	7.08
40	2.22	5.79
60	2.05	4.70
80	1.87	3.79
100	1.70	3.02
120	1.52	2.38
140	1.35	1.84
160	1.17	1.38
180	1	1

parameter, Q, as the *stress intensification*: $Q = \sigma_{11(max)}/2\tau_Y$. In the following section, we examine how Q depends on the size of the plastic zone, before general yield. Tresca's yield criterion is assumed, so that $2\tau_Y$ may be equated to the uniaxial yield stress, σ_Y.

2.13 Stress Intensification

The increase of $\sigma_{11\,(\text{max})}$ with extent of plastic zone, d_Y, ahead of a semi-circular notch of radius a is given from equation 2.11.3 as

$$\sigma_{11\,(\text{max})} = \sigma_Y \left[1 + \ln\left(1 + \frac{d_Y}{a}\right)\right] \tag{2.13.1}$$

The deformations producing this elevation of tensile stress were discussed in Section 2.11.

For acute 'V'-notches of the types used in fracture testpieces, it has not proved possible to devise simple slip-line fields for local yield zones, but detailed studies have been made for the 45° V-notch, containing a root radius of 0.25 mm: the

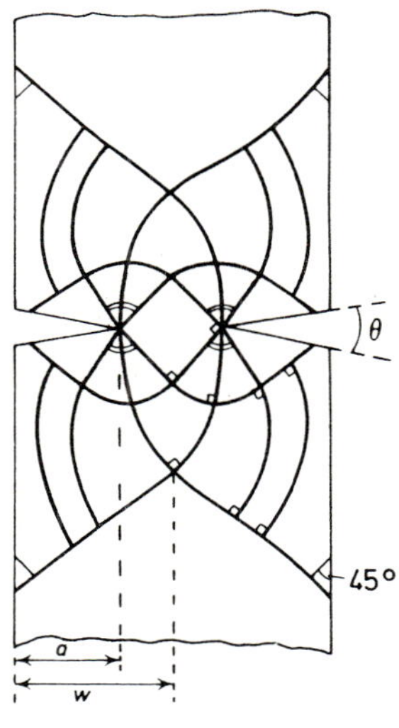

Figure 2.20 Lower-bound slip-line field for the general yield of a shallow-notched bar in plane strain tension (after Ewing and Hill[17])

particular geometry used in Izod and Charpy specimens (see Section 1.7). A simple approach[18] starts by calculating the elastic stress concentration factor of the notch, to determine the load at which yielding begins. The plastic zone is then assumed to spread as the logarithmic spiral slip-line field from the notch root radius. Eventually, as the applied load is increased, the extreme slip lines meet the straight sides of the notch. Although the zone grows larger with further increases in load, the angle turned through by the slip-lines remains constant, so that there is no corresponding increase in the value of $\sigma_{11\,(\text{max})}$. The situation is depicted in *Figure 2.21a* and *2.21b*. The highest value of $\sigma_{11\,(\text{max})}$ is of course identical to that given by the general yield slip-line field. The discontinuous nature of the graph of stress intensification $[Q = \sigma_{11\,(\text{max})}/\sigma_Y]$ versus applied load/general yield load (P/P_{GY}) seems rather unreasonable in terms of the physical model of incompatible deformations, which tends to indicate that σ_{22}, and hence σ_{11}, should continue to increase with plastic zone size.

Detailed calculations have been made for a more realistic elastic/plastic material for a rather similar specimen geometry, using finite element analysis, and will be discussed in Section 3.18. In general, the curve agrees with that predicted

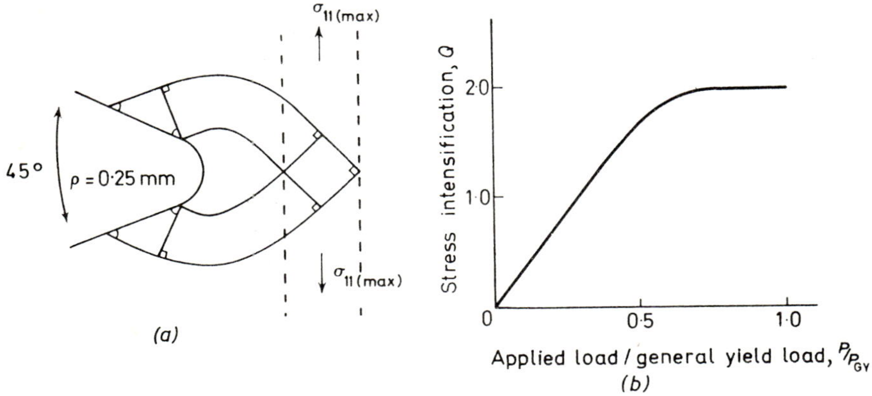

Figure 2.21 Stress intensification in Charpy specimen. (a) Notch tip slip-line field; (b) variation of Q with applied load (after Wilshaw, Rau and Tetelman[18]) using Tresca's yield criterion

by the slip-line field method at low applied loads, but at higher loads, it shows a continuing, although small, increase of $\sigma_{11\,(\max)}$ with applied load.

Similar results have also been obtained analytically for a tensile specimen containing two deep external notches possessing hyperbolic profiles[19].

No analytical solution for the increase in Q with extent of yielding has been developed for sharp cracks in plane strain deformation. Results obtained, using finite element techniques, are also discussed in Section 3.18.

We shall discover, in Chapter 7, that the magnitude of $\sigma_{11\,(\max)}$ has a critical role to play in the occurrence of brittle cracking ahead of a stress concentrator. For a given yield stress, $\sigma_{11\,(\max)}$ is determined by the value of Q, which is, in turn, a function of the amount of plastic deformation present in a testpiece. The interrelation of local fracture events [reacting to $\sigma_{11\,(\max)}$] and macroscopic ductility depends critically on the way in which Q increases with applied load. The detailed computation of curves such as *Figure 2.21b* therefore represents a vital step in this interrelation.

2.14 Conclusions

It has been shown how the development of elasticity theory, involving stress equilibrium, compatibility and stress–strain relations, leads to a final equation in terms of the second derivatives of the stresses σ_{11} and σ_{22}, in plane deformation. Elastic problems are usually solved using an appropriate stress function, which must satisfy all boundary conditions. The derivation of the solution for the stress distribution around a circular hole not only provides an example of the method but also gives a useful starting-point, from which many interesting semi-quantitative predictions may be made.

We have discussed situations in which some yielding has occurred, both to show how general yield loads (limit loads) are calculated and to draw attention to the most important 'intensification' of the tensile stress, $\sigma_{11(\max)}$ ahead of the stress concentrator.

In the following chapter, we shall study the stress distributions near stress concentrators of crack-like geometry.

REFERENCES

1. Frocht, M. M., *Photoelasticity*, Wiley, New York (1941)
2. Airy, G. B., *Brit. Assoc. Advance., Sci. Rept.* (1862)
3. Tresca, H., *Mém. Sci. Savants Acad. Sci., Paris,* **18**, 733 (1868); **20**, 75, 281 (1872)
4. Bridgman, P. W., *Studies in Large Plastic Flow and Fracture*, McGraw-Hill, New York (1952)
5. Ewing, D. J. F., *J. Mech. Phys. Solids*, **16**, 81 (1968)
6. Orowan, E., *Trans. Inst. Engrs Shipbuilders Scotland*, **89**, 165 (1945)
7. von Mises, R., *Gottinger Nachrichten, math.-phys. Klasse*, 582 (1913)
8. Prager, W., 'James Clayton Memorial Lecture', *Inst. Mech. Engrs*, 14th Jan. (1955)
9. Green, A. P., *Q. Jl. Mech. appl. Maths*, **6**, 223 (1953)
10. Green, A. P. and Hundy, B. B., *J. Mech. Phys. Solids*, **4**, 128 (1956)
11. Knott, J. F. and Cottrell, A. H., *J. Iron Steel Inst.*, **201**, 249 (1963)
12. Lianis, G. and Ford, H., *J. Mech. Phys. Solids*, **7**, 1 (1958)
13. Ewing, D. J. F., *J. Mech. Phys. Solids*, **16**, 205 (1968)
14. Knott, J. F., 'Fracture 1969', *Proceedings of the Second International Conference on Fracture*, Brighton, 205, Chapman and Hall, London (1969)
15. Green, A. P., *J. Mech. Phys. Solids*, **6**, 259 (1956)
16. Lianis, G. and Ford, H., *J. appl. Maths. Phys.* (ZAMP), **8**, 360 (1957)
17. Ewing, D. J. F. and Hill, R., *J. Mech. Phys. Solids*, **15**, 115 (1967)
18. Wilshaw, T. R., Rau, C. A. and Tetelman, A. S., *Engineering Fracture Mechanics*, **1**, 191 (1968)
19. Hendrickson, J. A., Wood, D. S. and Clark, D. S., *Trans. Am. Soc. Metals*, **51**, 629 (1959)

Three

COMPLEX FUNCTIONS AND STRESSES AROUND CRACKS

3.1 Introduction

In essence, the stress function method of solving elasticity problems, described in the previous chapter, has involved the finding of a suitable algebraic or trigonometrical function of two variables (x_1, x_2 or r, θ) which will satisfy the compatibility equation $\nabla^2(\nabla^2\phi) = 0$ and from which stresses which satisfy the boundary conditions can be obtained. To extend the method so that it can be used for crack-like geometries, it is mathematically convenient to express the stress function as a complex function of two variables, as described below.

3.2 Complex Variables

We make use of a complex number z, which may be expressed in Cartesian co-ordinates x_1 and x_2 as

$$z = x_1 + \mathrm{i}x_2 \tag{3.2.1}$$

or, in polar co-ordinates, by $z = r\mathrm{e}^{\mathrm{i}\theta}$, where $\mathrm{i} = \sqrt{-1}$ and x_1, x_2 or r, θ are real numbers. An analytic function f(z) is one whose derivatives depend on z only, being the same for all directions (of dz) at the point z. It has an indefinite integral, defined as the function having f(z) as its derivative with respect to z, and written $\int \mathrm{f}(z)\,\mathrm{d}z$. It is possible to regard f(z) as having partial derivatives with respect to both x_1 and x_2:

$$\begin{aligned} \frac{\partial}{\partial x_1}\mathrm{f}(z) &= \frac{\partial}{\partial z}\mathrm{f}(z)\frac{\partial z}{\partial x_1} = \mathrm{f}'(z)\frac{\partial z}{\partial x_1} = \mathrm{f}'(z) \\ \frac{\partial}{\partial x_2}\mathrm{f}(z) &= \mathrm{f}'(z)\frac{\partial z}{\partial x_2} = \mathrm{i}\mathrm{f}'(z) \end{aligned} \tag{3.2.2}$$

If f(z) is written in the form

$$\mathrm{f}(z) = \alpha + \mathrm{i}\beta \tag{3.2.3}$$

where α and β are real functions of x_1 and x_2, we have:

$$\frac{\partial f(z)}{\partial x_1} = \frac{\partial \alpha}{\partial x_1} + i\frac{\partial \beta}{\partial x_1} = f'(z)$$
$$\frac{\partial f(z)}{\partial x_2} = \frac{\partial \alpha}{\partial x_2} + i\frac{\partial \beta}{\partial x_2} = if'(z) \tag{3.2.4}$$

$$i\left(\frac{\partial \alpha}{\partial x_1} + i\frac{\partial \beta}{\partial x_1}\right) = \frac{\partial \alpha}{\partial x_2} + i\frac{\partial \beta}{\partial x_2} \tag{3.2.5}$$

Equating real and imaginary parts, remembering that α and β are real and that $i^2 = -1$, gives

$$\frac{\partial \alpha}{\partial x_1} = \frac{\partial \beta}{\partial x_2} \qquad \frac{\partial \alpha}{\partial x_2} = -\frac{\partial \beta}{\partial x_1} \tag{3.2.6}$$

These are called the *Cauchy–Riemann* equations. We may eliminate β by differentiating the first equation with respect to x_1; the second with respect to x_2; and adding. This gives:

$$\frac{\partial^2 \alpha}{\partial x_1^2} + \frac{\partial^2 \alpha}{\partial x_2^2} = 0 \qquad \text{or} \qquad \nabla^2 \alpha = 0 \tag{3.2.7}$$

which is *Laplace's* equation; ∇^2 is known as the Laplacian operator. A solution to Laplace's equation is called an harmonic function. We could also have eliminated α from the Cauchy–Riemann equations to obtain:

$$\frac{\partial^2 \beta}{\partial x_1^2} + \frac{\partial^2 \beta}{\partial x_2^2} = \nabla^2 \beta = 0 \tag{3.2.8}$$

Thus, both the real and imaginary parts of any analytic function (derivative depending on z only) will separately provide solutions to Laplace's equation. The functions α and β are said to be conjugate harmonic functions.

Now, if ψ is any function of x_1 and x_2, we have, by differentiation:

$$\left(\frac{\partial^2}{\partial x_1^2} + \frac{\partial^2}{\partial x_2^2}\right)(x_1\psi) = x_1\left(\frac{\partial^2 \psi}{\partial x_1^2} + \frac{\partial^2 \psi}{\partial x_2^2}\right) + 2\frac{\partial \psi}{\partial x_1} \tag{3.2.9}$$

If ψ is harmonic, the term $\left(\frac{\partial^2 \psi}{\partial x_1^2} + \frac{\partial^2 \psi}{\partial x_2^2}\right)$ is zero. Also, $(\partial\psi/\partial x_1)$ is an harmonic function, because

$$\left(\frac{\partial^2}{\partial x_1^2} + \frac{\partial^2}{\partial x_2^2}\right)\left(\frac{\partial \psi}{\partial x_1}\right) = \frac{\partial}{\partial x_1}\left(\frac{\partial^2 \psi}{\partial x_1^2} + \frac{\partial^2 \psi}{\partial x_2^2}\right) = 0 \tag{3.2.10}$$

So, if we apply the Laplacian operator $\nabla^2 \equiv (\partial^2/\partial x_1^2 + \partial^2/\partial x_2^2)$ again to equation 3.2.9, we obtain

$$\nabla^2[\nabla^2(x_1\psi)] = \nabla^2\left(0 + 2\frac{\partial \psi_1}{\partial x_1}\right) = 0$$

i.e.
$$\nabla^4(x_1\psi) \equiv \left(\frac{\partial^4}{\partial x_1^4} + 2\frac{\partial^4}{\partial x_1^2 \partial x_2^2} + \frac{\partial^4}{\partial x_2^4}\right)(x_1\psi) = 0 \tag{3.2.11}$$

This equation is identical to the compatibility equation (2.9.2) and so comparison shows that $x_1\psi$ may be used as a *stress function*, when ψ is harmonic. By similar methods, it is possible to show that both $x_2\psi$ and $(x_1^2 + x_2^2)\psi \equiv r^2\psi$ can also be used as stress functions, as, of course, can ψ itself.

In general, it can be shown that *any* stress function may be expressed in the form

$$\phi = \text{Re}\ [(x_1 - \text{i}x_2)\,\psi(z) + \chi(z)] \tag{3.2.12}$$

where 'Re' means 'real part of' and $\psi(z)$ and $\chi(z)$ are suitably chosen analytic functions. Conversely, equation 3.2.12 gives a stress function (i.e. an admissible solution for the compatibility equation) for any choice of $\psi(z)$ and $\chi(z)$. The analytic functions $\psi(z)$ and $\chi(z)$ are termed complex potentials and provide a very convenient way of determining stresses and displacements from the stress function.

3.3 Stresses and Displacements Determined from Complex Potentials

It is necessary to define one further function of a complex variable. This is the *conjugate function* (not to be confused with conjugate harmonic functions), defined as follows. If $\mathrm{f}(z)$ is a complex function, it may be expressed as $\mathrm{f}(z) = \alpha + \mathrm{i}\beta$, where both α and β are real. By the conjugate function, written $\bar{\mathrm{f}}(\bar{z})$ we mean the value taken by $\mathrm{f}(z)$, when i is replaced, wherever it occurs in $\mathrm{f}(z)$, by $-$i. Thus,

if $\quad \mathrm{f}(z) = \alpha + \mathrm{i}\beta, \quad \bar{\mathrm{f}}(\bar{z}) = \alpha - \mathrm{i}\beta,$

i.e. $\quad$ if $\quad \mathrm{f}(z) = \mathrm{e}^{\mathrm{i}z}, \quad \bar{\mathrm{f}}(\bar{z}) = \mathrm{e}^{-\mathrm{i}\bar{z}} = \mathrm{e}^{-\mathrm{i}(x_1 - \mathrm{i}x_2)} = \mathrm{e}^{-\mathrm{i}x_1}\,.\,\mathrm{e}^{-x_2}$

Note that $\mathrm{f}(\bar{z})$ would be $\mathrm{e}^{\mathrm{i}\bar{z}} = \mathrm{e}^{\mathrm{i}(x_1 - \mathrm{i}x_2)} = \mathrm{e}^{\mathrm{i}x_1}\,.\,\mathrm{e}^{x_2}$.

It is apparent that $\mathrm{f}(z) + \bar{\mathrm{f}}(\bar{z}) = 2\alpha = 2\ \text{Re}\ \mathrm{f}(z)$. Similarly, we may rewrite equation 3.2.12:

$$\phi = \text{Re}[(x_1 - \text{i}x_2)\,\psi(z) + \chi(z)] = \text{Re}[\bar{z}\psi(z) + \chi(z)] \tag{3.3.1}$$

and write

$$2\phi = [\bar{z}\psi(z) + z\bar{\psi}(\bar{z}) + \chi(z) + \bar{\chi}(\bar{z})] \tag{3.3.2}$$

If we obtain $\partial\phi/\partial x_1$ and $\mathrm{i}\partial\phi/\partial x_2$ by partial differentiation, add, and collect terms, we have:

$$\frac{\partial\phi}{\partial x_1} + \mathrm{i}\frac{\partial\phi}{\partial x_2} = \psi(z) + z\bar{\psi}'(\bar{z}) + \bar{\chi}'(\bar{z}) \tag{3.3.3}$$

To obtain σ_{11}, σ_{22} and σ_{12}, we differentiate again with respect to x_1 or x_2 ($\sigma_{11} = \partial^2\phi/\partial x_2{}^2$, etc., equation 2.9.1) add as necessary and produce:

$$\sigma_{11} + \sigma_{22} = 4 \text{ Re } \psi'(z)$$

$$\sigma_{22} - \sigma_{11} + 2\mathrm{i}\sigma_{12} = 2[\bar{z}\psi''(z) + \chi''(z)] \tag{3.3.4}$$

Separation of the real and imaginary parts of this last equation enables $(\sigma_{22} - \sigma_{11})$ and σ_{12} to be determined separately. It is therefore possible to determine the stresses in a relatively straightforward manner from the complex potentials which form the stress function. It is also possible to determine displacements from these potentials. If the displacement in the x_1 direction is u_1 and in the x_2 direction is u_2, the relationships are

$$2\mu(u_1 + \mathrm{i}u_2) = \left(\frac{3-\nu}{1+\nu}\right)\psi(z) - z\bar{\psi}'(\bar{z}) - \bar{\chi}'(\bar{z}) \text{ in plane stress} \tag{3.3.5}$$

$$2\mu(u_1 + \mathrm{i}u_2) = (3 - 4\nu)\,\psi(z) - z\bar{\psi}'(\bar{z}) - \bar{\chi}'(\bar{z}) \text{ in plane strain} \tag{3.3.6}$$

The individual values of u_1 and u_2 may be obtained by equating real and imaginary parts.

We now examine how complex potentials are used to determine stress distributions around stress concentrators of crack-like geometry, and consider first the elliptical notch. Use is made of the curvilinear co-ordinate system described earlier, in Section 2.3.

3.4 Curvilinear Co-ordinates

For a curvilinear co-ordinate system, we choose a complex variable $p = \alpha + \mathrm{i}\beta$ where α and β are co-ordinates in the curvilinear system. The relationship between p and the variable $z = x_1 + \mathrm{i}x_2$ in a Cartesian co-ordinate system is given by

$$z = c \cosh p \tag{3.4.1}$$

where c is a constant. If the real and imaginary parts of each side are separated, i.e.

$$x_1 + \mathrm{i}x_2 = \frac{c}{2}\,(\mathrm{e}^{\alpha + \mathrm{i}\beta} + \mathrm{e}^{-\alpha - \mathrm{i}\beta})$$

$$= \frac{c}{2}\,(\mathrm{e}^{\alpha}\cos\beta + \mathrm{i}\,\mathrm{e}^{\alpha}\sin\beta + \mathrm{e}^{-\alpha}\cos\beta - \mathrm{i}\,\mathrm{e}^{-\alpha}\sin\beta)$$

$$= c\,(\cosh\alpha\cos\beta + \mathrm{i}\sinh\alpha\sin\beta)$$

we obtain: $$x_1 = c\cosh\alpha\cos\beta, \quad x_2 = c\sinh\alpha\sin\beta \tag{3.4.2}$$

by equating real and imaginary parts.

If we eliminate β we obtain:

$$\frac{x_1{}^2}{\cosh^2\alpha} + \frac{x_2{}^2}{\sinh^2\alpha} = c^2 \tag{3.4.3}$$

For a constant value of α this equation represents an ellipse drawn in the X_1X_2 plane. Similarly, lines of constant β produce a series of confocal hyperbolae which intersect the ellipses at right angles (see *Figure 2.5*). A stress $\sigma_{\alpha\beta}$ is defined as acting on a face whose normal is orthogonal to the tangent to a curve $\alpha =$ constant, in the direction of the normal to the curve $\beta =$ constant. If the anti-clockwise angle between x_1 and $\hat{n}_\alpha$ is θ (*Figure 3.1a*) we have the equilibrium shown in *Figure 3.1b*.

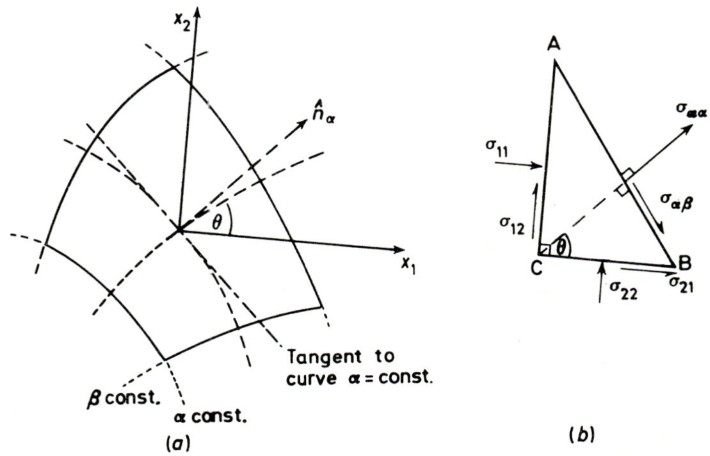

Figure 3.1 (a) Relationship of elliptical to Cartesian co-ordinates. (b) Relationship between components of stress (see also Figure 2.5)

The force parallel to $\hat{n}_\alpha$ is given by:

$$\sigma_{\alpha\alpha}\,\mathrm{AB} = \sigma_{11}\,\mathrm{AB}\cos^2\theta + \sigma_{22}\,\mathrm{AB}\sin^2\theta + 2\sigma_{12}\,\mathrm{AB}\sin\theta\cos\theta$$

i.e.

$$\sigma_{\alpha\alpha} = \sigma_{11}\cos^2\theta + \sigma_{22}\sin^2\theta + 2\sigma_{12}\sin\theta\cos\theta \tag{3.4.4}$$

Similarly it may be shown that:

$$\begin{aligned}\sigma_{\beta\beta} &= \sigma_{11}\sin^2\theta + \sigma_{22}\cos^2\theta - 2\sigma_{12}\sin\theta\cos\theta\\ \sigma_{\alpha\beta} &= (\sigma_{22} - \sigma_{11})\sin\theta\cos\theta + \sigma_{12}(\cos^2\theta - \sin^2\theta)\end{aligned} \tag{3.4.5}$$

These relationships simply provide a particular case of the general transformation of stress components, when referred to a new pair of orthogonal axes, inclined at an angle θ to OX_1 and OX_2, in two-dimensional deformation.

If these expressions are written in terms of functions of 2θ, i.e.

$$\begin{aligned}\cos^2\theta &= \tfrac{1}{2}(1 + \cos 2\theta)\\ \sin^2\theta &= \tfrac{1}{2}(1 - \cos 2\theta)\\ 2\sin\theta\cos\theta &= \sin 2\theta\end{aligned} \tag{3.4.6}$$

we may add terms and re-write the expressions to give:

$$\begin{aligned} \sigma_{\alpha\alpha} + \sigma_{\beta\beta} &= \sigma_{11} + \sigma_{22} \\ \sigma_{\beta\beta} - \sigma_{\alpha\alpha} + 2i\sigma_{\alpha\beta} &= (\sigma_{22} - \sigma_{11} + 2i\sigma_{12})e^{2i\theta} \end{aligned} \tag{3.4.7}$$

remembering that $e^{2i\theta} = \cos 2\theta + i \sin 2\theta$. The sum $(\sigma_{\alpha\alpha} + \sigma_{\beta\beta}) = (\sigma_{11} + \sigma_{22})$ is, in fact, the first invariant of the stress tensor (equation 2.5.6) in two dimensions.

The advantage of an expression in this form is that it is relatively easy to relate $e^{2i\theta}$ to the derivatives of $f(z)$ and $\bar{f}(\bar{z})$. We have

$$e^{2i\theta} = \frac{f'(z)}{\bar{f}'(\bar{z})} = \frac{\sinh p}{\sinh \bar{p}} \tag{3.4.8}$$

In terms of the complex potentials, the stresses are given by:

$$\begin{aligned} \sigma_{\alpha\alpha} + \sigma_{\beta\beta} &= 2[\psi'(z) + \bar{\psi}'(\bar{z})] = 4 \operatorname{Re} \psi'(z) \\ \sigma_{\beta\beta} - \sigma_{\alpha\alpha} + 2i\sigma_{\alpha\beta} &= 2e^{2i\theta} [\bar{z}\psi''(z) + \chi''(z)] \end{aligned} \tag{3.4.9}$$

see equation 3.3.4. We can now examine the solution for the stress distribution around an elliptical hole subjected to a uniform stress σ.

3.5 The Elliptical Hole in a Uniformly Stressed Plate: the Inglis Solution[1]

The hole is assumed to have semi-major axis a, semi-minor axis b, and to lie in an infinite plate with its major axis normal to a uniform tensile stress σ, as indicated in *Figure 3.2.* For these semi-axes we may write

$$c \cosh \alpha_0 = a, \quad c \sinh \alpha_0 = b$$

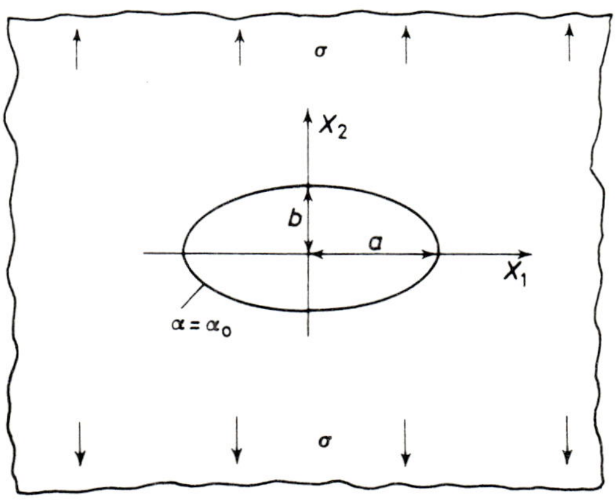

Figure 3.2 The elliptical hole in an infinite plate. (Note that the uniform stress, σ, is of the form σ_{22})

and therefore obtain for the equation of the ellipse, using equation 3.4.3

$$\frac{x_1^2}{a^2}+\frac{x_2^2}{b^2}=1 \tag{3.5.1}$$

In the limit, as $\alpha_0 \to 0$, the ellipse becomes a crack of length $2c = 2a$. When $a = b$, the ellipse becomes a circle.

A point goes once round the ellipse as β varies from 0 to 2π, so the continuity of stress and displacement demands that they be periodic also in β with period 2π, so that they have the same value at 0 and at 2π.

At infinity, suppose that $\sigma_{22} = \sigma$, $\sigma_{11} = \sigma_{12} = 0$, so that

$$\left.\begin{aligned} 4\,\mathrm{Re}\,\psi'(z) &= \sigma \\ 2[\bar{z}\psi''(z) + \chi''(z)] &= \sigma \end{aligned}\right\} \text{ at infinity} \tag{3.5.2}$$

At the boundary of the hole $\alpha = \alpha_0$, $\sigma_{\alpha\alpha} = \sigma_{\beta\alpha} = 0$.

Inglis[1] found complex potentials which satisfied these boundary conditions and which were periodic in β with period 2π, as follows:

$$4\psi(z) = \sigma c[(1 + e^{2\alpha_0})\sinh p - e^{2\alpha_0}\cosh p]$$
$$4\chi(z) = -\sigma c^2\left[(\cosh 2\alpha_0 - \cosh \pi)p + \frac{1}{2}e^{2\alpha_0} - \cosh 2\left(p - \alpha_0 - i\frac{\pi}{2}\right)\right] \tag{3.5.3}$$

Since $\sigma_{\alpha\alpha}$ is zero at the surface of the hole, we may obtain $\sigma_{\beta\beta}$ on the surface, from equation 3.4.9 as

$$\sigma_{\beta\beta(\alpha=\alpha_0)} = \frac{\sinh(2\alpha_0) - 1 + e^{2\alpha_0}\cos(2\beta)}{\cosh(2\alpha_0) - \cos(2\beta)}\,\sigma \tag{3.5.4}$$

The maximum values of $\sigma_{\beta\beta}$ correspond to those at the ends of the major axis where $\beta = 0\ (2\pi)$ or π. Here, $\cos(2\beta) = 1$ and we obtain for $\sigma_{\beta\beta}$, which is now equal to σ_{22} because the tangent to the ellipse is parallel to X_2:

$$\sigma_{\beta\beta(\beta=\substack{0\\\pi})} \equiv \sigma_{22} = \frac{\sinh(2\alpha_0) - 1 + e^{2\alpha_0}}{\cosh(2\alpha_0) - 1}\,\sigma \tag{3.5.5}$$

Since $c^2 = (a^2 - b^2)$; from equations 3.4.3 and 3.5.1, we write:

$$\sinh 2\alpha_0 = \frac{2ab}{c^2}, \qquad \cosh 2\alpha_0 = \frac{a^2+b^2}{c^2}$$

and obtain, at the crack tip:

$$\sigma_{\beta\beta(\beta=\substack{0\\\pi})} \equiv \sigma_{22} = \sigma\left(1 + 2\frac{a}{b}\right) \tag{3.5.6}$$

This increases without limit as the hole becomes longer and more slender; and becomes identical to the result given in Section 2.9 for a circular hole, when $a = b$: $\sigma_{\beta\beta} = 3\sigma$.

Taking the 'radius of curvature' at the tip of the ellipse (i.e. the radius of the circle which passes through the tip and two adjacent points) to be ρ, we have, by definition:

$$\rho^2 = \frac{(dx_1{}^2 + dx_2{}^2)^3}{(dx_1 d^2x_2 - dx_2 d^2x_1)^2} \tag{3.5.7}$$

where

$$\frac{x_1{}^2}{a^2} + \frac{x_2{}^2}{b^2} = 1$$

as for equation 3.5.1.

If we let $x_1 = a \cos t$, $x_2 = b \sin t$ it is possible to substitute in equation 3.5.7 to obtain:

$$\rho^2 = \frac{1}{a^2 b^2} (a^2 \sin^2 t + b^2 \cos^2 t)^3 \tag{3.5.8}$$

At the tip of the ellipse, $x_2 = 0, x_1 = \pm a$, i.e. $\sin t = 0$, $\cos t = 1$. We obtain, at the tip $\rho = b^2/a$ and may finally substitute in equation 3.5.6 above to produce the stress at the tip as

$$\sigma_{\beta\beta} = \sigma \left(1 + 2\sqrt{\frac{a}{\rho}}\right) \tag{3.5.9}$$

Similar methods have been used by Neuber[2] to calculate stress concentrations near a variety of internal and external notch profiles which approximate to the elliptical or hyperbolic form. Such solutions, substantiated by results obtained using photo-elastic models, provide valuable information for engineers who wish to calculate permissible design stresses in pieces containing stress concentrators. The information is used mainly to allow for effects of fillets or keyways in shafts subjected to fatigue stresses and is tabulated in several handbooks, e.g. Peterson[3].

The application of equation 3.5.9 to fracture in brittle solids will be discussed further in Section 4.3. Attention is also drawn to Section 2.10, where the present result for an elliptical hole is compared with a rough estimate of the stress concentration produced by a series of holes.

3.6 Stresses near a Crack: Westergaard's Solution[4]

The solution for the stresses near an elliptical hole was obtained by Inglis by writing down the equations for stresses in terms of complex potentials; writing further equations to specify limits in terms of boundary conditions; and then using a knowledge of complex functions and their properties to find the appropriate expression which satisfied all requirements.

The Westergaard treatment of stresses near a sharp crack[4] is best approached by examining the properties of a particular type of complex function and seeing what sort of boundary conditions would be compatible with these properties.

Consider a function $\phi(z)$, which is harmonic, and denote the first and second derivatives of $\phi(z)$, and the first and second integrals with respect to z by $\phi'(z)$, $\phi''(z)$, $\bar{\phi}(z)$ and $\bar{\bar{\phi}}(z)$ respectively. We are intending to devise some expressions for stresses which satisfy the appropriate equations, and hence for a stress function, Φ, which satisfies the biharmonic equation 2.9.2. We then examine the boundary conditions implied.

Let
$$\Phi = \text{Re}\,\bar{\bar{\phi}}(z) + x_2\,\text{Im}\,\bar{\phi}(z) \tag{3.6.1}$$

Then
$$\sigma_{22} = \frac{\partial^2\Phi}{\partial x_1{}^2} = \frac{\partial}{\partial z}\left(\frac{\partial\Phi}{\partial z}\right) = \text{Re}\,\phi(z) + x_2\,\text{Im}\,\phi'(z)$$
$$\sigma_{11} = \frac{\partial^2\Phi}{\partial x_2{}^2} = \text{Re}\,\phi(z) - x_2\,\text{Im}\,\phi'(z) \tag{3.6.2}$$
$$\sigma_{12} = \frac{-\partial^2\Phi}{\partial x_1\partial x_2} = -x_2\,\text{Re}\,\phi'(z)$$

To obtain $\partial^2\Phi/\partial x_2{}^2$ it is convenient to make use of the Cauchy–Riemann equations (3.2.6)

$$\frac{\partial(\text{Re})}{\partial x_1} = \frac{\partial(\text{Im})}{\partial x_2};\qquad \frac{\partial(\text{Im})}{\partial x_1} = -\frac{\partial(\text{Re})}{\partial x_2};\qquad \text{with}\qquad \frac{\partial}{\partial x_1} = \frac{\text{d}}{\text{d}z}$$

It may be shown that the stresses satisfy both equilibrium and compatibility equations, e.g. from equation 2.8.1 in two-dimensional deformation, $(\partial\sigma/\partial x_3 = 0)$ consider the first equation

$$\frac{\partial\sigma_{11}}{\partial x_1} + \frac{\partial\sigma_{12}}{\partial x_2} = 0$$

We have

$$\frac{\partial\sigma_{11}}{\partial x_1} = \frac{\partial}{\partial x_1}\,[\text{Re}\,\phi(z) - x_2\,\text{Im}\,\phi'(z)] = \text{Re}\,\phi'(z) - x_2\,\text{Im}\,\phi''(z)$$
$$\frac{\partial\sigma_{12}}{\partial x_2} = \frac{\partial}{\partial x_2}\,[-x_2\,\text{Re}\,\phi'(z)] = -\text{Re}\,\phi'(z) + x_2 . \frac{\partial}{\partial x_1}\,[\text{Im}\,\phi'\,(z)]$$
$$= -\frac{\partial\sigma_{11}}{\partial x_1}$$

in agreement with equation 2.8.1.

In plane strain, the displacements u_1 and u_2 are given by:

$$\begin{aligned} 2\mu u_1 &= (1-2\nu)\,\text{Re}\,\bar{\phi}(z) - x_2\,\text{Im}\,\phi(z) \\ 2\mu u_2 &= 2(1-\nu)\,\text{Im}\,\bar{\phi}(z) - x_2\,\text{Re}\,\phi(z) \end{aligned} \tag{3.6.3}$$

As a check, we may calculate the principal strain $\epsilon_{11} = \partial u_1/\partial x_1$ and show that it conforms to the usual expression:

$$E\epsilon_{11} = E\epsilon_1 = \sigma_{11} - \nu(\sigma_{22} + \sigma_{33}) \tag{2.6.12}$$

In plane strain,

$$\epsilon_{33} = 0, \quad \sigma_{33} = \nu(\sigma_{11} + \sigma_{22})$$

$$\therefore \quad E\epsilon_{11} = (1 - \nu^2)\sigma_{11} - \nu(1 + \nu)\sigma_{22} \tag{3.6.4}$$

By differentiating equation 3.6.3 with respect to x_1, we have:

$$\begin{aligned} 2\mu\epsilon_{11} &= (1 - 2\nu) \text{ Re } \phi(z) - x_2 \text{ Im } \phi'(z) \\ &= E\epsilon_{11}/(1 + \nu) \text{ (from equation 2.6.10)} \\ &= (1 - \nu) [\text{Re } \phi(z) - x_2 \text{ Im } \phi'(z)] - \nu [\text{Re } \phi(z) + x_2 \text{ Im } \phi'(z)] \\ &= (1 - \nu)\sigma_{11} - \nu\sigma_{22} \text{ (from equation 3.6.2)} \end{aligned}$$

$\therefore \quad E\epsilon_{11} = (1 - \nu^2)\sigma_{11} - \nu(1 + \nu)\sigma_{22}$ in agreement with equation 3.6.4.

In this way, we can deduce that equilibrium, compatibility and stress–strain relationships are all satisfied by the stress function Φ defined in equation 3.6.1. What we must now examine are the boundary conditions implied by the formulation and those which we may be able to set independently.

The main condition implied from the stress equation 3.6.2 is that when $x_2 = 0$, $\sigma_{12} = 0$ and $\sigma_{22} = \sigma_{11}$. The displacements are then:

$$\begin{aligned} u_2 &= \frac{2(1 - \nu^2)}{E} \text{Im } \bar{\phi}(z) \\ u_1 &= \frac{(1 - 2\nu)(1 + \nu)}{E} \text{Re } \bar{\phi}(z) \end{aligned} \tag{3.6.5}$$

3.7 The Crack under Tension

We use the stress function first to examine the problem of a crack of length $2a$ in an infinite body under an applied uniform biaxial tension $\sigma_{22} = \sigma_{11} = \sigma$ (*Figure 3.3*). The model can later be modified to represent uniaxial tension, with $\sigma_{22} = \sigma$, $\sigma_{11} = 0$. We select as $\phi(z)$ a suitable *harmonic* function such that σ_{22}, when $x_2 = 0$, is zero for $-a \leqslant x_1 \leqslant +a$ and is σ when x_1 tends to $\pm$ infinity. We expect the crack to exert a stress concentrating effect, so that σ_{22} is greater than σ close to the crack tip. We have $\sigma_{22} = \text{Re } \phi(z) + x_2 \text{ Im } \phi'(z)$, from equation 3.6.2, or, when $x_2 = 0$, $\sigma_{22} = \text{Re } \phi(z)$.

For $\sigma_{22} \to \sigma$ as $x_1 \to \infty$, and to be greater than σ when $|x_1|$ is just greater than a, the simplest function would be of the form $\sigma_{22} = \sigma/(1 - a/x_1)$ or, to make it symmetrical for $\pm x_1$, $\sigma_{22} = \sigma/\left(1 - \dfrac{a^2}{x_1^2}\right)$.

However, we need also to specify that when $x_2 = 0$ and $-a < x_1 < +a$, $\sigma_{22} = 0$. This can be achieved if the function $\phi(z)$ becomes imaginary when $-a < x_1 < +a$, since $\sigma_{22} = \text{Re } \phi(z)$. The simplest form for a denominator to give $\sigma_{22} \to \sigma$ as $x_1 \to \infty$ is $(1 - a^2/x_1^2)$ and if this is put under a square root sign, the function becomes imaginary (i.e. $\sigma_{22} = 0$), when $-a < x_1 < +a$.

Thus, when $x_2 = 0$,

$$\sigma_{22} = \frac{\sigma}{\sqrt{(1 - a^2/x_1^2)}} \tag{3.7.1}$$

as the simplest form, and we are led to investigate the complex function

$$\phi(z) = \frac{\sigma}{\sqrt{(1 - a^2/z^2)}}, \tag{3.7.2}$$

for the general solution. This is found to be compatible with all equations, and satisfies the boundary conditions for the crack in the infinite plate.

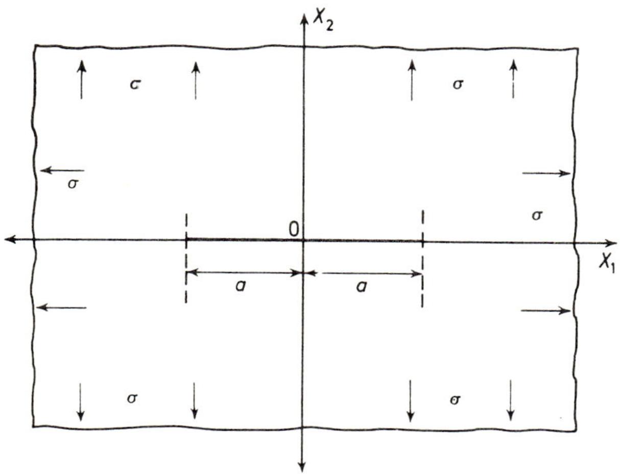

Figure 3.3 Westergaard's model of a crack under biaxial tension in an infinite sheet. (The crack is of length 2a)

The stress normal to the crack plane is given by equation 3.6.2 above and the displacement, u_2, in the x_2 direction, is given, from equation 3.6.5, by

$$u_{2(x_2=0)} = \frac{2(1 - \nu^2)\sigma}{E} \sqrt{(a^2 - x_1^2)} \tag{3.7.3}$$

which shows the shape of the crack to be elliptic:

$$\frac{u_2^2}{\text{const.}} + \frac{x_1^2}{1} = a^2 \tag{3.7.4}$$

If equation 3.7.1 is written in terms of the distance ahead of the crack tip $r = (x_1 - a)$, we obtain for the stress close to the tip $(r/a \ll 1)$:

$$\sigma_{22} = \sigma\sqrt{\left(\frac{a}{2r}\right)} \tag{3.7.5}$$

or, to the next degree of approximation:

$$\sigma_{22} = \frac{\sigma(a + r)}{\sqrt{(2ar)}} \left(1 - \frac{r}{4a}\right) \tag{3.7.6}$$

Taking this equation as an accurate expression it is seen that, if $r/a = 0.02$, equation 3.7.5 underestimates the local stress by some 1.5% (see Section 5.10).

Similarly, if σ_{22}, σ_{11} and σ_{12} are calculated from the full forms of equation 3.6.2 it is possible to calculate the angular dependencies of the stresses near the crack tip, by substituting the variable $\eta = (z - a)$, rather than simply $r = (x_1 - a)$, and writing $\eta = re^{i\theta}$. For example, from equation 3.6.2, σ_{22} is given by:

$$\sigma_{22} = \text{Re}\,\phi(z) + x_2\,\text{Im}\,\phi'(z)$$

Writing $\eta = z - a$, where $|\eta|/a \ll 1$, the first term becomes:

$$\text{Re}\,\phi(z) = \text{Re}\,\sigma\sqrt{\left(\frac{a}{2re^{i\theta}}\right)} = \sigma\sqrt{\left(\frac{a}{2r}\right)}.\cos\frac{\theta}{2}$$

and, with the relationship, $x_2 = r\sin\theta$, the second term becomes:

$$x_2\,\text{Im}\,\phi'(z) = r\sin\theta\,\text{Im}\frac{\sigma}{2}\sqrt{\left[\frac{a}{2(re^{i\theta})^3}\right]} = \sigma\sqrt{\left(\frac{a}{2r}\right)}.\sin\frac{\theta}{2}\cos\frac{\theta}{2}\sin\frac{3\theta}{2}$$

Therefore, the angular dependency of the stress σ_{22} is given by:

$$\sigma_{22} = \sigma\sqrt{\left(\frac{a}{2r}\right)}.\cos\frac{\theta}{2}\left(1 + \sin\frac{\theta}{2}\sin\frac{3\theta}{2}\right) + \ldots$$

and, similarly,

$$\sigma_{11} = \sigma\sqrt{\left(\frac{a}{2r}\right)}.\cos\frac{\theta}{2}\left(1 - \sin\frac{\theta}{2}\sin\frac{3\theta}{2}\right) + \ldots$$
$$\sigma_{12} = \sigma\sqrt{\left(\frac{a}{2r}\right)}.\sin\frac{\theta}{2}\cos\frac{\theta}{2}\cos\frac{3\theta}{2} + \ldots \qquad (3.7.7)$$

(see references 5 and 6).

Again, these are the first terms in series expansions. As θ tends to zero ($x_2 \to$ zero), so σ_{22} resumes the form of equation 3.7.5.

This solution pertains to a crack subjected to a uniform biaxial tensile stress at infinity $\sigma_{11} = \sigma_{22} = \sigma$, and produces biaxial stress at the crack tip: $\sigma_{22} = \sigma_{11}$ when $\theta = 0$. The uniaxial case may be reproduced simply by superimposing a pressure in the X_1 direction equal to $-\sigma$, which is the magnitude of σ_{11} at infinity. This superimposition does not affect the crack tip values of σ_{22} and σ_{11} given by equations 3.7.7.

It is, perhaps, unfortunate that the original Westergaard model put the crack in the X_1X_3 plane with its length running along $\pm x_1$ and therefore normal to x_2. It has been retained here in this form to facilitate the understanding of the original papers. In the application of Westergaard's results to fracture mechanics, it is more usual to site the crack along $\pm x_2$, normal to x_1, and to apply a uniform tensile stress $\sigma_{11} = \sigma$, as shown in *Figure 3.4a*, as drawn in previous sections (2.2, 2.7). The subscripts 1 and 2 in equations 3.7.1,

3.7.7 and 3.7.3 are simply interchanged so that the crack tip displacement (equation 3.7.3), becomes

$$u_{1(x_1=0)} = \frac{2(1-\nu^2)\sigma}{E}\sqrt{(a^2-x_2{}^2)} \tag{3.7.8}$$

This opening is in the X_1 direction, whence derives the definition of the configuration as a Mode I Opening (*Figure 3.4b*). It is hoped that this simple change of axes will not prove to be too confusing. The new axes, drawn in *Figure 3.4a*, will be used throughout the rest of this book, apart from sections in which the quoting of complex stress functions necessitates Westergaard's configuration to prevent confusion.

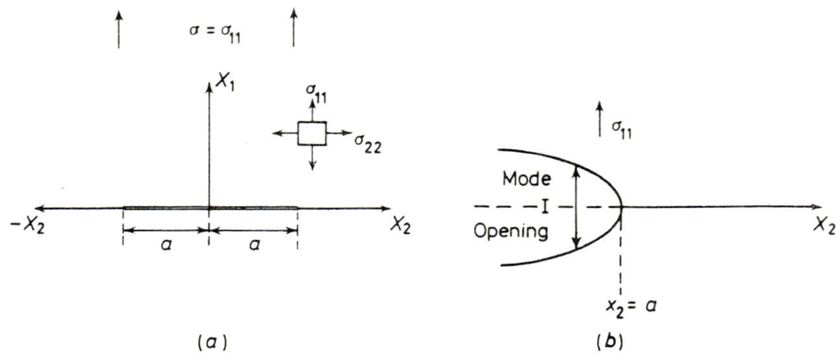

Figure 3.4 (a) Conventional choice of axes used for fracture mechanics terminology (note that X_1 and X_2 are interchanged by comparison with Figure 3.3). (b) Definition of Mode I opening (displacement u_1 in the X_1 direction)

The Westergaard approach may also be applied to cracks in shear, as described in the following section.

3.8 The Crack in Shear

We consider a crack of length $2a$ subjected to a shear stress $\sigma_{21} = \tau$ as referred to the axes in *Figure 3.3.* A suitable stress function for this situation is given by

$$\Phi_{\mathrm{II}} = -x_2 \ \mathrm{Re}\ \bar{\phi}_{\mathrm{II}}(z) \tag{3.8.1}$$

where $\phi_{\mathrm{II}}(z)$ is a function of z which is chosen to satisfy the boundary conditions. Differentiation of Φ_{II} to yield σ_{11}, σ_{22} and σ_{12} using equations (2.9.1) now gives:

$$\begin{aligned} \sigma_{11} &= 2\ \mathrm{Im}\ \phi_{\mathrm{II}}(z) + x_2\ \mathrm{Re}\ \phi'_{\mathrm{II}}(z) \\ \sigma_{22} &= -x_2\ \mathrm{Re}\ \phi'_{\mathrm{II}}(z) \\ \sigma_{21} = \sigma_{12} &= \mathrm{Re}\ \phi_{\mathrm{II}}(z) - x_2\ \mathrm{Im}\ \phi'_{\mathrm{II}}(z) \end{aligned} \tag{3.8.2}$$

Again, we note that, along the line $x_2 = 0$, σ_{22} vanishes and that σ_{21} can be made to vanish within the crack if $\phi_{II}(z)$ is imaginary when $-a < x_1 < +a$. We are led to choose the function

$$\phi_{II} = \frac{\tau}{\sqrt{(1 - a^2/z^2)}} \tag{3.8.3}$$

and find that, close to the crack tip,

$$\sigma_{21} = \tau\sqrt{\left(\frac{a}{2r}\right)} \tag{3.8.4}$$

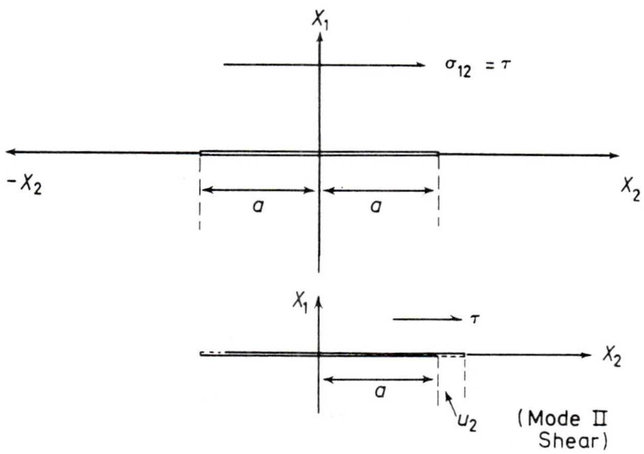

Figure 3.5 Definition of Mode II shear (displacement u_2 in X_2 direction)

With the change of axes given in *Figure 3.4* this situation is referred to as Mode II Shear, because the crack is given a shearing displacement u_2, by an applied stress σ_{12} (see *Figure 3.5*).

3.9 Antiplane Strain

An alternative mode of shearing a crack is possible if the crack has length $2a$ in the X_1 direction, lies in the X_1X_3 plane, and is sheared by a stress $\sigma_{23} = q$ in the X_3 direction (see *Figure 3.3*). This situation is called antiplane strain. Displacements occur only in the X_3 direction, which simplifies matters substantially, since u_1 and u_2 may be set as zero in equation 2.8.2. Equations 2.6.7 then reduce to

$$\frac{\partial u_3}{\partial x_1} = \frac{\sigma_{13}}{\mu}, \qquad \frac{\partial u_3}{\partial x_2} = \frac{\sigma_{23}}{\mu} \tag{3.9.1}$$

The equilibrium equations become, simply:

$$\frac{\partial \sigma_{13}}{\partial x_1} + \frac{\partial \sigma_{23}}{\partial x_2} = 0 \tag{3.9.2}$$

and so we obtain:

$$\frac{\partial^2 u_3}{\partial x_1^2} + \frac{\partial^2 u_3}{\partial x_2^2} = \nabla^2 u_3 = 0 \tag{3.9.3}$$

Choosing

$$u_3 = \frac{1}{\mu} \operatorname{Im} \phi_{\text{III}} (z)$$

gives

$$\sigma_{13} = \operatorname{Im} \phi'_{\text{III}} (z); \qquad \sigma_{23} = \operatorname{Re} \phi'_{\text{III}} (z) \tag{3.9.4}$$

for the shear stresses. At infinity, $\sigma_{23} = q$; within the crack, $\sigma_{23} = 0$. We choose the function

$$\phi'_{\text{III}} (z) = \frac{q}{\sqrt{(1 - a^2/z^2)}} \tag{3.9.5}$$

and again obtain, near the crack tip, a form for the local stress $\sigma_{23} = q\sqrt{(a/2r)}$. This situation is referred to as Mode III Antiplane Shear. With the conventional change of axes, the shear stress is σ_{13}, rather than σ_{23} (see *Figure 3.6*).

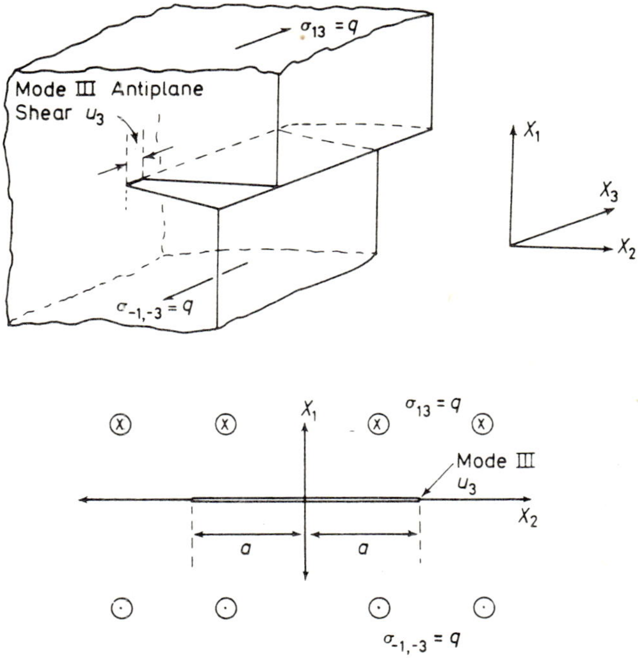

Figure 3.6 Definition of Mode III antiplane shear (displacement u_3 in direction X_3)

3.10 The Stress Intensity Factor

If one examines the expressions for the stress σ_{11} (with the change of axes described in Section 3.7) close to the tip of a crack, one has, from Westergaard, by altering equation 3.7.6

$$\sigma_{11\,(x_1\,=\,0)} = \frac{\sigma(a+r)}{\sqrt{(2ar)}}\left(1 - \frac{r}{4a} + \ldots\right) \tag{3.7.6}$$

or, from equation 3.7.7

$$\sigma_{11} = \sigma\sqrt{\left(\frac{a}{2r}\right)}.\cos\frac{\theta}{2}\left(1 + \sin\frac{\theta}{2}\sin\frac{3\theta}{2}\right) + \ldots \tag{3.7.7}$$

It should be emphasised that these are the initial terms in series forms. When, and only when, $r \ll a$, we may write

$$\sigma_{11\,(x_1\,=\,0)} = \sigma\sqrt{\left(\frac{a}{2r}\right)} = \frac{K'}{\sqrt{(2r)}} \tag{3.10.1}$$

where K' is a factor, given by $K' = \sigma\sqrt{a}$, and originally termed the stress intensity factor[5,6]. More usually, the *Stress Intensity Factor is defined as* $K = \sigma\sqrt{(\pi a)}$ and equation 3.10.1 then becomes

$$\sigma_{11} = \frac{K}{\sqrt{(2\pi r)}} \tag{3.10.2}$$

The units of K are MN $m^{-3/2}$ (or ksi $\sqrt{\text{in}}$).

The advantage in writing the stress close to the crack tip in terms of a single parameter K arises from the fact that the $1/\sqrt{(2\pi r)}$ dependency is followed, whatever the applied stress system. Compare the forms of the crack tip stresses in Modes I, II and III situations given in the three previous sections. In Mode I, combinations of applied stresses, tension, torsion, point loading, etc., each give rise to their own specific contributions to σ_{11}, and the resultant may be calculated simply by adding the individual stress intensities. Thus, in *Figure 3.7*

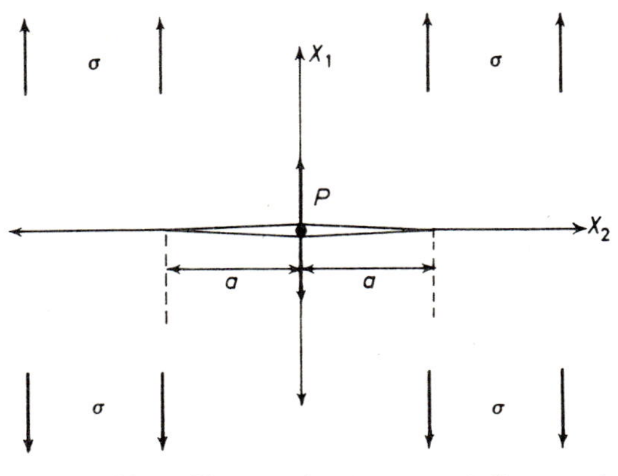

Figure 3.7 Crack loaded by uniform tension ($\sigma_{11} = \sigma$ at infinity) and point loading (P, at $x_1 = 0, x_2 = 0$)

a configuration is shown in which the cracked plate is loaded by a uniform tensile stress, σ, and a point wedging force, P, at $x_1 = 0, x_2 = 0$. If σ gives rise to a stress $\sigma_{11(\sigma)} = K_\sigma/\sqrt{(2\pi r)}$ and P to a stress $\sigma_{11(p)} = K_p/\sqrt{(2\pi r)}$, the stress σ_{11} is given by $\sigma_{11} = (K_\sigma + K_p)/\sqrt{(2\pi r)}$.

The form of K_p is of course different from that of K_σ. For point loading, the appropriate Westergaard stress function[4] is given by:

$$\phi(z) = \frac{Pa}{\pi z(z^2 - a^2)^{\frac{1}{2}}} \tag{3.10.3}$$

where P is the load per unit thickness, and x_1 and x_2 are defined as in *Figure 3.3*. With reference to *Figure 3.7*, when x_1 is zero, σ_{11} is given by [Westergaard's $\sigma_{22} \equiv \text{Re}\,\phi(z)$ when $x_2 = 0$, see equation 3.6.2].

$$\sigma_{11} = \frac{Pa}{\pi x_2(x_2{}^2 - a^2)^{\frac{1}{2}}} \tag{3.10.4}$$

Taking $x_2 = a + r$, where $r \ll a$, we may obtain:

$$\sigma_{11} = \frac{P}{\pi\sqrt{a}}\sqrt{\left(\frac{1}{2r}\right)} = \frac{K_p}{\sqrt{(2\pi r)}} \tag{3.10.5}$$

i.e.

$$K_p = \frac{P}{\sqrt{(\pi a)}} \tag{3.10.6}$$

Since P is the load per unit thickness, the dimensions of K_p are identical to those of K_σ.

The forms for K have been calculated for a number of loading configurations[7] and some of the more common expressions are listed in *Table 3.1*. For specimens of finite dimensions, it is often the practice to regard the solution as that for an infinite plate, modified by an algebraic or trigonometric function which is chosen to make the appropriate surface forces zero. Thus, for a central crack of length $2a$ in a plate of width W, we find that[8]:

$$K = \sigma\left[W \tan\left(\frac{\pi a}{W}\right)\right]^{\frac{1}{2}} \tag{3.10.7}$$

as compared with $K = \sigma\sqrt{(\pi a)}$ for the central crack in an infinite plate. If equation 3.10.7 is expanded, we have:

$$K = \sigma W^{\frac{1}{2}}\left[\left(\frac{\pi a}{W}\right) + \left(\frac{\pi^3 a^3}{3W^3}\right) + \ldots\right]^{\frac{1}{2}} \tag{3.10.8}$$

$$= \sigma\sqrt{(\pi a)}\left(1 + \frac{\pi^2 a^2}{3W^2} + \ldots\right)^{\frac{1}{2}}$$

$$= \sigma\sqrt{(\pi a)}\left(1 + \frac{\pi^2 a^2}{6W^2} + \ldots\right) \tag{3.10.9}$$

Thus, for the infinite solid, where $a/W = 0$, $K = \sigma\sqrt{(\pi a)}$. For an edge crack of length a in a semi-infinite plate, K is given by $1.12\,\sigma\sqrt{(\pi a)}$. The factor 1.12

Table 3.1 SOME FORMULAE FOR STRESS INTENSITY FACTORS

Type of crack	*Applied stress*	*Crack tip displacement*	*Mode*	*Stress intensity*
Central, of length $2a$ ($\pm a$ along $\pm X_2$) in infinite plate	$\sigma_{11} = \sigma$	u_{11}	I, opening (plane stress or plane strain)	$K_{\mathrm{I}} = \sigma\sqrt{(\pi a)}$
	$\sigma_{12} = \tau$	u_{12}	II, shear	$K_{\mathrm{II}} = \tau\sqrt{(\pi a)}$
	$\sigma_{13} = q$	u_{13}	III, antiplane strain	$K_{\mathrm{III}} = q\sqrt{(\pi a)}$
	P, load per unit thickness	u_{11}	I, opening	$K_{\mathrm{p}} = P/\sqrt{(\pi a)}$
Central, of length $2a$ in plate of width W	$\sigma_{11} = \sigma$	u_{11}	I, opening	$K_{\mathrm{I}} = \sigma\left[W \tan\left(\frac{\pi a}{W}\right)\right]^{\frac{1}{2}}$
Central, penny-shaped, of radius a in infinite body	$\sigma_{11} = \sigma$	u_{11}	Opening (radially symmetric)	$K = \frac{2}{\sqrt{\pi}}\sigma\sqrt{a}$
Edge, of length a in semi-infinite plate	$\sigma_{11} = \sigma$	u_{11}	I, opening	$K_{\mathrm{I}} = 1.12\sigma\sqrt{(\pi a)}$
	$\sigma_{13} = q$	u_{13}	III, antiplane strain	$K_{\mathrm{III}} = q\sqrt{(\pi a)}$
Two symmetrical edge cracks, each of length a, in a plate of total width W	$\sigma_{11} = \sigma$	u_{11}	I, opening	$K_{\mathrm{I}} = \sigma W^{\frac{1}{2}}\left[\tan\left(\frac{\pi a}{W}\right) + 0.1 \sin\left(\frac{2\pi a}{W}\right)\right]^{\frac{1}{2}}$

arises because stresses normal to the free surface must be put equal to zero and this affects the form of the stress function. The general methods for modifying stress functions in this way are termed 'Boundary Collocation', and will be discussed further in Section 3.15.

The K parameter itself sometimes produces confusion, particularly because its dimensions of stress $\sqrt{}$ length are not commonly encountered in other physical formulae. It should be emphasised that it is a factor which characterises the *intensity* of the stress field ahead of a crack. For any length of crack or combination of forces applied to the body, it is known that the local stress will diminish as the inverse of the square root of distance, and K gives the precise magnitude of the stress by embracing in one convenient form both the crack geometry and the net contribution of the applied forces.

For the purposes of analogy, we may compare the stress intensity factor with field intensities encountered in magnetostatics or electromagnetism, but there is one major difference in the form of definition. If, for example, we have a short bar magnet of length $2l$ and pole strength m (see *Figure 3.8*) we know that the force exerted on a unit pole situated at a point P on the extended axis of the magnet and at a distance r from its centre is given by:

$$F = \frac{2 \times 2\,ml \times 1}{r^3}$$

and that this is defined as the intensity, H, of the field at P. Note that the fact that the force becomes infinite as r approaches zero (as it would for the inverse square attraction of a single pole) does not give cause for concern, because we know that, in practical terms, we shall always be concerned with events some distance from the pole. In detail, the limiting distance below which any such formula cannot hold will be about the size of a few magnetic dipoles. Similarly, equation 3.10.2 tells us all we need to know about stresses near the crack tip and must be replaced by atomic force laws for distances less than a few atomic bond lengths.

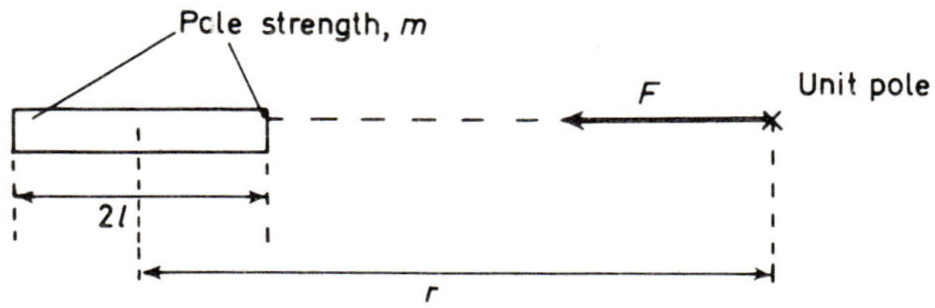

Figure 3.8 Force due to bar magnet (analogy between stress intensity and magnetic moment)

$$F = \frac{m}{(r-l)^2} - \frac{m}{(r+l)^2} \cong \frac{4ml}{r^3}$$

The difference in definition between *stress intensity K* and *magnetic field intensity H* results from the fact that the $1/\sqrt{r}$ distance term is excluded when defining K, but the $1/r^3$ distance term is included when defining H. Neither procedure is, however, fundamentally different in principle from the other.

If we take the *magnetic moment*, M, as an 'intensity' analogous to stress intensity, we see certain similarities. For the case above, M is given by $M = 2\,ml$, showing that the 'intensity' can be altered by the independent variation of both pole strength and magnetic length. Different combinations of m and l may give the same value of M and hence of magnetic field; similarly, different combinations of σ and a can produce the same value of K and hence of crack tip stresses. Further analogies may be sought in the differing relationships between field and magnetic moment found for different geometrical configurations and this point may, perhaps, be made even more clearly by considering the magnetic fields produced by currents flowing in straight wires as compared with solenoids.

These analogies have been emphasised because it is important to make the concept of stress intensity generally acceptable. Magnetic fields tend to be familiar, and so parameters associated with them are felt to be respectable, particularly, because they have been shown to provide convenient quantities for calculating the net effects of magnetic forces. Once the K concept has been accepted, it can be shown that it is equally useful in calculating the effects of applied mechanical stresses.

3.11 Plasticity in Cracked Pieces

Up to now, we have treated only the elastic deformation of bodies which contain sharp cracks. Indeed, until fairly recently, the problems of fracture in cracked pieces could be related only to solutions for stress systems which were elastic or fully plastic. Analytical solutions which treat mixed elastic/plastic stress fields have generally been obtained only for plane stress or antiplane strain configurations and numerical methods have to be employed for plane strain. The most physically reasonable plane strain solution to be derived is one which takes account of the changes in crack tip geometry as the plastic 'enclave' adjacent to the crack tip increases in size (see Section 3.18).

The next sections describe briefly the features of particularly relevant examples of mixed elastic/plastic stress analysis in cracked pieces.

3.12 The Equivalent Elastic Crack

The effect of a small amount of plastic yielding on the stresses and displacements ahead of a sharp crack may be estimated crudely as follows[9]. Consider the elastic stress distribution in plane stress as shown in *Figure 3.9a* and how it is altered by an amount of yielding (*Figure 3.9b*) which is small in comparison both with the crack length and with the dimensions of the body in which it is contained. The plastic zone is supposed to extend a (small) distance, d_Y, ahead of the crack tip ($x_2 = a$) and, within this zone, the tensile stress σ_{11} is equal to the uniaxial yield stress, σ_Y (plane stress, no work-hardening, Tresca criterion). The problem is to estimate the size of d_Y for a given applied stress, σ or stress intensity $K = \sigma\sqrt{(\pi a)}$.

As a first, lower, estimate, we could say that the zone extended to the point, r_Y, at which the elastic stress, σ_{11}, was equal to the yield stress, σ_Y (*Figure 3.9c*). However, it is immediately apparent that this will be very much an underestimate,

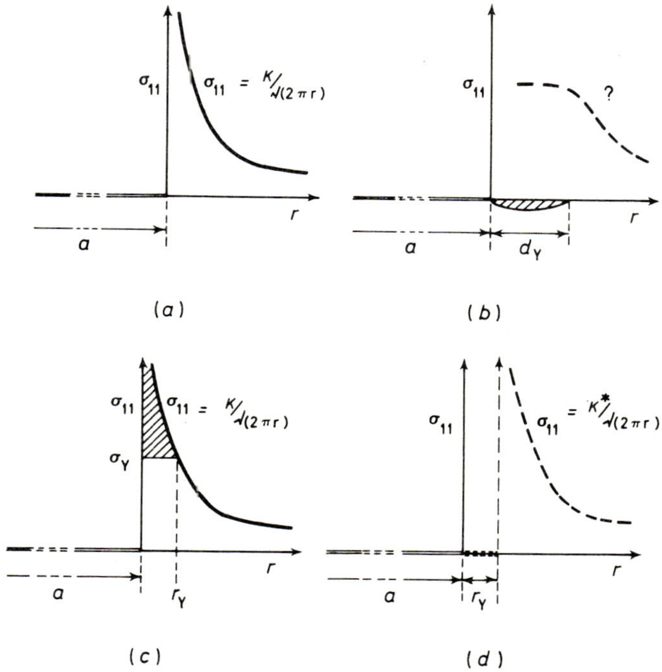

Figure 3.9 Local yielding at a crack tip. (a) Elastic stress distribution. (b) Formation of plastic zone. (c) First estimation of plastic zone size. (d) 'Notional' elastic crack

because the stress present in the shaded area between the lines $\sigma_{11} = \sigma_Y$, $\sigma_{11} = K/\sqrt{(2\pi r)}$ and the stress ordinate is available to produce further yielding. It is convenient to calculate the magnitudes of these stresses. At r_Y, the *total* area under the curve $\sigma_{11} = K/\sqrt{(2\pi r)}$ is given by

$$\int_0^{r_Y} \frac{K}{\sqrt{(2\pi r)}}\,\mathrm{d}r = \sqrt{\frac{2}{\pi}}.Kr_Y^{\frac{1}{2}} \qquad (3.12.1)$$

and, at $r = r_Y$, $\sigma_{11} = K/\sqrt{(2\pi r_Y)} = \sigma_Y$,

i.e.
$$r_Y = \frac{K^2}{2\pi\sigma_Y{}^2} \qquad (3.12.2)$$

Thus
$$\sqrt{\frac{2}{\pi}}.\,Kr_Y^{\frac{1}{2}} = \sqrt{\frac{2}{\pi}}.\,\sigma_Y\sqrt{(2\pi r_Y)}\,r_Y^{\frac{1}{2}}$$
$$= 2\sigma_Y r_Y \qquad (3.12.3)$$

Now the area below the line $\sigma_{11} = \sigma_Y$ up to $r = r_Y$ is obviously $\sigma_Y r_Y$ and so the shaded area must also be equal to $\sigma_Y r_Y$. It has been conventional to suppose, as a first approximation, that the plastic zone extends a distance

$$d_Y = 2r_Y = K^2/\pi\sigma_Y^2 \tag{3.12.4}$$

ahead of the crack.

Use has been made, for purposes of stress analysis, of a concept of a 'notional' elastic crack, the stress distribution ahead of which is equivalent to that of the configuration of a real crack and plastic zone at moderately large distances. The notional crack tip is supposed to be located at the point r_Y, and the new elastic stress distribution is then given by

$$\sigma_{11} = K^*/\sqrt{(2\pi r)}, \quad \text{where} \quad K^* = \sigma\sqrt{[\pi(a + r_Y)]} \tag{3.12.5}$$

The situation is drawn in *Figure 3.9d*. It is possible also to associate an opening at the original crack tip with this notional crack, as described in the following section.

3.13 The Spread of Plasticity from a Crack: the Dugdale Solution[10]

A more accurate relationship between applied stress, crack length and extent of plasticity in plane stress may be derived by making use of Westergaard's stress functions[4,11]. The solution was first derived by Dugdale[10], although he used somewhat different mathematical techniques and only referred to the Westergaard method in passing. It is, however, more convenient for our purposes, since we have already discussed the appropriate functions in Sections 3.7 and 3.10.

We imagine a situation in which a crack of length $2a$ is subject to a tensile stress σ. Let the plasticity spread from the ends of the crack for a further distance $|c - a|$ at each end as in *Figure 3.10*. If we now consider the elastic stress distribution immediately ahead of the plastic zone, we can derive

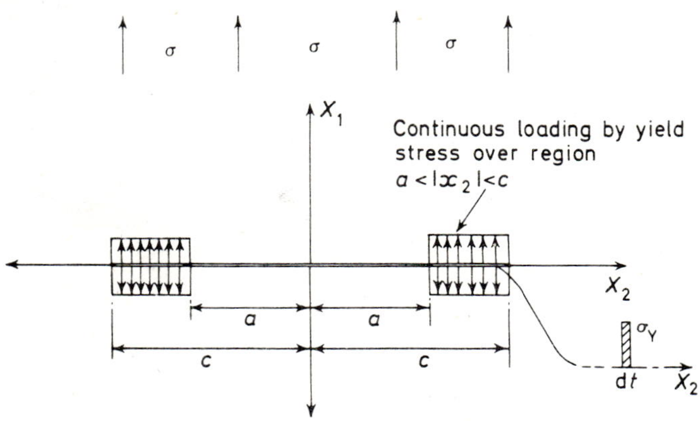

Figure 3.10 Spread of plasticity from a crack: Dugdale's model, based on Westergaard's stress functions

the relationship between σ, a and c as follows. First, we take the distribution produced when a crack of length $2c$ is loaded by a pair of point forces $\sigma(t) \times \mathrm{d}t \times$ unit thickness, where $x_2 = \pm t$ and $|t| < |c|$. It will be recalled that the Westergaard function for a single such force, P, at $x_1 = 0$ gives (by reference to equation 3.10.4).

$$\sigma_{11} = \frac{Pc}{\pi x_2 \sqrt{(x_2^2 - c^2)}}$$

For a pair of forces, $\sigma_{(t)}\, \mathrm{d}t$, at $x_2 = \pm t$, this takes the form

$$\sigma_{11} = \frac{2\sigma(t)\, \mathrm{d}t}{\pi} \frac{x_2 \sqrt{(c^2 - t^2)}}{(x_2^2 - t^2)\sqrt{(x_2^2 - c^2)}} \tag{3.13.1}$$

We now take the complete loading between a and c as arising from a complete set of forces $\sigma_Y \,.\, \mathrm{d}t$ integrated between a and c. Considering the region immediately ahead of the plastic zone, we take $x_2 = c + r$, where $r \ll c$. Then equation 3.13.1 gives for a pair of forces $\sigma_Y \mathrm{d}t$

$$\sigma_{11} = 2\frac{\sigma_Y}{\pi} \sqrt{\left(\frac{c}{2r}\right)} \frac{\mathrm{d}t}{\sqrt{(c^2 - t^2)}} \tag{3.13.2}$$

i.e.

$$K = \frac{2\sigma_Y}{\sqrt{\pi}} \sqrt{c} \frac{\mathrm{d}t}{\sqrt{(c^2 - t^2)}} \tag{3.13.3}$$

For a set of splitting forces, we integrate between $t = a$ and $t = c$, to obtain:

$$K = \frac{2\sigma_Y}{\sqrt{\pi}} \sqrt{c} \cos^{-1}\left(\frac{a}{c}\right) \tag{3.13.4}$$

This then gives the stress intensity for point loading equivalent to yield loads across the plastic zone.

To obtain the same stress intensity for an elastic crack of length $2c$ we would have to apply a stress σ such that

$$K = \sigma\sqrt{(\pi c)} \tag{3.13.5}$$

The relationship between σ, a and c may now be obtained by equating 3.13.4 and 3.13.5.

$$\sigma\sqrt{(\pi c)} = \frac{2\sigma_Y}{\sqrt{\pi}} \sqrt{c} \cos^{-1}\left(\frac{a}{c}\right) \tag{3.13.6}$$

This 'cancels out' the stress singularity for an approach along the x_2 axis *only*: not in the plane.

Hence

$$\left(\frac{a}{c}\right) = \cos\left(\frac{\pi\sigma}{2\sigma_Y}\right) \tag{3.13.7}$$

From this, we can see clearly that, as $\sigma \to \sigma_Y$, the value of $(a/c) \to 0$, i.e. $c \to \infty$ and yielding spreads across the infinite plate. As (σ/σ_Y) becomes small, so

$$\left(\frac{a}{c}\right) \to 1 - \frac{\pi^2\sigma^2}{8\sigma_Y^2} + \ldots \tag{3.13.8}$$

using the cosine expansion, and if we denote the spread of plasticity by $d_Y = (c - a)$

we have
$$\left(\frac{a}{a + d_Y}\right) = \left(1 + \frac{d_Y}{a}\right)^{-1} = \left(1 - \frac{d_Y}{a} + \ldots\right) \tag{3.13.9}$$

by the binomial expansion. Hence, at very low applied stress levels,

$$d_Y = \frac{\pi^2 \sigma^2}{8 \sigma_Y{}^2} a = \frac{\pi K^2}{8 \sigma_Y{}^2} \tag{3.13.10}$$

This may be compared with the value $d_Y = 2r_Y = K^2/\pi\sigma_Y{}^2$ obtained previously in equation 3.12.4. Given the forms of the Westergaard stress functions appropriate to the yield loads across the plastic zones and to the applied stress acting on a crack of length $2c$, it is possible to calculate the value of the displacement, u_1, at the ends of the real crack, $x_2 = \pm a$. The detailed derivation is somewhat complicated and is given in Burdekin and Stone[11]. This displacement will feature largely in our later consideration of fracture criteria (see Chapter 6). It is called the Crack Opening Displacement and is usually denoted by the symbol, δ.

Then, δ is given by:

$$\delta = \frac{8}{\pi} \frac{\sigma_Y}{E} a \ln \left[\sec \left(\frac{\pi \sigma}{2 \sigma_Y} \right) \right] \tag{3.13.11}$$

At very low stresses, this may be simplified to give

$$\delta = \frac{\sigma^2 \pi a}{\sigma_Y E} = \frac{K^2}{\sigma_Y E} \tag{3.13.12}$$

or using 4.5.6
$$G = \sigma_Y \delta \tag{3.13.13}$$

or
$$\delta = \frac{\sigma_Y}{E} . 2\pi \frac{K^2}{2\pi\sigma_Y{}^2} \tag{3.13.14}$$

If we return to the crude model of the 'notional' crack (*Figure 3.9*) which has its crack tip in the centre of the real plastic zone, we arrive at a similar answer for δ, by considering that the zone is circular and that δ is given by the yield strain, σ_Y/E, spread over a 'gauge length' given by the circumference of the zone, $2\pi r_Y$ (see Wells[12]).

In Mode III (antiplane strain) deformation, the appropriate form for the spread of plasticity is

$$\frac{a}{c} = \cos \left(\frac{\pi q}{2 \tau_Y} \right) \tag{3.13.15}$$

where q and τ_Y are the applied and yield stresses in shear. The (shear) displacement, u_{13}, at the crack tip is given by

$$u_{13} = \frac{4\tau_Y}{\pi\mu} a . \ln \left[\sec \left(\frac{\pi q}{2\tau_Y} \right) \right] \tag{3.13.16}$$

This result agrees exactly with that described in the following section, which was derived originally for the same Mode III deformation.

3.14 The Bilby–Cottrell–Swinden (BCS) Model[13]

An alternative method[13] for calculating the sizes of plastic zones and displacements in Mode III deformation is of interest, firstly, because it provided the initial analytical stimulus for focusing attention on crack tip displacements and, secondly, because the method has been used widely to develop solutions for problems involving more than one crack, and for others involving the presence of free surfaces.

The yielded zone and crack are modelled as arrays of dislocations (*Figure 3.11*). In Mode III deformation, the dislocations are of screw character and are distributed along the x_2 axis, so that each dislocation, if discrete, would have a line vector

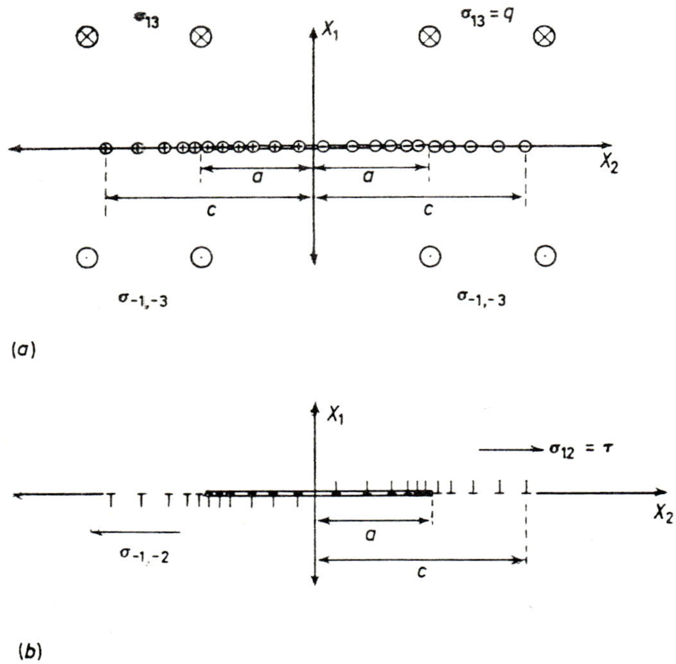

Figure 3.11 Representation of yielded zones by arrays of dislocations (a) Mode III antiplane shear ($\sigma_{13} = q$). (b) Mode II shear ($\sigma_{12} = \tau$)

$l = l_3$ and a Burger's vector $\boldsymbol{b} = b_3$. To accommodate the displacement gradient, from the high value, $S(a)$, at the crack tip, to the low value at the plastic/elastic interface, the dislocation array must take the form of an 'inverse' pile-up, with the dislocations closely spaced near the crack tip and widely spaced at the far extent of the yield zone. In the model, the dislocations are not treated as discrete but the theory of continuous distributions of dislocations is applied.

The applied shear stress at infinity, σ_{13} is equated to q as in Section 3.9. The resistance to motion of a dislocation is taken to be τ_i ($<q$) in the region $-a < x_2 < +a$ (τ_i is zero for a freely-slipping crack) and τ_Y in the yield zones, i.e.

$$-c < x_2 < -a \qquad \text{and} \qquad +a < x_2 < +c$$

There are taken to be $f(x_2)$. dx_2 dislocations, each of Burger's vector $\boldsymbol{b} > 0$, in the distance dx_2. Then, for equilibrium over the total region $2c$, along the x_2 axis, in which the dislocations are distributed, we sum the net repulsive forces on each screw dislocation. At a point x_2, the shear stress due to the dislocations at x_2' is given by:

$$\sigma_{13} = \frac{\mu b}{2\pi} \frac{f(x_2')\,dx_2'}{(x_2 - x_2')} \tag{3.14.1}$$

Hence, for equilibrium

$$\frac{\mu b}{2\pi} \int_{2c} \frac{f(x_2')\,dx_2'}{(x_2 - x_2')} = P(x_2) \tag{3.14.2}$$

where $P(x_2)$ is the resultant external shear stress, σ_{13}, at x_2. The value of $P(x_2)$ is given one discrete value, $P = q - \tau_i$, within the crack ($|x| < a$) and another, $P = q - \tau_Y$ within the plastic zone ($a < |x| < c$). The solution of equation 3.14.2 for these conditions leads to an expression for plastic zone size as:

$$\frac{a}{c} = \cos\left(\frac{\pi q}{2\tau_Y}\right) \tag{3.14.3}$$

and for the relative sliding displacement at the crack tip, $S(a)$, as

$$S(a) = \frac{4\tau_Y}{\pi\mu}\, a \ln\left[\sec\left(\frac{\pi q}{2\tau_Y}\right)\right] \tag{3.14.4}$$

This result is identical to that obtained in the previous section (equation 3.13.16). If we regard the dislocations in the BCS model as edge dislocations with Burger's vectors $\boldsymbol{b}$, producing displacements in the x_2 direction, *Figure 3.11b*, we modify equation 3.14.1 by inserting a factor of $(1 - \nu)$ in the denominator and derive:

$$u_{12}(a) = \frac{4\tau_Y(1-\nu)}{\pi\mu}\, a \ln\left[\sec\left(\frac{\pi\tau}{2\tau_Y}\right)\right] \tag{3.14.5}$$

or, from equation 2.6.10, putting $\mu = E/2(1 + \nu)$

$$u_{12}(a) = \frac{8\tau_Y(1-\nu^2)}{\pi E}\, a \ln\left[\sec\left(\frac{\pi\tau}{2\tau_Y}\right)\right] \tag{3.14.6}$$

in plane strain Mode II shear. In plane stress Mode I tension, the Burgers vectors are chosen to produce displacements in the x_1 direction and the crack tip displacement is given by:

$$u_1(a) = \frac{8\sigma_Y}{\pi E} a \ln\left[\sec\left(\frac{\pi\sigma}{2\sigma_Y}\right)\right]$$

exactly as before (equation 3.13.11).

The spread of plasticity in plane strain tension is a far more difficult problem to treat, because allowance must be made for the effects of constraint as the yielded zone becomes larger. The basic principles underlying the numerical solutions of such problems will be described in Sections 3.16 and 3.17. Alternative stress function methods of determining elastic stress distributions will be described in the following section, to indicate briefly how general boundary conditions may be satisfied.

3.15 Polynomial Stress Functions: Boundary Conditions

An alternative technique for solving the basic elasticity equation (2.9.2): $\nabla^2(\nabla^2\Phi) = 0$; is to make use of stress functions which are written as polynomial series, rather than as single algebraic functions of the type employed by Westergaard. Application of this method directly to calculate stress distributions around cracks was made initially by Williams[14, 15].

The method starts from the point that the term $\nabla^2\Phi$ must be harmonic (see Section 3.2) since $\nabla^2(\nabla^2\Phi) = 0$. We are therefore led to look for Laplace-type solutions for $\nabla^2\Phi$. The biharmonic equation may be satisfied if Φ is expressed in terms of two harmonic functions, f and g (i.e. $\nabla^2 f = 0, \nabla^2 g = 0$) through an equation of the type:

$$\Phi = r^2 f + g \tag{3.15.1}$$

where $r^2 = x_1^2 + x_2^2$. Essentially, f and g are expressed as simple polynomial functions of the complex variable: $z = x_1 + ix_2 = re^{i\theta}$. To conform with Williams's co-ordinate system, we take x_1 from the crack tip, x_2 normal to the crack and θ as the angle rotating anticlockwise from $-x_1$, towards $-x_2$. The open end of the crack is at $x = -x_0$ (see *Figure 3.12*). The harmonic functions f and g are chosen as the real parts of two polynomial series. Simply raising z to a power, say $(\lambda - 1)$, gives:

$$f = \Sigma \operatorname{Re} z^{(\lambda-1)} = \Sigma \operatorname{Re} [(re^{i\theta})^{(\lambda-1)}]$$

$$f = \Sigma r^{(\lambda-1)} \cos(\lambda - 1)\theta \tag{3.15.2}$$

Because we have chosen a function $\Phi = r^2 f + g$, it is appropriate to choose for g a polynomial which has r raised to the power $(\lambda + 1)$. This then gives the coefficient of θ in the cosine term and we expect to see, for the total stress function, Φ, something of the form:

$$\Phi = \Sigma r^{(\lambda+1)}[A_1 \cos(\lambda - 1)\theta + B_1 \cos(\lambda + 1)\theta] = \Sigma r^{(\lambda+1)} f(\theta) \tag{3.15.3}$$

where values of A_1, B_1, etc., have to be determined, according to the boundary conditions. Williams, in fact, uses a stress function, given by

$$\Phi = \Sigma r^{(\lambda+1)}[b_1 \sin(\lambda+1)\theta + b_2 \cos(\lambda+1)\theta + b_3 \sin(\lambda-1)\theta + b_4 \cos(\lambda-1)\theta] = \Sigma r^{(\lambda+1)} F(\theta) \tag{3.15.4}$$

which is derived using generally similar principles. If we take the standard expressions for two-dimensional stresses in polar co-ordinates (equation 2.9.5) we have, for each λ, the stress contributions:

$$\begin{aligned} \sigma_{rr} &= \frac{1}{r^2}\frac{\partial^2\Phi}{\partial\theta^2} + \frac{1}{r}\frac{\partial\Phi}{\partial r} = r^{(\lambda-1)}[F''(\theta) + (\lambda+1)F(\theta)] \\ \sigma_{\theta\theta} &= \frac{\partial^2\Phi}{\partial r^2} = r^{(\lambda-1)}[\lambda(\lambda+1)F(\theta)] \\ \sigma_{r\theta} &= \frac{1}{r^2}\frac{\partial\Phi}{\partial\theta} - \frac{1}{r}\frac{\partial^2\Phi}{\partial r\partial\theta} = r^{(\lambda-1)}[-\lambda F'(\theta)] \end{aligned} \tag{3.15.5}$$

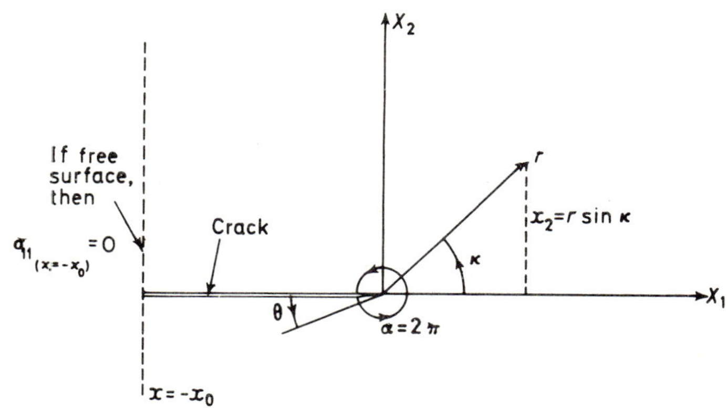

Figure 3.12 Co-ordinate system employed by Williams (see text)

For crack edges which are free from any imposed stresses, we obtain the conditions that $\sigma_{\theta\theta}$ (stress normal to the crack face) = 0 when $\theta = 0$ or 2π, and that $\sigma_{r\theta}$ (shear stress on the crack face) = 0 when $\theta = 0$ or 2π. From the equations above, this implies that $F(0) = F(2\pi) = 0$. Also $F'(0) = F'(2\pi) = 0$, because $\sigma_{r\theta} = 0$ at $\theta = 0$ and $\theta = 2\pi$. There are four unknown constants $b_i(b_1, b_2, b_3, b_4)$ in Williams' expression for $F(\theta)$ and the boundary conditions give four equations. For a solution to exist, the determinant of the coefficients of b_i in the four equations must vanish. In general, for a V-shaped notch of angle α, this equating of the determinant to zero yields the 'eigen equation':

$$\sin \lambda \alpha = \pm \lambda \sin \alpha \tag{3.15.6}$$

where λ must be greater than zero (to allow for the continuity of displacements).

For a crack, the eigen equation is particularly simple, i.e.

$$\sin (2\pi\lambda) = 0 \tag{3.15.7}$$

requiring that $\lambda = n/2$, where n is a positive integer. Putting in the condition that $F(0) = 0$ gives $b_2 = -b_4$, and that $F'(0) = 0$ gives $b_1(\lambda + 1) = -b_3(\lambda - 1)$, so that we may write for the total stress function Φ, with $\lambda = n/2$:

$$\Phi = \Sigma r^{\left(\frac{n}{2}+1\right)} \left\{ b_3 \left[\sin \left(\frac{n}{2} - 1\right)\theta - \frac{n-2}{n+2} \sin \left(\frac{n}{2} + 1\right)\theta \right] + b_4 \left[\cos \left(\frac{n}{2} - 1\right)\theta - \cos \left(\frac{n}{2} + 1\right)\theta \right] \right\} \tag{3.15.8}$$

where the summation is implied for all values of n. The values of the constants b_3 and b_4 may vary for the different values of n: we shall denote the 'b_3's by a_n and the 'b_4's by b_n, where the subscripts refer to the particular n value. Thus:

$$\Phi_{(n=1)} = r^{\frac{3}{2}} \left\{ a_1 \left[-\sin \frac{\theta}{2} + \frac{1}{3} \sin \frac{3\theta}{2} \right] + b_1 \left[\cos \frac{\theta}{2} - \cos \frac{3\theta}{2} \right] \right\}$$

$$\Phi_{(n=2)} = r^2 \left\{ b_2 [1 - \cos 2\theta] \right\} \tag{3.15.9}$$

by substitution of the appropriate figure for n in equation 3.15.8. We may then determine the stress, σ_{rr}, from equation 3.15.5.

It is more usual to refer the stresses to the bisector angle, κ, given by:

$$\kappa = \theta - \pi \tag{3.15.10}$$

(see *Figure 3.12*). Making this substitution, we obtain for the stress function:

$$\Phi_{(r,\kappa)} = r^{\frac{3}{2}} \left[a_1 \left(-\cos \frac{\kappa}{2} - \frac{1}{3} \cos \frac{3\kappa}{2} \right) + b_1 \left(-\sin \frac{\kappa}{2} - \sin \frac{3\kappa}{2} \right) \right] + b_2 r^2 [1 - \cos 2\kappa] + \ldots \tag{3.15.11}$$

and for the stress σ_{rr}:

$$\sigma_{rr(r,\kappa)} = \frac{1}{4r^{\frac{1}{2}}} \left[a_1 \left(-5 \cos \frac{\kappa}{2} + \cos \frac{3\kappa}{2} \right) + b_1 \left(-5 \sin \frac{\kappa}{2} + 3 \sin \frac{3\kappa}{2} \right) \right] + 4b_2 \cos^2 \kappa + \ldots \tag{3.15.12}$$

The stresses, $\sigma_{\kappa\kappa}$ and $\sigma_{r\kappa}$ may be determined similarly.

We now consider the term $b_2 r^2 [1 - \cos 2\kappa]$ in equation 3.15.11. Writing this in Cartesian form, we have, for this component of the stress function:

$$\Phi_{(n=2)} = b_2 r^2 [1 - \cos 2\kappa] = 2b_2 r^2 \sin^2 \kappa = 2b_2 x_2^2 \tag{3.15.13}$$

because x_2 is given by $r \sin \kappa$ (see *Figure 3.12*). Now, we may derive the stresses, referred to Cartesian axes, by use of equations 2.9.1, i.e.

$$\sigma_{11} = \frac{\partial^2 \Phi}{\partial {x_2}^2}, \qquad \sigma_{22} = \frac{\partial^2 \Phi}{\partial {x_1}^2}, \qquad \sigma_{12} = -\frac{\partial^2 \Phi}{\partial x_1 \partial x_2}$$

and obtain, for this component of the stress function:

$$\sigma_{11} = 4b_2, \qquad \sigma_{22} = 0, \qquad \sigma_{12} = 0 \tag{3.15.14}$$

For edge cracks, σ_{11} is usually zero at $x_1 = -x_0$ (*Figure 3.12*), so that b_2 is zero, and it is seen that the radial stress then has the characteristic inverse square root dependency: $\sigma_{rr} \propto r^{-\frac{1}{2}}$. If loadings are superimposed at the boundary $x_1 = -x_0$, the second term contributes to the crack tip stress field and suitable choice of the constant b_2 leads to a method of representing the effects of these loadings, not on K, which characterises the first term in the series, but on the crack tip stress distribution.

If the symmetrical solution ($b_i = 0$) is examined, σ_{rr} is given by:

$$\sigma_{rr} = \frac{a_1}{4r^{\frac{1}{2}}}\left(-5 \cos\frac{\kappa}{2} + \cos\frac{3\kappa}{2}\right) \tag{3.15.15}$$

and, if the stress state is referred to Cartesian co-ordinates, we have, for the stress across the cracking plane:

$$\sigma_{22} = -\frac{a_1}{r^{\frac{1}{2}}} \cos\frac{\kappa}{2}\left(1 + \sin\frac{\kappa}{2} \sin\frac{3\kappa}{2}\right) \tag{3.15.16}$$

using only the first term (n = 1).

Comparison of this with Westergaard's results (equation 3.7.7) gives for the relationship with stress intensity:

$$K_{\mathrm{I}} = -\sqrt{(2\pi)}\, a_1 \tag{3.15.17}$$

Calculation of the principal tensile stress gives:

$$\sigma_2 = -\frac{a_1}{r^{\frac{1}{2}}} \cos\frac{\kappa}{2}\left(1 \mp \sin\frac{\kappa}{2}\right) = \sigma_1 \tag{3.15.18}$$

which shows that the maximum principal stress occurs, not across the line of crack extension, but at ±60° to it. The direction of the maximum principal stress is in the X_2 direction. The maximum shear stresses occur at ±90° and the maximum octahedral stress at ± 70°. It is therefore expected that yield zones should spread out at some ± 70–90° to the line of crack extension, rather than at ± 45° as they would in rigid/plastic material. *Figure 3.13* shows that the real-life compromise is that the zones form angles of about 60–65°.

The advantage of the polynomial stress function method is that it allows the addition of loadings on the boundaries of the crack or specimen. Westergaard's method is restricted to the particular situation of stresses along the line of the crack and cannot treat non-uniform loadings on the specimen boundary. In Williams' stress function further terms can be included in the polynomial expression to allow for the extra stresses, e.g. the constant b_2 in equations 3.15.11 and

3.15.12 can be adjusted suitably if σ_{11} is non-zero on the line $x_1 = -x_0$. One disadvantage of Williams' method is that it is not very suitable for dealing with internal cracks. It has been used (e.g. by Gross *et al.*[16]) to derive stress intensity factors for geometries, such as the single-edge-cracked tensile specimen, where free surfaces are not infinitely removed from the crack tip region (see Section 3.10). Here, appropriate constants in the polynomial expansion were

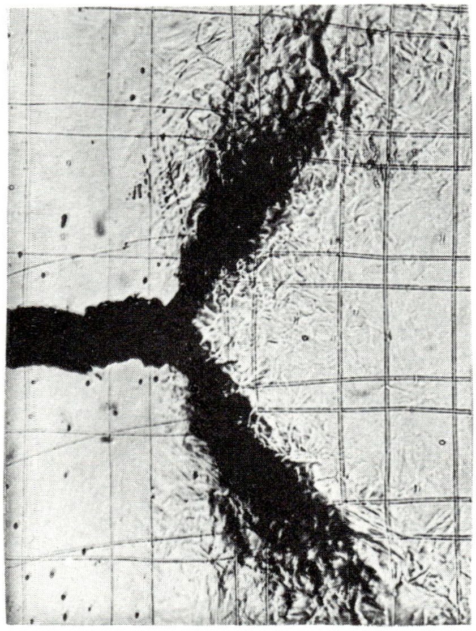

Figure 3.15 Slip deformation ahead of a sharp crack stressed in tension. The crack opening is accommodated by plastic flow at approximately 60° to the line of crack extension. (Note the shear displacements shown by the grid lines) (×150)

obtained, as indicated above, by matching the stress distribution given by the function to the imposed boundary conditions (e.g. uniform stress, zero tensile stress normal to a free surface, etc.). The technique is known as '*boundary collocation*'. The single-edge-cracked specimen required collocation at some twenty boundary points, but the fact that the functions are polynomials permits the use of computer routines.

Williams' method may be stated in general terms by saying that polynomial forms for the stress function, Φ, arise when Φ is expressed in terms of complex potentials (e.g. equation 3.3.1) $\psi(z)$ and $\chi(z)$, which are written as polynomial series:

$$\begin{aligned} \psi(z) &= \psi_0 + \psi_1 z + \psi_2 z^2 + \ldots + \psi_n z^n \\ \chi(z) &= \chi_0 + \chi_1 z + \chi_2 z^2 + \ldots + \chi_n z^n \end{aligned} \qquad (3.15.19)$$

and in which the unknown constants ψ_i and χ_i can be determined by matching stresses at discrete points on the boundary. Suitable modification of the relationship between Φ, ψ and χ enables internal crack problems to be treated. Analysis of the convergence of the various series chosen has not always been carried out rigorously and accuracy is usually assumed if the addition of more terms to the series does not produce any marked change in the results. Comparison with similar geometrical configurations is often used as a test of validity.

A further powerful method of incorporating the effects of free surfaces is to use polynomial functions in combination with mapping techniques. In mapping, the complex variable, z, corresponding to the crack geometry, is expressed as a function of a further complex variable, ξ, which corresponds to the geometry of a unit circle or a half-plane, in an infinite body. A neat illustration of the method is given by Paris and Sih[7], in considering the situation of a single concentrated force, F, applied at arbitrary inclination to a crack surface. To represent the tension and shear fields at the crack tip due to the force, they employ a complex stress intensity factor: $K = K_{\mathrm{I}} - \mathrm{i}K_{\mathrm{II}}$ and by deriving the stresses σ_{11} and σ_{22} formally from the full complex form of the Westergaard stress function, using the variable $\eta = (z - z_1)$, rather than the real distance variable $r = (x_1 - a)$, (as in the derivation of equation 3.7.7 from 3.7.2) they are able to write:

$$\sigma_{11} + \sigma_{22} = \mathrm{Re}\left[\frac{\sqrt{2}}{(\pi\eta)^{\frac{1}{2}}} K\right] \qquad (3.15.20)$$

From 3.3.4, we have: $\sigma_{11} + \sigma_{22} = 4\,\mathrm{Re}\,[\psi'(z)]$ where $\psi'(z)$ is the derivative of the appropriate harmonic function; in the present case, that corresponding to point loading (cf. 3.10.3). It is then possible to write:

$$K \equiv K_{\mathrm{I}} - \mathrm{i}K_{\mathrm{II}} = 2(2\pi)^{\frac{1}{2}} \lim_{z \to z_1} (z - z_1)^{\frac{1}{2}} \psi'(z) \qquad (3.15.21)$$

where $(z - z_1)$ is the variable η.

The mapping is accomplished by a functional relationship of the form: $z = w(\xi)$ whence equation 3.15.21 becomes:

$$K = 2(2\pi)^{\frac{1}{2}} \lim_{\xi \to \xi_1} [w(\xi) - w(\xi_1)]^{\frac{1}{2}} \frac{\psi'(\xi)}{w'(\xi)} \qquad (3.15.22)$$

The mapping of a crack of length $2a$ into a circular hole of unit radius is given by:

$$z = w(\xi) = \frac{a}{2}\left(\xi + \frac{1}{\xi}\right) \qquad (3.15.23)$$

and, for this mapping, equation 3.15.22 becomes:

$$K = 2\left(\frac{\pi}{a}\right)^{\frac{1}{2}} \psi'(1) \qquad (3.15.24)$$

When the appropriate stress function (equivalent to the form of equation 3.10.3 for the mapping ξ-plane) is found for $\psi(\xi)$, K_{I} and K_{II} may be obtained directly from equation 3.15.24 by separation of the real and imaginary parts. The very important result that derives from this mapping procedure is that the effects on K_{I} or K_{II} of any imposed tensile or shear stress put on the boundary of a crack in the infinite solid can be calculated. In a general situation, the mapping function (e.g. equation 3.15.23 above) may not be of simple form: it is, however, possible to express the function $w(\xi)$ as a polynomial, or even as a ratio of polynomials, and thence to obtain the stress function (from which the crack tip stresses are calculated) also in a polynomial form.

A final method of assessing the effects of free surfaces or loaded boundaries relies on the modelling of the crack singularities in terms of a continuous distribution of dislocations. In simple cases, the stress field outside the crack is completely elastic and the dislocations are distributed within the crack itself. A singular integral equation of the form given in 3.14.2 applies: if we wish to model a crack of length $2a$ subjected to a tensile stress σ_{11}, we must choose edge dislocations, modify the term outside the integral by $1/(1-\nu)$, integrate between $-a$ and $+a$, and substitute σ_{11} for $P(x_2)$. If the displacements are to be single-valued, the total Burger's vector around a circuit enveloping the whole crack must be zero. If boundaries are present, superposition may be used: singularities are represented by dislocation configurations and non-singular stress distributions by the techniques described previously, functions being chosen to

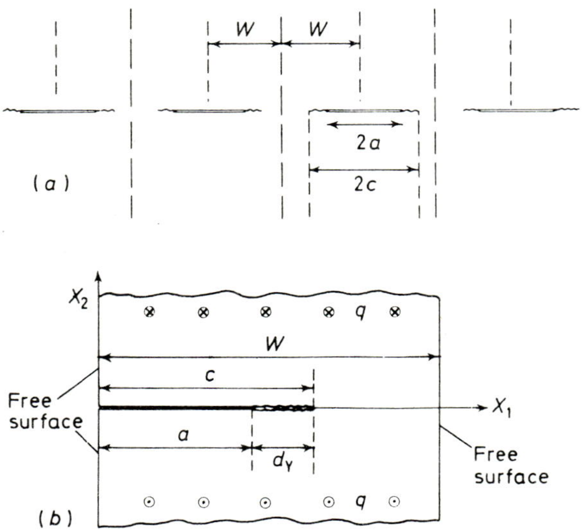

Figure 3.14 Effects of free surfaces on elastic/plastic stress distribution in antiplane strain. (a) Infinite array of cracks in antiplane strain: crack lengths 2a; lengths of cracks + plastic zones 2c ($c = a + d_Y$). Spread of yield given by:

$$\frac{\sin\left(\dfrac{\pi a}{2W}\right)}{\sin\left(\dfrac{\pi c}{2W}\right)} = \cos\left(\frac{\pi q}{2\tau_Y}\right)$$

(b) This situation can be used, in antiplane strain, to model the presence of free surfaces at the crack centres and ± W from the centres

reduce to zero the stresses produced on free surfaces by singularity stress fields. In more complicated models, the effects of surfaces on elastic/plastic stress and strain fields may be calculated by the dislocation distribution method. Smith[17,18], particularly, has published results on a variety of crack configurations. That for an infinite array of collinear cracks, which may be used to represent free surfaces, is shown in *Figure 3.14*.

3.16 Numerical Methods

The use of numerical methods to assist in the determination of stress distributions in notched or cracked pieces may be placed in three categories. The first was described in the previous section: where polynomial functions are used, it is obviously important to evolve good numerical techniques to evaluate the functions rapidly.

The second use of numerical techniques involves finite difference equations[19]. Here, if we have a function which is defined, not in analytic form, but by a series of values corresponding to regularly-spaced values of the appropriate variable, we can approximate differentials by finite differences, e.g. if the spacing along the x direction is d, and the values of the function at x_0, x_1, x_2, etc., are $F_0, F_1, F_2, \ldots$, etc., we have:

$$\left(\frac{\mathrm{d}F}{\mathrm{d}x}\right)_0 \simeq \frac{F_0 - F_{-1}}{d} \qquad \left(\frac{\mathrm{d}F}{\mathrm{d}x}\right)_1 \simeq \frac{F_1 - F_0}{d}$$

$$\left(\frac{\mathrm{d}^2F}{\mathrm{d}x^2}\right)_0 \simeq \frac{1}{d}\left[\left(\frac{\mathrm{d}F}{\mathrm{d}x}\right)_1 - \left(\frac{\mathrm{d}F}{\mathrm{d}x}\right)_0\right] \simeq \frac{F_1 - 2F_0 + F_{-1}}{d^2} \tag{3.16.1}$$

If F is a function of two variables, x_1 and x_2, we may evaluate the partial derivatives of F with respect to x_1 and x_2 independently in a similar manner.

The Laplace equation is given by:

$$\nabla^2\Phi = \frac{\partial^2\Phi}{\partial x_1{}^2} + \frac{\partial^2\Phi}{\partial x_2{}^2} = 0 \tag{3.16.2}$$

and so, if Φ is a function of x_1, and x_2, having values $\Phi_{0,0}$ at $x_1 = 0, x_2 = 0$; $\Phi_{1,0}$ at $x_1 = 1, x_2 = 0$; $\Phi_{0,1}$ at $x_1 = 0, x_2 = 1$, etc., we may approximate the ∇^2 operator at $x_1 = 0, x_2 = 0$ by the finite difference equation:

$$\nabla^2\Phi \simeq \frac{\Phi_{1,0} + \Phi_{0,1} + \Phi_{-1,0} + \Phi_{0,-1} - 4\Phi_0}{d^2} = 0 \tag{3.16.3}$$

assuming that the spacing of points in the x_1 and x_2 directions is identical. A method of solution of Laplace's equation for the stress distribution in a body therefore involves the 'guessing' of values for the stress function Φ at each of the nodes of a square grid into which the body is subdivided. The individual values of Φ at each node are then altered until the Laplace equation, as for node $x_1 = 0, x_2 = 0$ above, is satisfied for each node and, simultaneously, all boundary conditions are satisfied. At the boundaries, of course, the stresses $\sigma_{11}, \sigma_{22}, \sigma_{21}$ ($\sigma_{11} = \partial^2\Phi/\partial x_2{}^2$, etc., see equations 2.9.1), must be resolved parallel and orthogonal to the surface normal to calculate the surface forces.

It is possible to extend the finite difference equations to cater for the solution of the biharmonic equation $\nabla^2(\nabla^2\Phi) = 0$. The expression for $\nabla^4\Phi_{0,0}$ at

$x_1 = 0, x_2 = 0$ is rather complicated, but if we denote the sum of the values of nodes symmetrically disposed about 0,0 by Σ (e.g. $\Sigma_4\Phi_{1,0}$ would represent the first four terms in the numerator of equation 3.16.3), we may write:

$$\nabla^4\Phi_{0,0} = \frac{1}{d^4}(\Sigma_4\Phi_{2,0} - 8\Sigma_4\Phi_{1,0} + 2\Sigma_4\Phi_{1,1} + 20\Phi_{0,0}) \qquad (3.16.4)$$

i.e. if $\nabla^4\Phi_{0,0}$ is found to be, say, 1 rather than zero, a change in -1 on $\nabla^4\Phi_{0,0}$ to reduce it to zero would produce 'residual' effects on the value of Φ: at $x_1 = 0$, $x_2 = 0$ of -20; at $x_1 = \pm 1, x_2 = 0$ and at $x_1 = 0, x_2 = \pm 1$ of $+8$, and so on. These changes then obviously affect the value of $\nabla^4\Phi$ calculated at $x_1 = 1, x_2 = 0$ and at each of the other nodes contained in the $\nabla^4\Phi$ formula. The method has been used to calculate stress distributions in notched pieces by calculating consistent values of Φ in this way at each nodal point and deriving the appropriate values of stress from each value of stress function. The method is quite useful for the Laplace or Poisson equation ($\nabla^2\Phi$ = constant) but converges rather slowly for the purposes of determining stress functions. Special techniques have to be used at the boundaries of stress concentrators, because the grid size is so large ($x_1 = -2d$ to $x_1 = +2d, x_2 = -2d$ to $x_2 = +2d$) that values could otherwise not be calculated within $2d$ of the surface.

The method provided, however, the first *elastic*-plastic (as opposed to rigid-plastic, see Section 2.13) solutions for stress distributions around stress concentrators[20]. The elastic stress distribution conforms to the biharmonic equation. The plastic strains are equivoluminal (Poisson's ratio is 0.5) and the main assumption made is that the *incremental* principal shear strains are proportional to the magnitude and *instantaneous* directions of the principal shear stresses,

i.e. $$\Delta\gamma_1 : \Delta\gamma_2 : \Delta\gamma_3 = \tau_1 : \tau_2 : \tau_3 \qquad (3.16.5)$$

or, in equivalent manner:

$$(\Delta\epsilon_2 - \Delta\epsilon_3) : (\Delta\epsilon_3 - \Delta\epsilon_1) : (\Delta\epsilon_1 - \Delta\epsilon_2) = (\sigma_2 - \sigma_3) : (\sigma_3 - \sigma_1) : (\sigma_1 - \sigma_2) \qquad (3.16.6)$$

where Δ's represent incremental quantities.

Mises' yield criterion is employed,

i.e. $$(\sigma_1 - \sigma_2)^2 + (\sigma_2 - \sigma_3)^2 + (\sigma_3 - \sigma_1)^2 = 6\tau_Y{}^2 \qquad (3.16.7)$$

where τ_Y is the shear yield stress.

If we write:

$$\frac{(\Delta\epsilon_2 - \Delta\epsilon_3)}{(\sigma_2 - \sigma_3)} = \frac{(\Delta\epsilon_3 - \Delta\epsilon_1)}{(\sigma_3 - \sigma_1)} = \frac{(\Delta\epsilon_1 - \Delta\epsilon_2)}{(\sigma_1 - \sigma_2)} = \frac{3}{2}\Delta\lambda \qquad (3.16.8)$$

where $\Delta\lambda$ represents an increment in the parameter λ, we may deduce that:

$$2\Delta\epsilon_1 - \Delta\epsilon_2 - \Delta\epsilon_3 = \tfrac{3}{2}\Delta\lambda(2\sigma_1 - \sigma_2 - \sigma_3) \qquad (3.16.9)$$

and from this: $$\Delta\epsilon_1 = \Delta\lambda[\sigma_1 - \tfrac{1}{2}(\sigma_2 + \sigma_3)] \qquad (3.16.10)$$

The effective value of 'Young's modulus' is thus $1/\Delta\lambda$. Poisson's ratio is $\frac{1}{2}$. The *total* strain, ϵ is then given by summing the elastic (see equations 2.6.12) and plastic components, i.e. for an incremental total strain

$$\Delta\epsilon_1 = \frac{1}{E}[\Delta\sigma_1 - \nu(\Delta\sigma_2 + \Delta\sigma_3)] + \Delta\lambda[\sigma_1 - \tfrac{1}{2}(\sigma_2 + \sigma_3)] \qquad (3.16.11)$$

In terms of applied stress, σ_{11}, σ_{22}, etc., rather than principal stresses, Mises' criterion becomes: in plane stress,

$$\sigma_{11}{}^2 - \sigma_{11}\sigma_{22} + \sigma_{22}{}^2 + 3\sigma_{12}{}^2 \leqslant 3\tau_Y^2 \qquad (3.16.12)$$

in plane strain, *assuming that Poisson's ratio is 0.5 in the elastic as well as in the plastic region:*

$$(\sigma_{11} - \sigma_{22})^2 + 4\sigma_{12}{}^2 \leqslant 4\tau_Y^2 \qquad (3.16.13)$$

These conditions hold in the plastic region; $\nabla^4\Phi = 0$ holds in the elastic region. In plane strain, problems arise regarding the value of Poisson's ratio and the value of σ_{33}; the substitutions:

$\sigma_{33} = \nu(\sigma_{11} + \sigma_{22})$ in the elastic parts of the expressions for $\Delta\epsilon_{11}$ and $\Delta\epsilon_{22}$, and $\sigma_{33} = \frac{1}{2}(\sigma_{11} + \sigma_{22})$ in the plastic parts, have been made, giving, in the plastic enclave:

$$\Delta\epsilon_{11} = \frac{1+\nu}{E}[(1-\nu)\Delta\sigma_{11} - \nu\Delta\sigma_{22}] + \tfrac{3}{4}\Delta\lambda(\sigma_{11} - \sigma_{22}) \qquad (3.16.14)$$

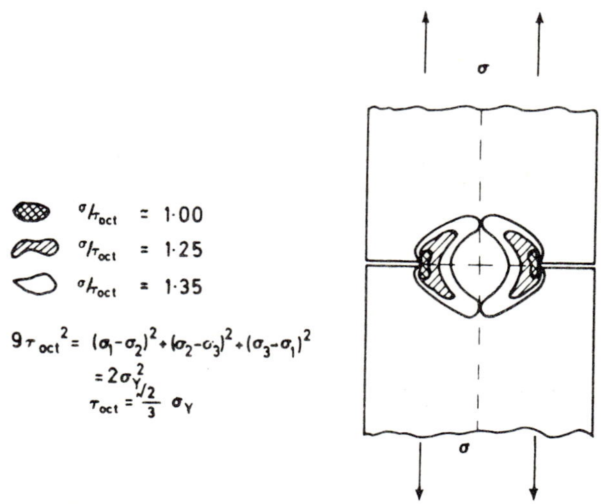

Figure 3.15 Development of plastic zones in precracked specimens in plane stress (elastic/plastic behaviour, after Stimpson and Eaton[21])

The finite difference equations are set up as follows. For the elastic deformation, we have: $\nabla^4\Phi = 0$. For the plastic deformation, using Mises' criterion:

$$(\sigma_{11} - \sigma_{22})^2 + 4\sigma_{12}^2 \leqslant 4\tau_Y^2$$

we write

$$\Phi_p = (\sigma_{11} - \sigma_{22})^2 + 4\sigma_{12}^2 - 4\tau_Y^2 \tag{3.16.15}$$

$$= \left[\left(\frac{\partial^2}{\partial x_2^2} - \frac{\partial^2}{\partial x_1^2}\right)\Phi\right]^2 + 4\left(\frac{\partial^2 \Phi}{\partial x_1 \partial x_2}\right)^2 - 4\tau_Y^2 \tag{3.16.16}$$

Thus, in solving a particular problem, we must select values of Φ such that $\nabla^4\Phi = 0$ in all the elastically-stressed regions and that $\Phi_p = 0$ in the plastic region. The partial differentials are determined from finite differences as before. Displacements must be obtained from strains, as given by equations of the form in 3.16.14 above. The method has been applied to determine elastic/plastic distributions in notched and cracked parts in plane strain and in plane stress; the predicted form[21] of the yield zone in a cracked specimen in plane stress is shown in *Figure 3.15.*

3.17 The Finite Element Method

In recent years, techniques based on finite difference equations have been largely superseded by finite element methods (see e.g. Dugdale and Ruiz[22]). The main principle involved in these methods is that the body, which we have hitherto regarded as a continuum, obeying certain types of stress-strain relationships, is considered as being replaced by a structural framework of elements, which are usually triangular or trapezoidal in shape for problems concerned with two-dimensional deformation. The framework forms a complete lattice, whose external form corresponds to that of the continuous body. The stress distribution throughout the body is calculated by considering the equilibrium of forces at the joining points, or nodes, of the lattice: the strain distribution is calculated in terms of the displacements suffered by these nodes.

In a conventional framework, we would normally consider the forces and displacements at the end of each beam, truss or strut. In general two-dimensional deformation, each end is subjected to net resolved forces in two orthogonal directions, and to a net bending moment. If the axes are X_1, X_2 and the beam lies parallel to one of these, we may write the conditions at end I of the beam in terms of a column vector, e.g.

$$\boldsymbol{F}^{\mathrm{I}} = \begin{bmatrix} F_1^{\mathrm{I}} \\ F_2^{\mathrm{I}} \\ M^{\mathrm{I}} \end{bmatrix} \tag{3.17.1}$$

The corresponding displacements and rotation may be written as another column vector,

$$u^{\mathrm{I}} = \begin{bmatrix} u_1{}^{\mathrm{I}} \\ u_2{}^{\mathrm{I}} \\ \omega^{\mathrm{I}} \end{bmatrix} \qquad (3.17.2)$$

It is assumed that, in general, each component of force is linearly proportional to each component displacement, so that a set of linear equations between F and u is produced. This is usually written in matrix form, e.g.

$$\mathbf{N}u = \begin{bmatrix} a_1 & b_1 & c_1 \\ a_2 & b_2 & c_2 \\ a_3 & b_3 & c_3 \end{bmatrix} \begin{bmatrix} u_1 \\ u_2 \\ \omega \end{bmatrix} = \begin{matrix} a_1u_1 + b_1u_1 + c_1u_1 \\ a_2u_2 + b_2u_2 + c_2u_2 \\ a_3\omega + b_3\omega + c_3\omega \end{matrix} \qquad (3.17.3)$$

A full analysis of the equilibrium of forces and the relationships between forces and displacements may readily be made for the components at each end of the member to give:

$$\begin{aligned} F^{\mathrm{I}} &= \mathbf{N}_{11}u^{\mathrm{I}} + \mathbf{N}_{12}u^{\mathrm{II}} \\ F^{\mathrm{II}} &= \mathbf{N}_{21}u^{\mathrm{I}} + \mathbf{N}_{22}u^{\mathrm{II}} \end{aligned} \qquad (3.17.4)$$

$$F = \begin{bmatrix} F^{\mathrm{I}} \\ F^{\mathrm{II}} \end{bmatrix} = \begin{bmatrix} \mathbf{N}_{11} & \mathbf{N}_{12} \\ \mathbf{N}_{21} & \mathbf{N}_{22} \end{bmatrix} \begin{bmatrix} u^{\mathrm{I}} \\ u^{\mathrm{II}} \end{bmatrix} = \mathbf{N}u \qquad (3.17.5)$$

where F and u are now six-component column vectors and $\mathbf{N}$ is a 6 x 6 matrix, composed of terms involving combinations of Young's modulus and the length, cross-sectional area and moment of inertia of the member. Some terms are zero.

This 6 x 6 matrix is known as the *stiffness matrix* (an analogy is to be found with the elastic coefficients in equations 2.6.2). The relationship between F and u may have to be further modified by a transformation matrix, if the beam does not lie initially along X_1 or X_2 and if the column vectors are referred to the new co-ordinates X_1' and X_2'. In a general framework, a member links two nodes. At each node, it will be possible to relate the net resolved nodal forces to nodal displacements through simultaneous linear equations of the form:

$$(F) = [\text{stiffness matrix}]\ (u) \qquad (3.17.6)$$

If two nodes, i and j, are connected by a single member (where i is equivalent to end I and j to end II), the contribution of the member to the total stiffness matrix is to add a term $[\mathbf{N}_{11}]_{ij}$ to the element situated on row i, column i; $[\mathbf{N}_{12}]_{ij}$ to row i, column j; $[\mathbf{N}_{21}]_{ij}$ to row j column i; and $[\mathbf{N}_{22}]_{ij}$ to row j, column j. A similar contribution is obtained from each component member of the framework.

The method for solving framework problems is divided into three stages:

(1) determining the stiffness matrix for an elementary member;
(2) assembling the members to form a complete framework, and determining the complete stiffness matrix;
(3) using the matrix to calculate unknown forces and displacements from known values at loading points and free surfaces.

For the triangular elements into which a continuous body may be divided, the total stress acting over the element is resolved as force components acting through the corners. Strains are represented by the displacements of the corners. Elements must deform so that the edges remain straight, although they may be rotated, displaced or changed in length. The general displacement of a point P within an element may be related to its position relative to the corners, i, j, k, and to the displacements of these corners: u_1^i, u_2^i, etc. The strains at P may then be found, using standard relationships (equation 2.4.5). If the strain column vector is

$$e = \begin{bmatrix} \epsilon_{11} \\ \epsilon_{22} \\ 2\epsilon_{12} \end{bmatrix} \tag{3.17.7}$$

and the corner displacement column vector is:

$$u = \begin{bmatrix} u_1^i \\ u_2^i \\ u_1^j \\ u_2^j \\ u_1^k \\ u_2^k \end{bmatrix} \tag{3.17.8}$$

we may write:

$$e = \mathbf{B}u \tag{3.17.9}$$

where **B** is the 3 x 6 matrix:

$$\mathbf{B} = \frac{1}{\Delta}\begin{bmatrix} (x_2^j - x_2^k) & 0 & (x_2^k - x_2^i) & 0 & (x_2^i - x_2^j) & 0 \\ 0 & (x_1^k - x_1^j) & 0 & (x_1^i - x_1^k) & 0 & (x_1^j - x_1^i) \\ (x_1^k - x_1^j) & (x_2^j - x_2^k) & (x_1^i - x_1^k) & (x_2^k - x_2^i) & (x_1^j - x_1^i) & (x_2^i - x_2^j) \end{bmatrix} \tag{3.17.10}$$

Δ is given by $\quad (x_1^j x_2^k + x_1^i x_2^j + x_1^k x_2^i) - (x_1^j x_2^i + x_1^i x_2^k + x_1^k x_2^j)$

and $x_1^i x_2^i$; $x_1^j x_2^j$; $x_1^k x_2^k$ are the co-ordinates of corners i, j and k respectively.

Stress is written as the column vector:

$$S = \begin{bmatrix} \sigma_{11} \\ \sigma_{22} \\ \sigma_{12} \end{bmatrix} \tag{3.17.11}$$

and the elastic stress–strain relationships (equations 2.6.7) are written in matrix form as:

$$S = \mathbf{D}e \tag{3.17.12}$$

where, for a cubic material,

$$\mathbf{D} = \frac{E}{(1-\nu^2)} \begin{bmatrix} 1 & \nu & 0 \\ \nu & 1 & 0 \\ 0 & 0 & \frac{1-\nu}{2} \end{bmatrix} \quad \text{in plane stress} \tag{3.17.13}$$

$$\mathbf{D} = \frac{E(1-\nu)}{(1+\nu)(1-2\nu)} \begin{bmatrix} 1 & \nu/(1-\nu) & 0 \\ \nu/(1-\nu) & 1 & 0 \\ 0 & 0 & \frac{1-2\nu}{2(1-\nu)} \end{bmatrix} \quad \text{in plane strain} \tag{3.17.14}$$

Note that the final term in the column vector for e has been taken as $2\epsilon_{12}$ (i.e. the engineering shear γ, rather than the pure shear $\gamma/2$, see Section 2.4).

The internal stresses in an element are related to the forces acting on the corners by equating the work done by both systems, and the stiffness matrix for a single element is found to be:

$$\mathbf{N} = (\mathbf{B}^t \, \mathbf{D} \, \mathbf{B}) \times (\text{area of triangle}) \times \text{thickness} \tag{3.17.15}$$

where $\mathbf{B}$ and $\mathbf{D}$ are defined by equations 3.17.10 and 3.17.13 or 3.17.14 and $\mathbf{B}^t$ is the transpose of $\mathbf{B}$ (i.e. the matrix $\mathbf{B}$, with columns and rows interchanged). The forces on the corners of the elements are then related to their displacements through the equation:

$$F = \mathbf{N}u \tag{3.17.16}$$

where F and u are columns of six elements and $\mathbf{N}$ is a 6 x 6 matrix. Given the co-ordinates of i, j and k, Young's modulus and Poisson's ratio, it is possible to calculate $\mathbf{N}$ directly given computer facilities to handle the linear equations.

The complete stiffness matrix is then assembled as for the open, structural framework. The contribution of a triangular element I, J, K, to the stiffness matrix consists, by analogy with the framework, of adding the term $[\mathbf{N}_{IJ}]_{IJK}$ to row I, column J., etc., where $[\mathbf{N}_{IJ}]$ is a 2 x 2 matrix. Many terms in the total stiffness matrix turn out to be zero. The solution of a problem involves the determination of unknown forces and displacements from known values on the boundaries of the body. Once these have been calculated, the strains and stresses may be determined from equations 3.17.9 and 3.17.12. The procedure followed is to find displacements which are compatible and which also minimise the total potential energy of the system.

We proceed to examine two important stress distributions calculated by finite element methods.

3.18 Plane Strain Stress Distributions in Notched and Cracked Specimens

Finite Element Methods (FEM) are being increasingly applied to a wide variety of elastic and elastic/plastic stressing problems, and one recent analysis has treated the elastic/plastic stress distribution (with linear work-hardening in the plastic region) in a notched bar subjected to pure bending in plane strain. The mesh was made particularly fine near the root of the notch, so that the local stress distribution could be examined in detail.

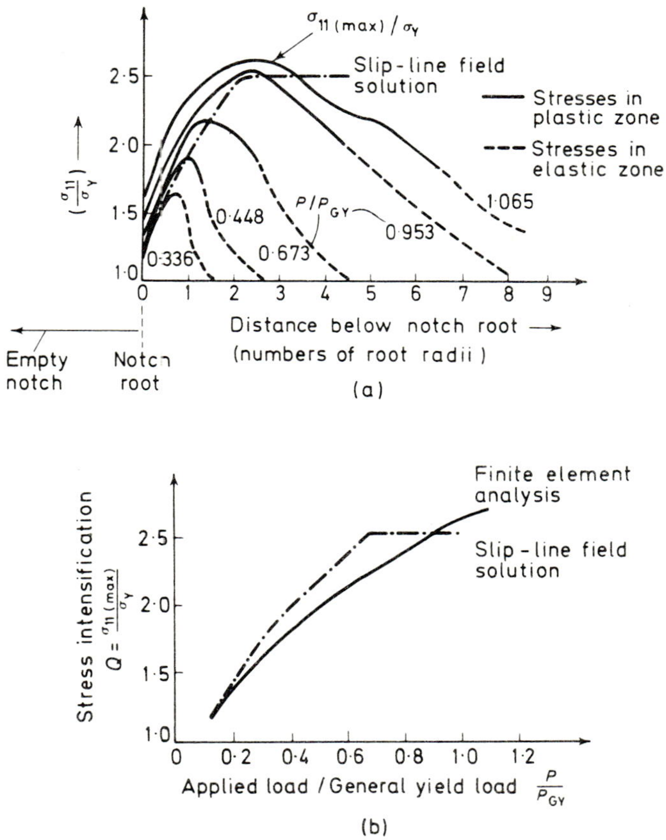

Figure 3.16 Finite element analysis of stress below a deep notch in pure bending. (a) Variation of maximum principal stress (σ_{11}) with distance below notch root for various applied loads (P/P_{GY}). (b) Dependence of stress intensification on applied load using Mises yield criterion (after Griffiths and Owen[23])

The distribution of the maximum principal stress σ_{11}, expressed as a multiple of the uniaxial yield stress σ_Y, is shown as a function of the distance below the notch for several different applied loads in *Figure 3.16a*. At low loads, the distribution is similar to that predicted by slip-line field theory (see Section 2.13), but the effect of the substantial amounts of work-hardening at higher loads is to raise the whole level of the curve. It is important to realise that at these high loads, the maximum value of σ_{11} occurs well behind the plastic/elastic interface. Values of stress intensification, $Q = \dfrac{\sigma_{11(\max)}}{\sigma_Y}$ are plotted *v.* load in *Figure 3.16b*. Again, the results are broadly similar to those given by slip-line field analyses, but show no sharp discontinuities. The mesh size is considered adequate, because refinement of the mesh gives no difference in the values of Q.

The method also enables notch root strain to be calculated as a function of applied load. The effective gauge length of the notch, over which the strain is constant, is found to be about 1.2 times the notch root radius. At about half the general yield load, the root strain is about 1.5%; at three-quarters, it is about 2.5%, and, at general yield, it is about 7%. These results agree well with earlier experimental values of notch strains[24].

The strain gradient below the notch is also calculable. For general yield loading, this decreases in an almost linear fashion from 7% at the notch root to some 1.5% at a distance equal to one root radius below the notch surface.

The plane strain stress distribution in a locally yielded zone around a crack must be calculated with extreme care, having in mind some physical model of the likely crack tip deformation field. The most appealing results are those due to Rice and Johnson[25] who, whilst drawing on finite element analyses for quantitative detail, ensure physical reality by devising a solution which is compatible, in the limit of a rigid/perfectly plastic material, with a local Prandtl slip-line field around the crack tip (see Section 2.12). The point is made particularly with regard to choice of the shape of the finite element.

If a Prandtl field is assumed for the limiting case, we can examine the situation drawn in *Figure 3.17a*. Here, in regions A and B, the stresses are constant because the slip lines are straight. The tension in region B is given by: $\sigma_{11} = 2\tau_Y(1 + \pi/2)$ (see equation 2.12.4) and it is clear that no intense strain concentration occurs as the crack tip is approached from directly ahead (i.e. moving from a positive position on the X_2 axis towards zero). In region C, however, the slip-lines form a fan which is centred on the crack tip: the shear strain, $\gamma_{r\theta}$, acting on faces normal to the radial lines in this fan region, becomes infinite as the crack tip is approached along a radial line:

$$\gamma = \gamma_0 \frac{R(\theta)}{r} \tag{3.18.1}$$

where γ_0 is the initial yield strain and $R(\theta)$ is an as yet undetermined function of θ. The point is that different shear strain contributions to the crack tip displacement are made for different directions of approach to the crack tip, so that a *discrete* opening displacement is predicted. Rice and Rosengren[26] had previously produced a dominant singularity solution for a power-law hardening material, which gives the maximum value of $R(\theta)$ as:

$$R_{max} = 0.286\left(\frac{K}{\sigma_Y}\right)^2 \tag{3.18.2}$$

where K and σ_Y have their usual meanings of stress intensity and yield stress (the 'plane stress plastic zone radius' r_Y is about $0.4\,(K/\sigma_Y)^2$ for comparison). The characteristic crack opening displacement is then

$$\delta_t = 2.5\frac{\sigma_Y}{E}R_{max} = 0.715\frac{K^2}{\sigma_Y E} \tag{3.18.3}$$

(cf. equation 3.13.12)

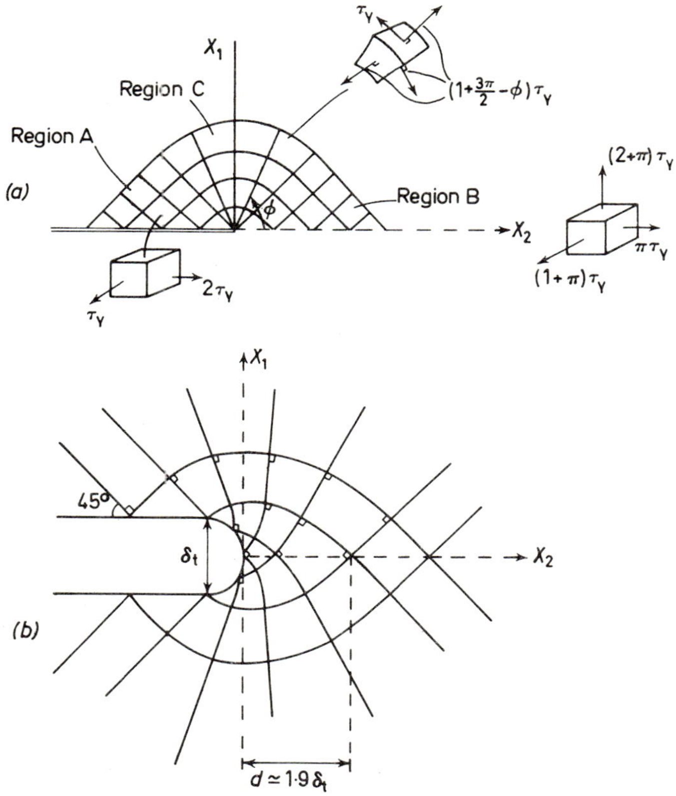

Figure 3.17 Local plane strain plastic fields around a crack tip. (a) Prandtl field in region of crack tip. (b) Modification of slip-line field due to crack-tip blunting (cf. Figure 2.21)

Rice makes the point that conventional numerical methods (finite differences or finite elements) have not been able to provide the required accuracy near the crack tip singularity. Boundary conditions near the crack tip are difficult to satisfy, particularly for the large finite difference grid required for the biharmonic equation (Section 3.16), but most analyses have, in fact, placed a single node at the crack tip, so that the displacement variations known to exist there were unable to be calculated. A modified four-sided finite element, bounded by lines r = const., θ = const., and therefore very suitable for the fan region, was developed and produced results giving the required $1/r$ singularity in shear strain. Two of the nodes of the element at the crack tip were situated at the same physical point but permitted different displacements, dependent on the radial path of approach.

The results then give:

$$R_{max} = 0.155(K/\sigma_Y)^2 \tag{3.18.4}$$

$$\delta_t = 2.7\frac{\sigma_Y}{E}R_{max} = 0.425\frac{K^2}{\sigma_Y E} \tag{3.18.5}$$

The computer results reveal that R_{max} is in fact, very similar to the extent of the plastic zone, which was calculated to extend a distance $0.175\,(K/\sigma_Y)^2$ along a line inclined at $\theta = 70°32'$ to X_2 (cf. 3.15.18 et seq.), and a distance 0.032 $(K/\sigma_Y)^2$ along X_2 (assuming $\nu = 0.3$). A more recent result[30] gives

$$\delta_t = 0.49\frac{K^2}{\sigma_Y E} \tag{3.18.6}$$

The stress distribution ahead of the crack, based on a power-law hardening material, would give rise to singularities at the crack tip (as for a linear elastic material), were it not for the blunting of the crack tip occasioned by its characteristic opening displacement. Supposing that the tip blunts to a semi-circle of diameter δ_t, the slip-line field shown in *Figure 3.17b* results. There is no longer a focusing of radial slip-lines at the tip and the stress distribution becomes similar locally to that for a circular notch (*Figure 2.21*). For a non-hardening material, σ_{11} increases from a value of σ_Y (Tresca's criterion) or $1.15\,\sigma_Y$ (Mises' criterion, as used by Rice) to the maximum value of approximately $3\sigma_Y$ at the fullest extent of the logarithmic spiral ($d \simeq 1.9\delta_t$).

For hardening materials, the situation is drawn in *Figures 3.18a and b*. Farther than about $2\delta_t$ from the crack tip, the stress distribution is given by the curve for the dominant singularity: this would become singular at the tip if the associated crack opening and blunting were ignored. It should be noted that the distance quoted above, for the spread of yield along the X_2 axis (at A), i.e. $0.032\,(K/\sigma_Y)^2$, is still an order of magnitude greater than the region influenced by the crack tip blunting. The maximum stress therefore occurs well behind the plastic/elastic interface in the X_2 direction. This does not conflict with the results given in *Figure 3.16* for the stress distribution in the notched bar. It is clear that, the higher the work-hardening exponent and the smaller the crack tip displacement (governed by σ_Y/E), the closer will this position be to the crack tip. In most cases, the maximum extent of the plastic zone, at some 70° to X_2, is over fifty times greater than the region affected by crack tip blunting.

The stress distributions ahead of a crack in plane strain, as shown in *Figure 3.18*,

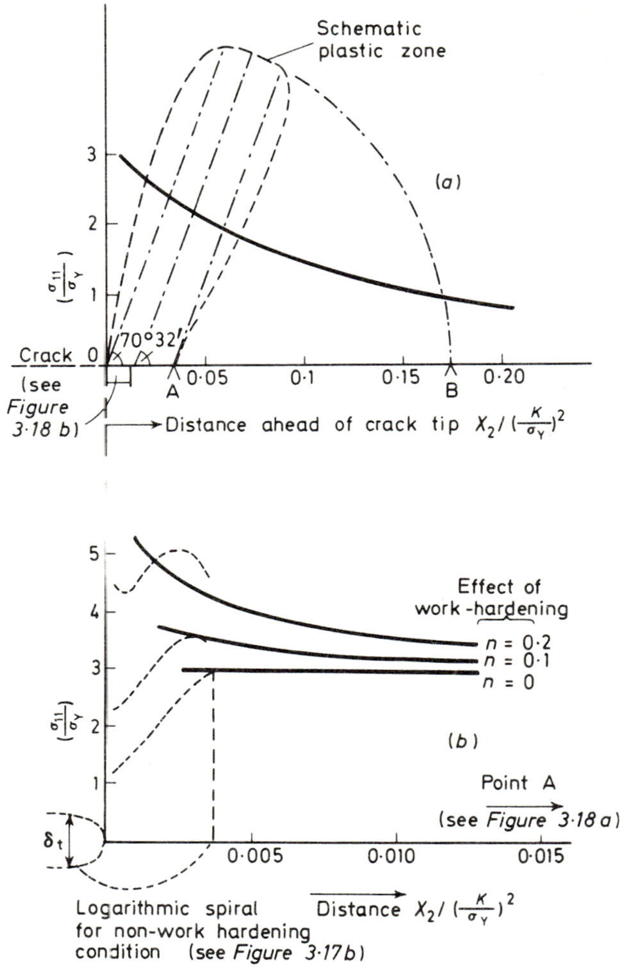

Figure 3.18 (*a*) *Variation of maximum stress ahead of crack with distance (point B represents the furthest extent of the plastic zone, at 70° 32′ to X_2. Point A represents its extent along X_2). Small-scale yielding model assumed.* (*b*) *As Figure 3.18*(*a*) *but for region dominated by crack tip singularity and including crack blunting (note very much smaller distances ahead of crack tip compared with Figure 3.18*(*a*). *Assuming* $\frac{\sigma_Y}{E} = 2.5 \times 10^{-3}$

therefore bear similarities to notched bar results, allow for hardening and provide plausible slip-line field analogies for non-hardening material. Detailed modifications may have to be made, but the general pattern seems to be the most convincing yet produced.

3.19 Conclusions

This chapter has described a variety of methods for determining stress distributions around sharp stress concentrators or cracks. The analytical solutions all involve the use of complex variables in one form or another. For the major features of crack tip stress fields, Westergaard's single stress functions are usually adequate, but, for more complicated situations, such as those pertaining to practical testpiece geometries, polynomial stress functions or mapping functions may have to be used. Alternatively, dislocation arrays may be used to model crack singularities. Finite difference equations have been generally superseded by finite element methods: computers are then essential to handle the large numbers of simultaneous linear equations represented by the stiffness matrix. Dislocation methods or superposition of stress functions can model elastic/plastic deformation in Mode I plane stress or Modes II (shear) or III (antiplane strain), but Mode I plane strain or work-hardening materials necessitate numerical methods. Recently, credible stress distributions in plane strain deformation have been presented for notched and for cracked pieces. The general situation is therefore an optimistic one.

Solutions have, however, been confined primarily to the limiting two-dimensional conditions of plane stress or plane strain. Many important practical situations involve three-dimensional deformation which cannot be modelled in terms of radial symmetry: for example, problems concerned with cracks in plates of moderate thickness and with 'part-through' thumbnail cracks. An analytical solution has been obtained for the elastic stress distribution around a circular hole in an infinite plate of arbitrary thickness[27], the plane strain solution for which was discussed in Section 2.11. This plane strain solution is obviously inappropriate for a plate of finite thickness, because it does not predict that the stress σ_{33} is zero on the top and bottom surfaces of the plate: $x_3 = \pm B/2$. The problem is solved generally by determining three polynomial stress functions. Choice of these is dependent on satisfaction of boundary conditions, symmetry requirements, the need to produce a finite total strain energy, and some anticipation of the general form of σ_{33} (i.e. taking a maximum value at $x_3 = 0$ and becoming zero at $\pm B/2$). The calculated variations of the maximum tensile stress, σ_{11} for a plate loaded by a tensile stress, σ, at infinity and for the corresponding stress, σ_{33} are shown in *Figure 3.19a* and *3.19b*. It can be seen that the elastic value of σ_{11} varies little with thickness, but that the variation of σ_{33} is

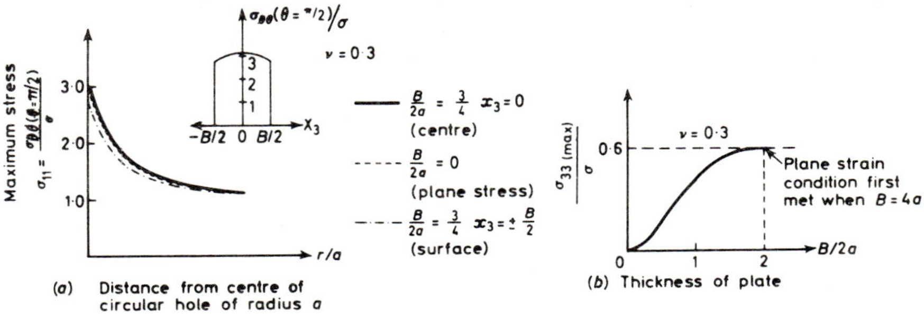

Figure 3.19 Variation of stresses in plate of finite thickness, B, containing hole of radius a (cf. Figure 2.9). (a) Variation of $\sigma_{11} = \sigma_{\theta\theta}\left(\theta = \frac{\pi}{2}\right)$ with distance. (b) Variation of σ_{33} with thickness

substantial, obtaining its plane strain value first when the ratio of plate thickness to hole radius is about 4.

A solution has been produced for the stress intensity ahead of a semi-elliptical crack in a plate of finite thickness[28]. Here, each part-through cracked section is treated as a line spring and the total spring constant of the piece (or its reciprocal, the compliance: see Section 4.4) is calculated by integrating the effect of each section through the width of the crack. The stress intensity may then be calculated by a virtual work argument (Section 4.5). In no case is the stress intensity as high as it would be if the crack had its maximum length throughout its width, but cracks which are wide and short compared with the plate thickness, approximate most closely to the crack of uniform length. The method of solution, however, is again essentially based on theories of two-dimensional deformation.

If three-dimensional solutions are sought, it is necessary to determine three stress functions, as for the problem of the circular hole in the plate of finite thickness. Neuber[2] has shown how the three stress functions may be applied to stress concentrators of hyperbolic or elliptical geometry and, recently, an attempt has been made to solve the three-dimensional crack problem, by determining the elastic stress distribution around a quarter-infinite crack in a half space[29]. A point of interest here is that, if σ_{ij} is expressed in terms of the spherical co-ordinates r, θ, χ by an equation of the form:

$$\sigma_{ij} = \text{const.}\, r^{-\lambda} \mathrm{f}_{ij}(\theta, \chi) \tag{3.19.1}$$

it is found that the exponent λ, which has the characteristic value 0.5 in all two-dimensional deformation modes, takes values depending on Poisson's ratio. For Poisson's ratio = 0.3, the exponent has a value of 0.45, thus replacing the familiar inverse square root singularity by the form:

$$\sigma_{11} = \text{const.}\, r^{-0.45}\, \mathrm{f}_{11}(\theta, \chi), \text{ etc.} \tag{3.19.2}$$

It is apparent that full three-dimensional analyses may somewhat modify the conclusions obtained from two-dimensional analyses.

If three-dimensional elastic solutions are rare, there is an even greater lack of knowledge concerning three-dimensional stress distributions in situations where there is a locally yielded zone ahead of a crack tip. As we shall see later, in Chapter 5, a knowledge of how the through-thickness stress, σ_{33}, varies with testpiece thickness is of great importance in explaining modes of deformation and fracture. An argument developed in Section 5.7 is concerned with stress-states, in test specimens, which can be classed neither as plane stress nor plane strain.

There is, therefore, still a need to produce good, three-dimensional elastic and elastic/plastic stress analyses. In principle, the finite element method (Section 3.17) can be extended directly to cope with such problems, although the necessary computing effort is vastly increased. Lacking such three-dimensional analyses, the present theory of fracture mechanics is based largely on the limiting states of plane stress or plane strain. We now proceed to follow the development of this theory and to examine the extent to which it can be used to explain experimental results.

REFERENCES

1. Inglis, C. E., *Trans. Instn Nav. Archit.*, **LV**, 1, 219 (1913)
2. Neuber, H., Kerbspannunglehre, Springer, Berlin (1938); translated as *Theory of Notch Stresses*, David Taylor Model Basin, Translation 74, (Nov., 1945)
3. Peterson, R. E., *Stress Concentration Design Factors*, Wiley, New York (1953)
4. Westergaard, H. M., *J. appl. Mech.*, A, 49 (June, 1939)
5. Irwin, G. R., 'Fracture', *Encyclopaedia of Physics*, Vol. VI, Springer, Berlin (1958)
6. Irwin, G. R., Kies, J. A. and Smith, H. L., *Proc. Am. Soc. Test. Mater.*, **58**, 640 (1958)
7. Paris, P. C. and Sih, G. C. M., *Proc. Am. Soc. Test. Mat.*, Spec. Tech. Pub. No. 381, 30 (1965)
8. Irwin, G. R., *J. appl. Phys.*, **24**, 361 (1957)
9. Irwin, G. R., *Appl. Mater. Res.*, **3**, 65 (1964)
10. Dugdale, D. S., *J. Mech. Phys. Solids*, 8, 100 (1960)
11. Burdekin, F. M. and Stone, D. E. W., *J. Strain Anal.*, **1**, 145 (1966)
12. Wells, A. A., *Br. Welding J.*, **10**, 563 (1963)
13. Bilby, B. A., Cottrell, A. H. and Swinden, K. H., *Proc. R. Soc.*, **A272**, 304 (1963)
14. Williams, M. L., *J. appl. Mech.*, **24**, 109 (1957)
15. Williams, M. L., *J. appl. Mech.*, **28**, 78 (1961)
16. Gross, B. *et al.*, *Technical Note D–2395 NASA August (1964);*
    Technical Note D–2396 NASA August (1964);
    Technical Note D–2603 NASA January (1965);
    Technical Note D–3092 NASA December (1965)
17. Smith, E., *Proc. R. Soc.*, **A285**, 46 (1965)
18. Smith, E., *Proc. R. Soc.*, **A282**, 422 (1964)
19. Southwell, R. V., *Relaxation Methods in Theoretical Physics*, Clarendon Press, Oxford (1946)
20. Allen, D. N. de G. and Southwell, R. V., *Phil. Trans. R. Soc.*, **A242**, XIV, 49 (1949–1950)
21. Stimpson, L. D. and Eaton, D. M., *Aero. Res. Lab. Report*, ARL 24, Wright-Patterson Air Force Base (July, 1961)
22. Dugdale, D. S. and Ruiz, C., *Elasticity for Engineers*, McGraw-Hill, New York (1971)
23. Griffiths, J. R. and Owen, D. R. J., *J. Mech. Phys. Solids,* **19,** 419 (1971)
24. Knott, J. F. and Cottrell, A. H., *J. Iron Steel Inst.*, **201**, 249 (1963)
25. Rice, J. R. and Johnson, M. A., *Inelastic Behaviour of Solids* (edited by M. F. Kanninen et al.), 641, McGraw-Hill, New York (1970)
26. Rice. J. R. and Rosengren, G. F., *J. Mech. Phys. Solids*, **16**, 1 (1968)
27. Sternberg, E. and Sadowsky, M. A., *J. appl. Mech.*, 27 (March, 1949)
28. Rice, J. R. and Levy, N., *J. appl. Mech.*, Paper No. 71–APM–20
29. Benthem, J. P., Report WTHO 74, Delft University of Technology, September, 1975
30. Rice, J. R., *Third International Congress on Fracture,* Munich (April, 1973)

Four

LINEAR ELASTIC FRACTURE MECHANICS

4.1 Introduction

The preceding two chapters have described the mathematical techniques by which the stress distributions near blunt stress concentrators or sharp cracks may be determined. These techniques have a two-fold purpose. First, for a structure of complicated geometry, they enable calculations to be made of the local stress in any particular region. Secondly, they can be used to determine the localised stresses around any crack-like defect in the structure, with the aim of relating the weakening effect of the defect on the material's resistance to fracture directly to the atomistic mechanisms by which the material ahead of the defect breaks. Although this final aim has been achieved for only a few very special situations, it is found in practice that fracture occurs at well-defined critical levels of stress intensity around a defect. The present chapter describes how crack-like defects reduce a material's fracture strength and how failure stress may be related to crack geometry through measurements of critical levels of stress intensity.

4.2 Ideal Fracture Strength

The tensile strength of an ideal crystalline body is the stress which must be applied to cause it to fracture across a particular crystallographic plane. We consider a cubic lattice of spacing, b_0 subjected to a tensile stress σ, as shown in *Figure 4.1.* To calculate the stress required to cause fracture across the plane XX, we suppose that σ is the sum of the forces, F, acting between pairs of atoms, such as CC$'$, on either side of the fracture plane, per unit area. To a first approximation, we may suppose that the value of F required to cause fracture is identical to that needed to separate an isolated pair of atoms such as C–C$'$. The calculation is approximate, because it ignores any interaction, that may exist between atom C and other atoms, such as A$'$, B$'$, D$'$, B$''$, C$''$, D$''$, etc. However, it serves to estimate the order of magnitude of fracture strength.

If we consider a pair of atoms such as C and C′, we can draw a curve to represent their energy of interaction as a function of distance of separation b. Such curves differ in detail, depending on the types of bonding associated with a particular material, but, for metals, they are generally of the form shown in

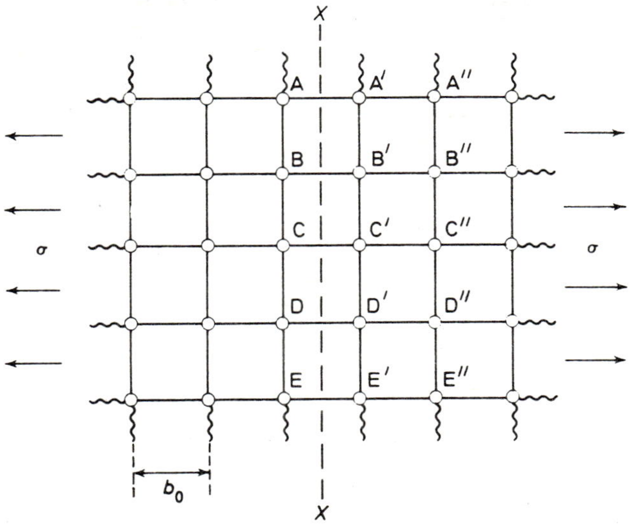

Figure 4.1 Cleavage across a lattice plane (XX)

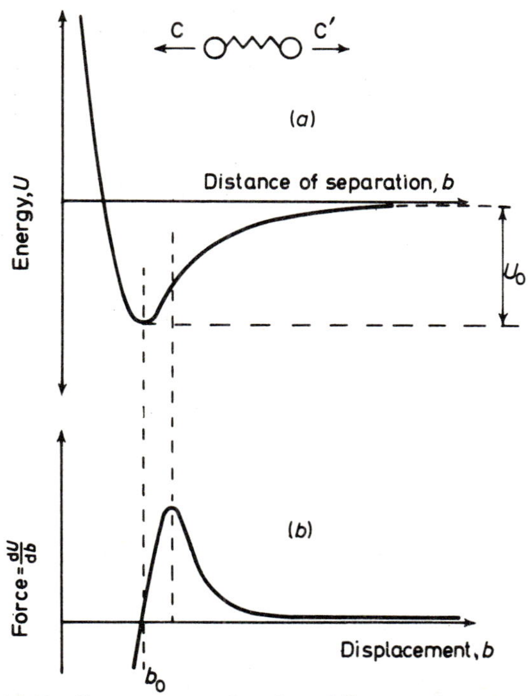

Figure 4.2 (a) Bonding energy as a function of distance of separation. (b) Force–displacement curve

Figure 4.2a. The resultant energy, U, exhibits a minimum at the equilibrium lattice spacing b_0. The total amount of energy which must be supplied to separate the two atoms to infinity is given by U_0. This 'work to fracture' in a crystalline solid is often equated to twice the surface tension, γ, of unit area of a free surface of the appropriate fracture plane of the solid, because the work done has provided enough energy to create two surfaces, each of energy

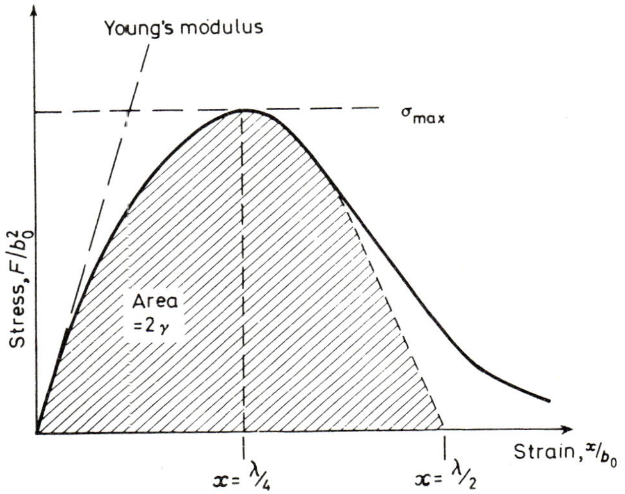

Figure 4.3 Atomic stress–strain curve

γ. Estimates of U_0 can therefore be made by extrapolating results obtained on the surface tensions of liquid droplets or by solid-state measurements made in 'zero-creep' experiments.

The force required to separate the atoms may be derived directly by differentiating the energy–distance curve with respect to distance to give,

$$F = \frac{dU}{db} \tag{4.2.1}$$

The resultant force–displacement curve is shown in *Figure 4.2b.*

The force is zero at the equilibrium spacing $b = b_0$ and attains its maximum at the point of inflection of the energy–distance curve. The initial slope of the force–displacement curve represents the stiffness of the atomic spring and is related directly to Young's modulus. This modulus therefore depends on the form of the energy–distance curve and so general relationships between modulus and type of atomic bonding may be readily deduced.

If we let $(b - b_0)$ be equal to x, we may write strain directly as x/b_0. We equate the stress, σ, to F/b_0^2. The atomic stress–strain curve is therefore as shown in *Figure 4.3.* The initial slope of this curve is Young's modulus. To determine the value of σ_{max}, the stress–displacement curve is usually approximated to that of half a sine wave, so that the relationship between σ and x is given by:

$$\sigma = \sigma_{max} \sin\left(2\pi\frac{x}{\lambda}\right) \tag{4.2.2}$$

where λ is the wavelength, i.e. when $x = \lambda/4$, $\sigma = \sigma_{max}$. The total area under this curve represents the work supplied when the plane is fractured and so we can equate this work to $+U_0 = +2\gamma$.

$$\frac{\lambda}{2\pi}\,\sigma_{max}\left[-\cos\left(\frac{2\pi x}{\lambda}\right)\right]_0^{\lambda/2} = 2\gamma \tag{4.2.3}$$

For very small displacements, we have

$$\sigma = \sigma_{max}\,2\pi\frac{x}{\lambda} = E\frac{x}{b_0} \tag{4.2.4}$$

and, substituting for λ in equation 4.2.3 we have

$$\frac{\sigma_{max}^2\,b_0}{E}\,[2] = 2\gamma$$

Hence, we derive as an expression for the ideal fracture strength,

$$\sigma_{max} = \sqrt{\left(\frac{E\gamma}{b_0}\right)} \tag{4.2.5}$$

The above expression was derived simply by considering the energies of interaction between pairs of atoms across the fracture plane. It is possible to derive similar expressions for fracture stress, by considering the interaction energies between a given atom, say, C, in *Figure 4.1* and its nearest, second-nearest, third-nearest, etc., neighbours and differentiating these to determine more accurate forms of atomic force–displacement curves. The calculations become very complicated and are suited to computer techniques. The interaction energies have been expressed as Morse potentials[1] of the form

$$U_{(b-b_0)} = U_0\left[e^{-2s(b-b_0)} - 2e^{-s(b-b_0)}\right] \tag{4.2.6}$$

where s is the reciprocal of the 'range' of the interatomic forces, and is hence dependent on the type of bonding. In other cases[2], the interatomic forces have been given as functions of the form:

$$F = \text{const.}\left[\left(\frac{b_0}{b}\right)^m - \left(\frac{b_0}{b}\right)^n\right] \tag{4.2.7}$$

where m and n are again dependent on the type of bonding. The exponent m is representative of the repulsive forces: for closed electron shells, m can be as high as 12. The value of n depends on the attractive forces: for ionic bonding, $n = 2$; for Van der Waal's bonding, $n = 7$. Often, the true interionic potentials are not known and it is then necessary to write pseudo-potential functions of the form given in equation 4.2.6 and to evaluate the various constants by calculating values of parameters such as the elastic constants (equations 2.6.2) or the coefficient of thermal expansion and comparing them with experimental values. The pseudo-potential may then be used to calculate the atomic stress–strain curve. The net conclusion of these results is that the fracture stress is of the order of $E/10$. This value may be estimated from equation 4.2.5, using the information that γ is of the order of 0.01 Eb_0 for many materials[3].

Equation 4.2.5 makes no predictions as to the strain associated with fracture, because the atoms have to be separated to infinity before they show no interaction energy. In practical terms, however, we may regard separation as having occurred once the force maximum has been attained. A rough estimate of this strain may be made by assuming that the stress–displacement curve (*Figure 4.3*) is triangular rather than sinusoidal, whence $\sigma_{max} = \sqrt{(2E\gamma/b_0)}$ and the strain at σ_{max} is $x/b_0 = \sigma_{max}/E = \sqrt{(2\gamma/Eb_0)}$. Substituting $\gamma \cong 0.01\ Eb_0$ as above, we have $x/b_0 \cong 0.14$. In fact, the strain to fracture for most materials lies roughly in the range 0.2 (for very stiff materials) to 0.4 (for rubbery materials).

4.3 Griffith Cracks

The theoretical fracture strength of a solid is of the order of $E/10$, but the strengths of crystals and glasses found in practice tend to be lower than this value by some two orders of magnitude. Griffith[4] first suggested reasons for this discrepancy between predicted and actual values and it is by developing his arguments that the present methods of measuring a material's fracture toughness have evolved. Griffith supposed that a macroscopically homogeneous test sample might contain small defects which enabled the stress to be concentrated sufficiently for the ideal fracture stress to be attained in small localised regions of the sample.

As a first step towards deciding what the fracture stress of a sample containing such a defect might be, we could imagine the defect as being elliptical in shape, with major axis $2a$. Then, if this were lying normal to the applied stress, σ, the concentrated value of stress at the ends of the ellipse would be, from the Inglis solution (equation 3.5.9)

$$\sigma_{11} = \sigma\left(1 + 2\sqrt{\frac{a}{\rho}}\right)$$

where ρ is the radius of the tip of the ellipse. If the defect were crack-like, we could take as a value for ρ the lattice spacing b_0. Then, since $2\sqrt{(a/b_0)} \gg 1$ we would obtain:

$$\sigma_{11} = 2\sqrt{\left(\frac{a}{b_0}\right)}\sigma \qquad (4.3.1)$$

and we could set, as a fracture criterion, $\sigma_{11} = \sigma_{max}$ as in equation 4.2.5,

i.e.

$$2\sigma\sqrt{\left(\frac{a}{b_0}\right)} = \sqrt{\left(\frac{E\gamma}{b_0}\right)} \qquad (4.3.2)$$

At fracture $\sigma = \sigma_F$, hence:

$$\sigma_F = \sqrt{\left(\frac{E\gamma}{4a}\right)} \tag{4.3.3}$$

It can be seen immediately that, if we put $\gamma \cong 0.01\ Eb_0$ as before, a defect length $2a$ of 5000 b_0, say 1 μm, is sufficient to lower the fracture strength by two orders of magnitude. We shall return to a consideration of the sizes of defects shortly, but it is appropriate first to point out the difficulties associated with the result obtained in equation 4.3.3. The Inglis solution for the stresses around an elliptical hole was derived strictly on the basis of linear elasticity and yet we predict from this result a crack-tip stress which we know to be well outside the realms of linear elastic behaviour. The left-hand side of equation 4.3.2 derives from linear elasticity: the right-hand side from a sinusoidal stress–strain curve. The left-hand side relates to the macroscopic applied stress: the right-hand side to the atomic force–displacement laws.

The main achievement of Griffith in providing a basis for the fracture strengths of bodies containing cracks was his realisation that it was possible to derive a thermodynamic criterion for fracture by considering the total change in energy of a cracked body as the crack length was increased. Only if the total energy decreased would the crack extend spontaneously under the applied stress. The value of the energetic approach is that, by considering energy changes in the body as a whole, it is possible to de-focus attention from the very highly strained region immediately surrounding the crack tip and yet derive a useful expression for the fracture stress. This general approach is followed in all later sections, but, at present, it is of interest briefly to trace the course of Griffith's own analysis.

The situation is simply as shown in *Figure 4.4* where a crack of length $2a$ is situated in an infinite body and lies normal to the applied tensile stress, σ. We

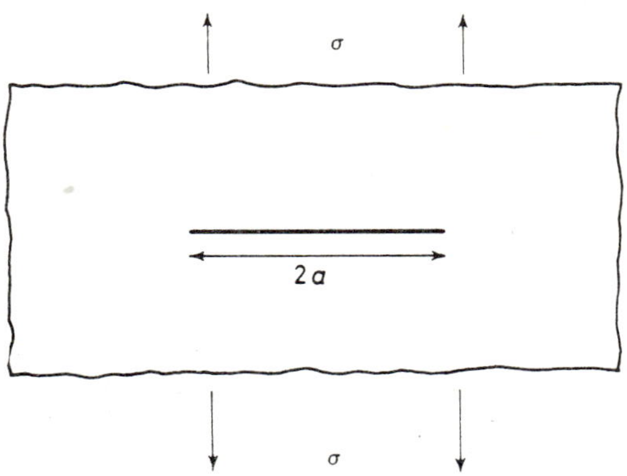

Figure 4.4 Griffith's crack: geometrical configuration

now consider the changes in energy which occur if the crack is extended by an infinitesimal amount. It is obvious, first, that new crack surfaces are created and since the stresses and displacements immediately ahead of the infinitesimally extended crack tip are to all intents and purposes identical to those ahead of the initial crack tip, the energy increase during crack extension is simply the 'work to fracture', 2γ, multiplied by the area of the new crack surfaces produced.

The other energy terms are best dealt with by supposing that it were possible to produce a macroscopic load–displacement curve for the cracked body for the two half-crack lengths, a, and $(a + \delta a)$. The resultant curves for elastic loading would then appear schematically as in *Figure 4.5.* Essentially, the body containing the longer crack is a weaker spring. Then under *fixed-grip* (i.e. fixed displacement $\equiv u_1$ fixed) loading, an *increase* in crack length from

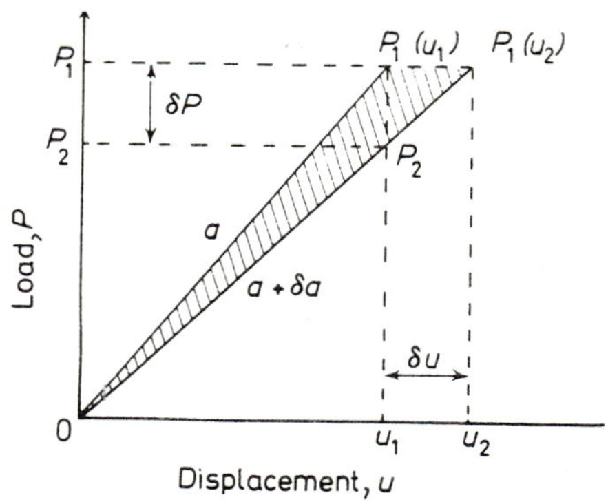

Figure 4.5 Elastic loading curves for crack lengths a and $(a + \delta a)$

a to $(a + \delta a)$ results in a *decrease* in stored elastic *strain energy* in the body, from $\frac{1}{2}P_1u_1$ to $\frac{1}{2}P_2u_1$, because the same displacement with a weaker spring implies less load and hence less strain energy stored. Under fixed grip conditions, crack extension therefore produces a *release* of stored elastic energy $\frac{1}{2}(P_1 - P_2)u_1$ in addition to the increase in surface energy.

Under *constant loading* conditions ($P = P_1$ = constant) the situation is slightly more complicated. Here, the weaker spring effect produces more displacement for a given load and so the *strain energy* stored for a crack of length $(a + \delta a)$ is greater $(\frac{1}{2}P_1u_2)$ than it is for a crack of length a $(\frac{1}{2}P_1u_1)$. The vitally important point, however, is that, during this increase of stored strain energy, the applied load has *moved* a distance $(u_2 - u_1)$ and so work, of magnitude $P_1(u_2 - u_1)$ has been done by the applied force on the system. The total *change in potential energy* is therefore a *decrease* of magnitude $P_1(u_2 - u_1) - \frac{1}{2}P_1(u_2 - u_1) = \frac{1}{2}P_1(u_2 - u_1)$. This magnitude is conveniently represented by the shaded area in *Figure 4.5* [triangle of base $(u_2 - u_1)$ and vertical height P_1].

Thus, under fixed grip conditions there is a decrease in strain energy of magnitude $\frac{1}{2}(P_1 - P_2)u_1$ [area $OP_{1(u_1)}P_2$ in *Figure 4.5*] as the crack extends from a to $(a + \delta a)$; under constant load, there is a decrease in potential energy of magnitude $\frac{1}{2}P_1(u_2 - u_1)$ [the area $OP_{1(u_1)}P_{1(u_2)}$ in the figure]. If we define

$(P_1 - P_2)$ as δP and $(u_2 - u_1)$ as δu, we are able to show that, as $\delta a \to 0$, the net energy release is the same for both situations. We write:

$$\text{strain energy release (fixed grips)} = -\tfrac{1}{2}u \,.\, \delta P \qquad (4.3.4)$$

$$\begin{aligned}\text{potential energy release (constant load)} &= -P\delta u + \tfrac{1}{2}P\delta u \\ &= -\tfrac{1}{2}P\delta u\end{aligned} \qquad (4.3.5)$$

Now, the relationship between u and P is given by:

$$u = CP \qquad (4.3.6)$$

where C is a constant for a given crack length, called the compliance of the system, by analogy with equation 2.6.3. As the change in crack length, δa, tends to zero, we may treat C as identical for crack lengths a and $(a + \delta a)$ and write, from equation 4.3.6

$$\delta u = C\delta P \qquad (4.3.7)$$

Using equation 4.3.6, equation 4.3.4 becomes:

$$-\tfrac{1}{2}u\delta P = -\tfrac{1}{2}CP\delta P \qquad (4.3.8)$$

Using equation 4.3.7, equation 4.3.5 becomes:

$$-\tfrac{1}{2}P\delta u = -\tfrac{1}{2}CP\delta P \qquad (4.3.9)$$

Alternatively, it may be shown geometrically that the shaded areas $OP_{1(u_1)}P_2$ and $OP_{1(u_1)}P_{1(u_2)}$ tend to the same value as the angle $P_1\hat{O}P_2$ tends to zero, (let $u_1\hat{O}P_2 = \theta$, $P_1\hat{O}P_2 = \delta\theta$ and draw $P_{1(u_1)}Q$ perpendicular to $OP_2P_{1(u_2)}$).

A most important statement follows from this result. *For an infinitesimally small amount of crack extension, the decrease in stored elastic energy of a cracked body under fixed grip conditions is identical to the decrease in potential energy under conditions of constant loading*[5].

Griffith recognised that the driving force for crack extension was the difference between the energy which could be released if the crack were extended and that needed to create new surfaces. His method for calculating energy release was rather complicated because he was considering energy changes in the body as a whole and therefore had to integrate the product of stress and strain over the whole infinite plate. We shall investigate a more straightforward method of calculation in the next section, but it may prove useful first to give an approximate solution for the problem, by considering two half-crack lengths a and $(a + \delta a)$ under fixed-grip conditions as in *Figure 4.6*. We suppose that the applied stress is σ and employ again the concept of stress-lines (Section 2.2). The stress-free regions above and below the crack are supposed to be roughly triangular in shape and to extend to a height βa. Then, for a length a, the strain energy release per unit thickness is given in plane stress by $\frac{1}{2}$ stress x strain x area,

i.e.
$$U = -\frac{1}{2}\sigma \,.\, \frac{\sigma}{E} \,.\, \beta a^2 \qquad (4.3.10)$$

or, for an increase da

$$dU = -\frac{\sigma^2}{E}.\ \beta a\, da \qquad (4.3.11)$$

In plane strain, the tensile strain associated with a tensile stress σ is $\sigma(1 - \nu^2)/E$ (see Section 2.6) and so the change in energy with crack length, $\partial U/\partial a$, becomes:

$$\frac{\partial U}{\partial a} = -\frac{\sigma^2 (1 - \nu^2)}{E}\beta a \qquad (4.3.12)$$

Griffith's accurate methods for calculation give:

$$U = -\frac{1}{2}\frac{\sigma^2 \pi a^2}{E} \qquad \frac{\partial U}{\partial a} = -\frac{\sigma^2 \pi a}{E} \text{ in plane stress} \qquad (4.3.13)$$

$$U = -\frac{1}{2}\frac{\sigma^2 \pi a^2}{E}(1 - \nu^2) \qquad \frac{\partial U}{\partial a} = -\frac{\sigma^2 \pi a}{E}(1 - \nu^2) \text{ in plane strain} \qquad (4.3.14)$$

By our previous arguments these values of $\partial U/\partial a$ also represent the decrease in potential energy when a crack extends by an infinitesimal amount da under constant load.

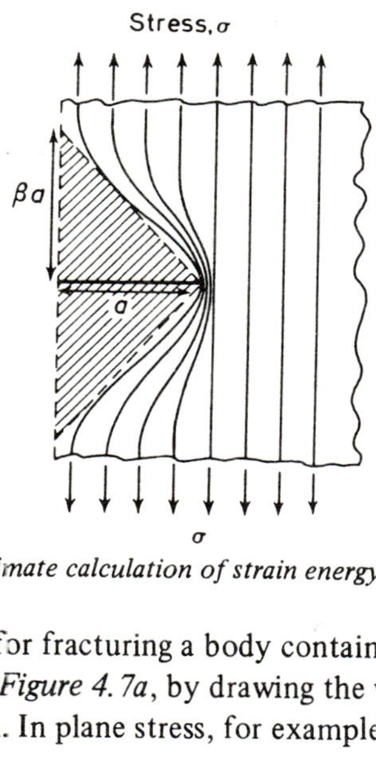

Figure 4.6 Model for approximate calculation of strain energy release rate (see also Figure 2.1)

The Griffith criterion for fracturing a body containing a crack of half length a may be visualised, as in *Figure 4.7a*, by drawing the way in which energy changes with crack length. In plane stress, for example, the total energy, W, is given by

$$W = U + S = -\frac{1}{2}\frac{\sigma^2 \pi a^2}{E} + 2\gamma a \qquad (4.3.15)$$

The maximum in the total energy curve is given by $\partial W/\partial a = 0$

i.e.

$$\frac{\sigma^2 \pi a}{E} = 2\gamma \qquad (4.3.16)$$

This situation is depicted in *Figure 4.7b*, where the intersection of the line $-(\partial U/\partial a) = \sigma^2 \pi a/E$ with that of 2γ is shown. The positive value of the slope $(\partial U/\partial a)$ is conventionally defined as the *strain energy release rate* (with respect

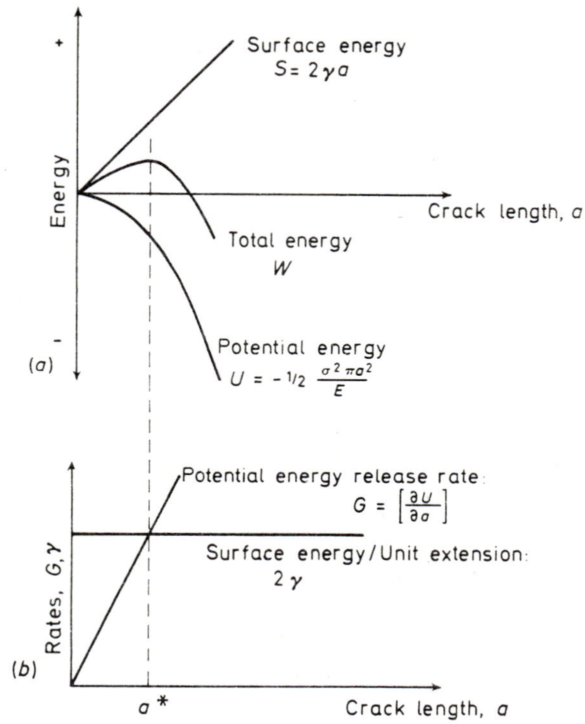

Figure 4.7 (a) Variation of energy with crack length. (b) Variation of energy rates with crack length. (a^* is the critical Griffith crack length)

to crack length) and is given the symbol G: more logically, it should be defined as the *potential energy release rate* to cater for crack extension under constant load. It should be noted that G has been defined for unit thickness.

For a given crack length a, the Griffith fracture stress is given, by equation 4.3.16, as:

$$\sigma_F = \sqrt{\left[\frac{2E\gamma}{\pi a}\right]} \quad \text{(in plane stress)} \tag{4.3.17}$$

$$\sigma_F = \sqrt{\left[\frac{2E\gamma}{\pi(1-\nu^2)a}\right]} \text{(in plane strain)} \tag{4.3.18}$$

These expressions show similarities in form to that given in equation 4.3.3, which was derived using the Inglis solution. The rationale of the Griffith approach is, however, very different in that it is concerned with changes in energy as a crack grows and can hence ignore the details of the fracture process at the crack tip. This thermodynamic approach must be emphasised.

Because it considers only the initial and final states and is not concerned with the details of fracturing over the length da, it represents a thermodynamic condition for fracture, which is necessary, but may not be sufficient. For example, the approach could treat the change in energy involved with the extension of a rather blunt elliptical hole and conclude that the energy of the system would decrease, yet the hole would not extend if there were not sufficient stress concentrated at its end to break atomic bonds.

Nevertheless, the Griffith theory swiftly achieved popularity and has since provided the basis for newer methods through the critical concept of treating fracture in terms of the change of energy remote from the immediate atomic environment of the crack tip.

The theory was proved initially by experiments on glass rods containing defects of various lengths[4]. The details of these experiments might not warrant too much further investigation, but the general role of defects in decreasing fracture strength was clearly proved as was the inverse square root relationship between crack length and fracture stress. Besides the values of γ obtained for glass, more recent experiments on brittle refractory metals[6] and on basal-plane cleavage of zinc, containing long initial cracks[7], provide good support for the Griffith theory in its original form when applied to materials which fracture in the classical, virtually elastic manner. Such materials are those where, by nature of their structure, much plastic flow is unlikely to occur before fracture. Typical examples would be:

(*a*) glasses, where no slip planes are present;

(*b*) {mica and other layer silicates; zinc (basal-plane cleavage)} these structures are very anisotropic and it is much easier to fracture by pulling the close-packed planes apart than to slip on other planes;

(*c*) {diamond; tungsten and other refractory metals} very high lattice resistance to dislocation movement (high Peierl's Nabarro force).

The Griffith theory does not explain the fracture strength of pure non-pre-cracked, crystals of metals such as zinc or iron. The cleavage strength of iron at 77 K is about 750 MN m^{-2}, i.e. approx. $E/300$. Taking

$$\frac{E}{300} \cong \sqrt{\left(\frac{2}{3} \cdot \frac{0.01\, E^2 b_0}{a}\right)}$$

we have:

$$2a \cong 1200\, b_0 \cong 0.3\ \mu\text{m}$$

An inherent defect, even as small as 0.3 μm, will not be present in a zone-refined crystal, but the case is more convincing for zinc where the critical defect size can be as great as several millimetres. Experiments have been performed, using zinc crystals which were smaller than, or comparable with, this critical defect size and yet which fractured at the low stress level[8]. In these pure crystals, it was necessary to produce crack nuclei and intense local stress concentrations by dislocation interactions and the fracture stress was related to the yield stress (mechanism control) rather than to Griffith's thermodynamic criterion.

Discrepancies also arise when attempts are made to use the Griffith theory in its original form to explain the fracture strengths of more ductile materials which contain cracks. This is the real aim of fracture theory because these are the materials commonly used in structural engineering, where fracture is a problem. The Griffith principle of relating crack extension to the change in energy of regions relatively remote from the crack tip is followed in all the methods of determining fracture toughness, but the changes in energy are derived differently. Two types of approach are used: one experimental; one theoretical.

4.4 Compliance Methods

If we refer back to the schematic load–extension curves drawn in *Figure 4.5* for an infinite body containing cracks of length a and $(a + \delta a)$ respectively, an experimental method for determining the strain (or potential) energy release rate, G, suggests itself immediately. That the energy release for an increase in crack length δa is given simply by $GB\delta a = \frac{1}{2} P\delta u$ where B is the thickness of the plate. It is obviously somewhat difficult to measure this (shaded area) quantity directly as $\delta a \rightarrow 0$, but it is relatively easy to rearrange terms to provide more easily measurable parameters.

In general, we may write:

$$u = CP \tag{4.4.1}$$

where C is the reciprocal of the slope of the load-deflection curve at a particular value of crack length, a, and, by direct analogy with stress–strain relationships (equation 2.6.3) is known as the *compliance* of the system. Then, for constant loading, $GB\delta a = \frac{1}{2} P^2\delta C$ or, in the limit as $\delta a \rightarrow 0$

$$G = \frac{1}{2}P^2 \left(\frac{\partial C}{\partial a}\right) \Big/ B \tag{4.4.2}$$

The problem then resolves itself into determining the compliance of the specimen as a function of crack length and measuring the gradient of the resultant curve at the appropriate initial crack length (see *Figures 4.8a* and *4.8b*)[9]. The method is clearly most useful for relatively small test specimens which can have precise measurements made on them in the laboratory. Under constant load, it is important to measure the movement of the loading points: in three-point-bend specimens, for example, the magnitude of the bending moment arm would not enter into any experimental calibration of a particular geometry. It would be included only if a theoretical compliance calibration, of application to different sized testpieces for example, were being developed. It is usually convenient and more sensitive practically to measure displacements close to the crack: the elastic strains in the rest of the specimen must then be determined before the 'compliance displacement' can be used to calculate G. In a small piece, the compliance method often provides the most direct method of allowing for free surfaces, and additional stress concentrators. It is important,

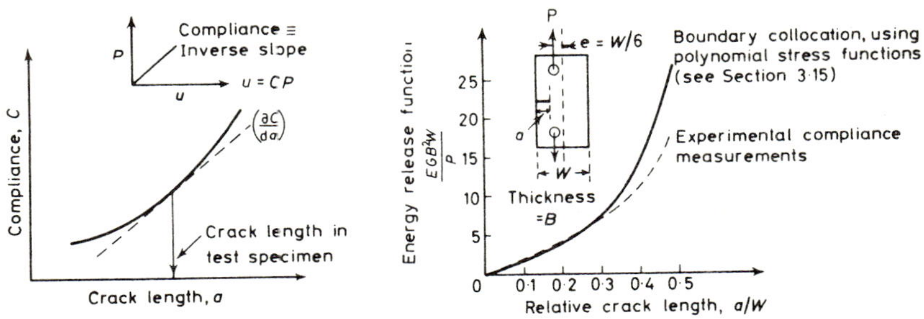

Figure 4.8 Compliance measurements. (a) Principle of measurement. (b) The compliance of an eccentrically-loaded SEN tension specimen (after Lubahn[9])

however, in the experimental calibration, to make allowance for effects such as plastic yielding near the loading points.

It is also apparent that the compliance method cannot be used for the large structures to which toughness measurements will be applied. Here, one must rely on theoretical methods for calculating the energy release as a crack advances: it is important also to ensure that these theoretical methods give good agreement with compliance measurements when both are dealing with test specimens, so that confidence in their application to the large pieces can be generated.

4.5 The Stress Intensity Approach[10]

The problem is to calculate the energy release when a crack of length a is extended to a length $(a + \delta a)$. Under constant loading, the release of potential energy is equal to the release of strain energy under fixed grips as $\delta a \rightarrow 0$. Rather than follow Griffith's methods, we concentrate on a crack tip region which is small compared with the body as a whole but sufficiently large with respect to atomic dimensions for us to be reasonably happy with the application of linear elasticity theory. The physical situation is shown in *Figure 4.9* where the stress distribution ahead of a crack of length a and the displacement distribution within a crack of length $(a + \delta a)$ are both shown. From Westergaard's solutions with change of axes (Section 3.7) we have:

$$\sigma_{11} = \frac{\sigma x_2}{(x_2^2 - a^2)^{\frac{1}{2}}} \text{ (cf. equation 3.7.1)}$$

and, in plane strain,

$$u_{11} = 2(1 - \nu^2)\frac{\sigma}{E} [(a + \delta a)^2 - x_2^2]^{\frac{1}{2}} \text{ (cf. equation 3.7.8)}$$

or, by writing $r = (x_2 - a)$

$$\sigma_{11} = \frac{\sigma\surd(\pi a)}{\surd(2\pi r)} = \frac{K}{\surd(2\pi r)} \tag{3.10.2}$$

$$u_{11} = 2(1 - \nu^2)\frac{\sigma}{E}\surd(2a)\,(\delta a - r)^{\frac{1}{2}} \tag{4.5.1}$$

We may evaluate the change in energy in the body as a whole by calculating the work done by the surface forces acting across the length δa when the crack is closed from length $(a + \delta a)$ to length a[10]. This virtual work principle is often

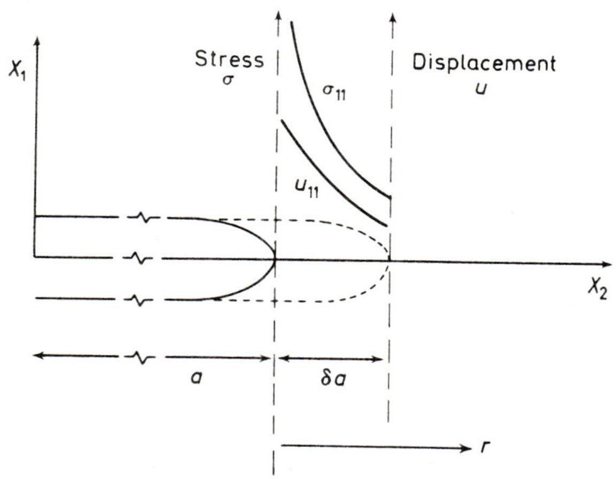

Figure 4.9 Calculation of G from work done by crack tip stress field

of great use in problems involving energy: another very good example is seen in the calculation of the strain energy of an edge dislocation[2]. The energy change may then be expressed as:

$$G\delta a = \int_0^{\delta a} \sigma_{11} u_{11}\, dr \tag{4.5.2}$$

where the thickness is taken as unity. If, for example, both σ_{11} and u_{11} were independent of r, we would simply recognise the integral as $\sigma \,.\, u \,.\, \delta a$ where

$\sigma \times \delta a \times 1$ (unit thickness) would be a constant force doing work by moving through a constant displacement. Physical situations in which both σ and u are constant will be considered later (Sections 5.4 and 6.2). As it is, we have, from the Westergaard solutions (3.7.5 and 4.5.1)

$$G\delta a = 2(1 - \nu^2)\frac{\sigma^2 a}{E}\int_0^{\delta a}\left(\frac{\delta a - r}{r}\right)^{\frac{1}{2}} \mathrm{d}r \tag{4.5.3}$$

Making the substitution $r = \delta a \sin^2 \omega$, this integral may be readily evaluated to give:

$$G\delta a = \frac{\sigma^2 \pi a}{E}(1 - \nu^2)\,\delta a \tag{4.5.4}$$

If, from equation 3.10.2, we now make the standard substitution $K = \sigma\sqrt{(\pi a)}$ we have: in plane strain

$$G = \frac{K^2}{E}(1 - \nu^2) \tag{4.5.5}$$

and, by an identical argument, in plane stress:

$$G = \frac{K^2}{E} \tag{4.5.6}$$

where K is the standard stress intensity, calculated by the techniques of elastic stress analysis described in Chapter 3. A useful way of remembering the result (and for this purpose only), is to recall that, if K is proportional to stress and (K/E) or $(K/E)(1 - \nu^2)$ is proportional to strain, then the strain (or potential) energy release rate is proportional to 'stress x strain', i.e., to $K(K/E)$ or $K(K/E)(1 - \nu^2)$.

Equations 4.5.5 and 4.5.6 are of very general application. It will be recalled that, whatever the macroscopic geometry of the piece in which a crack is contained, the stress singularity around the crack tip can always be represented in the form $\sigma = K/\sqrt{(2\pi r)}$, where K takes a value appropriate to the geometry and stress system. Similarly, the displacement singularity may be expressed uniquely in terms of K. Whatever the macroscopic geometry or applied stress system, application of the Colonetti theorem to calculate the work done in crack closure shows that the total change in energy in the body as a whole can be expressed directly in terms of the individual stress intensities which characterise the nature of the singularities in the crack tip stress- and displacement-fields. For example, suppose that a crack of half-length a, situated in an infinite body, were subjected to a tensile stress (Mode I opening), to a shear stress (Mode II sliding) and to an antiplane shear stress (Mode III). The local crack tip stresses which could do work would be σ_{11} (Mode I), the shear σ_{12} (Mode II) and the shear σ_{13} (Mode III). These would be characterised by the stress intensity parameters K_{I}, K_{II} and K_{III} respectively. Work would be done only when these stresses moved in appropriate directions and the expression for the strain energy (or the equivalent potential energy) release rate per unit thickness would be

$$G = \lim_{\delta a \to 0} \frac{1}{\delta a} \int_0^{\delta a} (\sigma_{11} u_{11} + \sigma_{12} u_{12} + \sigma_{13} u_{13})\, dr \tag{4.5.7}$$

Evaluating the integral as before, and assuming plane strain conditions for Modes I and II, we have:

$$EG = (1 - \nu^2) K_I^2 + (1 - \nu^2) K_{II}^2 + (1 + \nu) K_{III}^2 \tag{4.5.8}$$

or, in plane stress for Modes I and II:

$$EG = K_I^2 + K_{II}^2 + (1 + \nu) K_{III}^2 \tag{4.5.9}$$

Similarly, for a crack opened by uniform tension (K_I singularity) and also by wedging forces (K_w singularity) we would obtain, for plane strain deformation:

$$EG = (K_I^2 + K_w^2)(1 - \nu^2) \tag{4.5.10}$$

and for plane stress deformation:

$$EG = K_I^2 + K_w^2 \tag{4.5.11}$$

The power of this method of determining energy release rates lies in the fact that they are expressed directly in terms of standard values of stress intensity, which are now available and tabulated for a wide variety of geometries and stress systems. The application to large structures is particularly important because, in these, it is obviously impossible to make anything in the nature of an experimental compliance calibration to determine G.

Griffith's original concept of energy changes is well illustrated by this method. No mention of fracture has been made in the interrelation of K and G. What we calculate, and express as $G\,.\,da$ or $(K^2/E).\,da$ is the energy per unit thickness which would be released if the crack were to move forward by an amount da. Whether or not this is sufficient to cause catastrophic crack propagation depends on whether or not it reaches a critical magnitude. On the Griffith theory for perfectly elastic fractures, it would have to exceed the work needed to separate the two surfaces, 2γ.

As before, the Griffith criterion, in plane strain, would be written:

$$G_{crit} \cdot da = \frac{K_{crit}^2}{E} (1 - \nu^2)\, da = 2\gamma\, da \tag{4.5.12}$$

for unit thickness, i.e. by substituting $K = \sigma\sqrt{(\pi a)}$ and giving the value of σ at fracture the symbol σ_F, we obtain:

$$\sigma_F = \sqrt{\left[\frac{2E\gamma}{\pi(1 - \nu^2)a}\right]}$$

as before (equation 4.3.18). We shall now consider situations where the critical value of G is substantially greater than 2γ.

4.6 Quasi-brittle Fracture

Although the Griffith theory in its original form was considered to provide a good explanation for the fracture strengths of very brittle materials, it was realised that modifications to the theory would have to be made if it were going to be equally successful in explaining the fracture strengths of the more ductile materials commonly used in structural engineering. Experiments on the fracture at room temperature of large, thin sheets of aluminium containing central cracks gave cause for optimism that the theory could be of application, because it was found that the fracture stress was again of the form:

$$\sigma_F = \sqrt{\left(\frac{E\,.\,\text{const.}}{\pi a}\right)} \tag{4.6.1}$$

The constant was, however, found to be very much greater than the surface energy of the material[11].

These results led Orowan[12] and Irwin[13] independently to suggest that the energy release in the specimens was to a large extent dissipated by producing plastic flow around the crack tip, so that the critical value at fracture was apparently much greater than 2γ. However, it appeared that the amount of plastic work in the crack tip region which preceded unstable crack propagation was independent of the initial crack length and was hence as characteristic a measure of the material's resistance to fracture as would be its surface energy if it were breaking in a completely elastic manner. The important point was that the amount of plastic flow at instability was very much smaller in extent than either the crack length or the width of the sheet, so that, as far as the macroscopic energy release rate was concerned, methods of linear elasticity could still be used to relate crack tip events to the applied stress. If, for example, we referred back to the loading curves shown in *Figure 4.5*, it is reasonable to suppose that a small amount of crack tip plasticity will not alter significantly the slopes either for crack length a or $(a + \delta a)$ and so the relationship between G and applied load still holds. We shall examine shortly how the relationship between G and K is modified by rather larger amounts of crack tip plasticity.

Orowan re-wrote the Griffith relationship to give

$$\sigma_F = \sqrt{\left[\frac{E(2\gamma + \gamma_P)}{\pi a}\right]} \tag{4.6.2}$$

where γ_P represented the energy expended in the plastic work necessary to produce unstable crack propagation. Since it was found experimentally that $(\gamma_P + 2\gamma)$ was very much greater than 2γ, equation 4.6.2 could be rewritten

$$\sigma_F = \sqrt{\left(\frac{E\gamma_P}{\pi a}\right)} \tag{4.6.3}$$

and values of γ_P could then be determined directly from the fracture stresses of specimens containing cracks of known lengths.

Irwin's approach was similar, but he took more pains to justify the use of a linear elastic approach to relate fracture stress to crack length, even though crack-tip plasticity was preceding fracture. His results were expressed in terms of the critical value of strain (or potential) energy release rate at which unstable propagation occurred. This value, G_{crit}, provided a convenient parameter to include all supplementary energy-dissipating terms, such as plastic flow, which could in turn produce heat or sound, in addition to the work required to fracture the lattice. The constancy of G_{crit}, and hence its use as a measure of a material's resistance to fracture, will be found to depend critically on experimental testing conditions, but, for situations where small amounts of local plastic flow precede crack extension, which we shall call '*quasi-brittle*' behaviour, the critical value can always be related to the failure stress by linear elastic methods. Irwin's parameter, G_{crit}, became known as a material's 'fracture toughness', although this term is now generally reserved for the associated value of stress intensity, K_{crit}, as defined through equations 4.5.5 or 4.5.6. The development of fracture mechanics testing since Irwin's initial work has been to define experimental conditions under which reproducible toughness measurements, which have relevance to service application, can be measured in laboratory tests. The practical details of this development are discussed in the following chapter.

It is worth examining first how the linear elastic stress analysis is modified by the presence of the (small) plastic zone which is present before the crack extends in an unstable manner. In the simplest approach, use is made of the concept of the 'notional', or equivalent, elastic crack, described in Section 3.12. Here, in plane stress, the presence of a plastic zone of total extent $2r_Y = K^2/\pi\sigma_Y^2$ produces the same effect as that of a crack of half-length $(a + r_Y)$. The failure stress is then given by:

$$\sigma_F = \sqrt{\left[\frac{EG_{crit}}{\pi(a + r_Y)}\right]} \tag{4.6.4}$$

$$= \sqrt{\left[\frac{EG_{crit}}{\pi(a + \sigma_F^2\, a/2\sigma_Y^2)}\right]} \tag{4.6.5}$$

or, in terms of the critical stress intensity:

$$K_{crit} = \sigma_F \sqrt{[\pi a(1 + \sigma_F^2/2\sigma_Y^2)\ ]} \tag{4.6.6}$$

For fractures well below yield ($\sigma_F \ll \sigma_Y$), the correction factor is negligible. The difference between equation 4.6.6, and the simple relationship,

$$K_{crit} = \sigma_F \sqrt{(\pi a)} \tag{4.6.7}$$

(see equation 3.10.2) is about 4% when $(\sigma_F/\sigma_Y) = 0.4$; 8.5% when $(\sigma_F/\sigma_Y) = 0.6$; and 15% when $(\sigma_F/\sigma_Y) = 0.8$.

A similar conclusion is reached if the fracture stress is related to the plastic zone size calculated, using the Dugdale model, equation 3.13.7. Here, we obtained:

$$\frac{c}{a} = \frac{a + 2r_Y}{a} = \sec\left(\frac{\pi\sigma}{2\sigma_Y}\right)$$

where $2r_Y$ is the total extent of plasticity ahead of the crack. By expanding the secant term, we obtain:

$$1 + \frac{2r_Y}{a} = \left(1 - \frac{\pi^2\sigma^2}{8\sigma_Y^2} + \ldots\right)^{-1} \tag{4.6.8}$$

$$r_Y = \frac{\sigma^2\pi^2 a}{16\sigma_Y^2} \tag{4.6.9}$$

If this value is now used to calculate the size of the equivalent elastic crack, we have

$$K_{crit} = \sigma_F \sqrt{\left[\pi a\left(1 + \frac{\pi^2\sigma_F^2}{16\sigma_Y^2}\right)\right]} \tag{4.6.10}$$

An alternative approach is to define the stress intensity, K, as $(E\sigma_Y\delta)^{1/2}$, using equation 3.13.12. Then, if equation 3.13.11 is expanded to second-order terms, we obtain[14], with the substitution of K_{crit} for K and σ_F for σ:

$$K_{crit} = \sigma_F \sqrt{\left[\pi a\left(1 + \frac{\pi^2\sigma_F^2}{24\sigma_Y^2}\right)\right]} \tag{4.6.11}$$

Clearly, equations 4.6.6, 4.6.10 and 4.6.11 are all of similar form. It is also possible to substitute the full equation 3.13.11 into 3.13.12, to give:

$$K_{crit}^2 = \frac{8\sigma_Y^2}{\pi} a \ln\left[\sec\left(\frac{\pi\sigma_F}{2\sigma_Y}\right)\right] \tag{4.6.12}$$

If an 'apparent' K_{crit} value, K_{app}, is defined in terms of σ_F through equation 4.6.7 and equation 4.6.12 is rearranged, we obtain:

$$K_{app} = 2\sigma_Y\left(\frac{a}{\pi}\right)^{1/2} \cos^{-1}\left\{\exp\left(\frac{-\pi K_{crit}^2}{8\sigma_Y^2 a}\right)\right\} \tag{4.6.13}$$

which has been suggested[15] as a means of calculating toughness values, when extensive plasticity precedes instability.

In each case, however, it should be emphasised that these methods are approximate and that the relationship between energy release $(1/B)(dU/da) = G = K^2/E$, as proved for the linear elastic case, equation 4.5.6, has not been proved for the mixed elastic/plastic problem. The values of 'K_{crit}' may *characterise* fracture but cannot be used in a genuine Griffith energy balance if the amount of plasticity preceding instability is substantial.

4.7 Conclusions

This chapter has shown how the principles of fracture toughness testing have been developed from the original Griffith analysis. Unstable crack propagation occurs when the rate of release of strain energy (fixed grips) or potential energy (constant load) exceeds a critical value, which is equal to the surface energy for ideally elastic fractures. In practice, structural metals fracture under 'quasi-brittle' conditions and exhibit critical values of toughness, which are composed predominantly of the plastic work around the crack tip which precedes instability. The constancy of such toughness values in different testpiece geometries and under different conditions of test temperature and strain rate can be established only by experiment or by a full understanding of the factors which control the amount of plastic flow preceding instability. The following chapter describes how experimental techniques have been developed to enable consistent values of toughness to be measured and the micro-mechanisms of fracture at a crack tip are discussed in Chapters 7 and 8, to show how these may sometimes be used to predict the point of instability.

REFERENCES

1. Gilman, J. J., *Mechanical Behaviour of Crystalline Solids*, National Bureau of Standards Monograph, 59, 79 (1963)
2. Cottrell, A. H., *Dislocations and Plastic Flow*, Clarendon Press, Oxford (1953)
3. Cottrell, A. H., *Tewksbury Symposium on Fracture*, 1, University of Melbourne (1963)
4. Griffith, A. A., *Phil. Trans. R. Soc.*, **A221**, 163 (1920)
5. Paris, P. C. and Sih, G. C. M., *Am. Soc. Test. Mat.*, Spec. Tech. Pub. No. 381, 30 (1965)
6. Hull, D., Beardmore, P. and Valintine, A. P., *Phil. Mag.*, **12**, 1021 (1965)
7. Maitland, A. and Chadwick, G. A., *Phil. Mag.*, **19**, 645 (1969)
8. Deruyterre, G. and Greenhough, A. P., *J. Inst. Metals*, **84**, 337 (1955)
9. Lubahn, J. D., *Proc. Am. Soc. Test. Mat.*, **59**, 885 (1959)
10. Irwin, G. R., *Trans. Am. Soc. Mech. Engrs, Jnl. appl. Mech.*, **24**, 361 (1957)
11. Irwin, G. R., Kies, J. A. and Smith, H. L., *Proc. Am. Soc. Test. Mat.*, **58**, 640 (1958)
12. Orowan, E., *Trans. Inst. Engrs Shipbuilders Scotland*, **89**, 165 (1945)
13. Irwin, G. R., *9th Inter. Congr. appl. Mech.*, VIII, Paper 101 (II), University of Brussels, 245 (1957)
14. Cottrell, A. H., *Proc. Roy. Soc.*, A285, 10 (1965)
15. Heald, P. T., Spink, G. M. and Worthington, P. J., *Mater. Sci. and Eng.*, **10**, 129 (1972)

Five

FRACTURE TOUGHNESS TESTING IN PRACTICE

5.1 Introduction

The theory of fracture toughness testing, as described in the previous chapter, leads logically to the types of experiments used to determine critical values of strain energy release rate or stress intensity. Precracked specimens of standard geometry are loaded until they break and then, if the fractures are macroscopically brittle, the fracture loads can be used to calculate the toughness directly, using a set of standard tables for the compliance of the specimen. This procedure has now reached a state where it can be used in standard specifications (B.S. Draft for Development No. 3; ASTM Tentative Method E 399–70 T) as described in Section 5.9 et seq. To understand exactly what measurements are made in practice and why precise limitations are put on specimen dimensions to give 'valid' results, it is instructive to review the development of fracture toughness testing from the first experiments carried out by Irwin.

5.2 Testing of Thin Sheet

Irwin's experiments were devised to test his theory of fast fracture, in which the failure stress of an infinite body containing a central crack of length $2a$ is given by

$$\sigma_F = \sqrt{\left(\frac{EG_{crit}}{\pi a}\right)} \qquad (5.2.1)$$

in plane stress, where G_{crit} is the critical value of the strain energy release rate at fracture. The equivalent expression in terms of stress intensity is (equation 4.6.7):

$$K_{crit} = \sigma_F \sqrt{(\pi a)}$$

To simulate this situation as closely as possible, he chose to test large, centrally-cracked thin sheets of aluminium[1]. Irwin allowed for the fact that the stress-free boundaries were not at infinity by employing a relationship between stress intensity and applied stress of the form:

$$K = \sigma\left[W \tan\left(\frac{\pi a}{W}\right)\right]^{\frac{1}{2}}$$

where W is the total width of the specimen (see equation 3.10.7). It can be seen that, when (a/W) is very small, the function in the square brackets tends to the form $\sqrt{(\pi a)}$, as for an infinite body.

Irwin was then able to obtain the results shown in *Figure 5.1*, which were taken to indicate that σ_F is inversely proportional to $\sqrt{a}$. The value of G_{crit} calculated from the results is of the order of 130 kJ m^{-2}, i.e. about five orders of magnitude

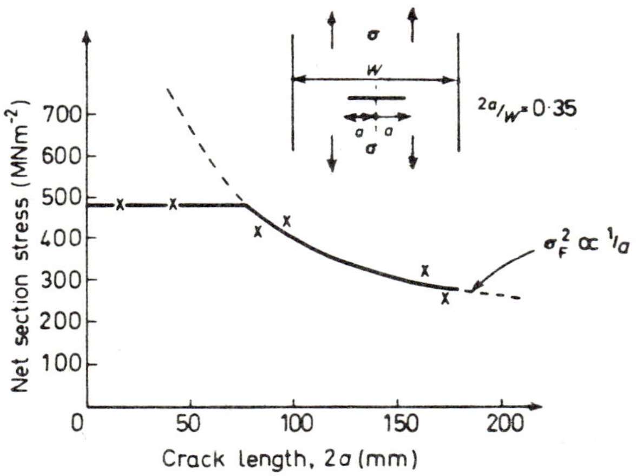

Figure 5.1 Net-section stress at instability v. crack length for 7075 Alloy (Al–Zn–Mg) – T6. *Face supports used to prevent buckling: plate thickness constant* (0.8 mm) (*after Irwin, Kies and Smith*[1])

greater than the surface energy of aluminium. The neglect of the surface energy term in deriving equation 4.6.3 from equation 4.6.2 is therefore justified. The whole analysis rests, of course, on the assumption that the behaviour is essentially elastic and so the specimen's dimensions must be much larger than the extent of plasticity which precedes fracture. The relationship illustrated by *Figure 5.1* showed that Irwin's treatment was correct in its essentials and led to an arousal of interest, particularly from the aircraft and aerospace industries, in the possibility of design against fracture using fracture mechanics techniques. Experiments were carried out on aluminium alloys, titanium alloys, maraging steels and other high-strength steels, mainly using specimens in the form of centrally-cracked thin sheets. One critical set of experiments will be discussed in detail in the following sections, because the conclusions reached have had a major effect on the specification of procedure and geometrical limits in modern testing practice.

5.3 The Variation of Fracture Toughness with Thickness

These experiments[2] were carried out, using an age-hardened aluminium alloy (7075–T6), to investigate the way in which toughness might be affected by the thickness of specimen tested. The results[3, 4], indicated in *Figure 5.2*, show that a large variation in toughness is produced and substantial doubt therefore appears to be cast on the possibility of measuring a single parameter which is capable of measuring a material's resistance to fast fracture in a way which can be applied universally in practice.

To understand the form of the toughness curve in *Figure 5.2a*, it is convenient to examine three regions: A, B, and C as indicated; and to take into account also the fracture profiles and the form of the stress–displacement curves obtained for each region as indicated in *Figure 5.2b*, A, B and C. The fractures

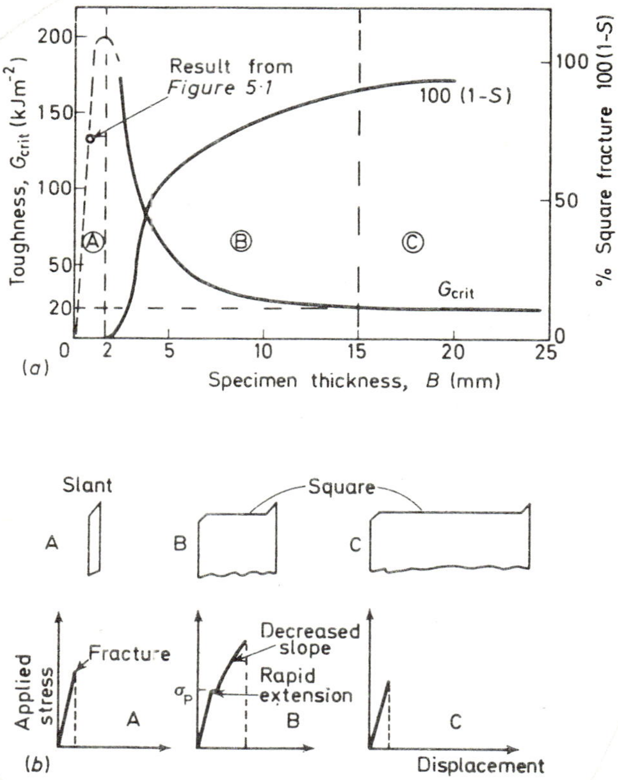

Figure 5.2 (a) Variation of toughness with thickness for 7075 Alloy (Al–Zn–Mg) – T6. *(b) Fracture profiles and stress–displacement curves typical of regions A, B and C*

are classified as 'slant' or 'square', depending on whether the macroscopic fracture surface is at 45° to the tensile axis or normal to it. The second curve in *Figure 5.2a* indicates how the proportion of square fracture varies with specimen thickness. It is apparent that, up to the maximum of the toughness curve (region A) the fractures are completely slant, in thick specimens (region C) they are almost completely square, and, for specimens of intermediate thickness, they are of 'mixed-mode'.

5.4 Fracture in Thin Sheet

We consider first the state of affairs in region A (*Figures 5.2a* and *5.2b*). Here, the specimens are very thin and tend to show increasing toughness with thickness. The load–displacement curve is linear up to fracture and the fracture is 100% slant. This behaviour can be explained as follows[5, 6]. In thin pieces, the stress in the thickness direction tends to zero and the stress-state is essentially that of plane stress. The specimen can be thought of as two free surfaces: in fact, it undergoes deformation (buckling) which relieves any stress produced in the thickness direction. From a previous argument (Section 2.11) we recognise that the yield criterion is $\sigma_{11} - \sigma_{33(\cong 0)} = 2\tau_Y$ and that yielding occurs on through-thickness planes at 45° to the X_1 and X_3 axes. Extension of the specimen therefore occurs by sliding which is not contained in the X_1X_2 plane. Since the crack extends normal to X_1 in any given X_1X_2 plane, the crack tip must move forward by an antiplane strain (K_{III}) mechanism. It can do this only because, in any practical testing configuration, some buckling and twisting allows the specimen's halves above and below the fracture plane to be displaced laterally. It is not clear that this K_{III} mode can occur if the specimen faces are constrained in such a way that they cannot so slide and the extent to which they can slide may be relevant when relating toughness values obtained on thin sheet specimens to failure in, say, a pressurised, thin-walled, precracked tube. Results obtained by Irwin, Kies and Smith[1] showed that the toughness of 0.8 mm thick specimens was increased by a factor of two when buckling was restrained.

In a representative model of this type of fracture, the movement of the surfaces ahead of the crack tip is indicated in *Figure 5.3a* and the sliding movement is supposed to be achieved by the movement of a number of screw dislocations on the 45° plane in the appropriate X_2 direction. As can be seen from the small inset diagram, each screw dislocation (MN) produces a relative shift of the top and bottom halves of magnitude $\boldsymbol{b}$, where $\boldsymbol{b}$ is the Burger's vector of the dislocation. The total number of dislocations, n, in the plastic zone is therefore given simply by $n = \sqrt{2}\, B/\boldsymbol{b}$ where $\sqrt{2}B$ is the relative displacement across the 45° plane required to give complete separation by a sliding-off mechanism. This situation of an 'inverse pile-up' of dislocations constituting the plastic zone in anti-plane strain deformation is exactly that treated by Bilby *et al.* (equation 3.14.4), who give the relationship between sliding displacement, S, and applied stress as:

$$S = \frac{4\tau_Y}{\pi\mu} a \ln\left[\sec\left(\frac{\pi q}{2\tau_Y}\right)\right]$$

where q and τ_Y are shear stresses.

As a criterion for crack extension, we put $S = \sqrt{2}\,B$ and then, for low stresses, obtain,

$$q = \sqrt{\left(\frac{2\mu\tau_Y\sqrt{2}B}{\pi a}\right)} \tag{5.4.1}$$

or, writing in terms of tensile stress ($\sigma = 2q$, $\sigma_Y = 2\tau_Y$) and Young's modulus [$E = 2\mu(1 + \nu)$].

$$\sigma = \sqrt{\left[\frac{2\sqrt{2}E\sigma_Y B}{\pi a(1 + \nu)}\right]} \tag{5.4.2}$$

This criterion for first crack extension is also the criterion for total instability, because the number of dislocations, and hence the plastic zone size, is constant for a specimen of constant thickness. As 'a' increases in equation 5.4.2, the stress needed to cause further increase falls, yet the stress available, at least in a 'soft' (load-controlled) system, increases because the same load is borne by a smaller specimen cross-sectional area. The crack therefore propagates at an accelerating rate once it has begun to extend. This behaviour has been classed as 'cumulative', because the contribution of each dislocation accumulates until catastrophe occurs[5]. The load–displacement curve (*Figure 5.2a*) is (macroscopically) linear up to the failure point, in agreement with this model for fracture.

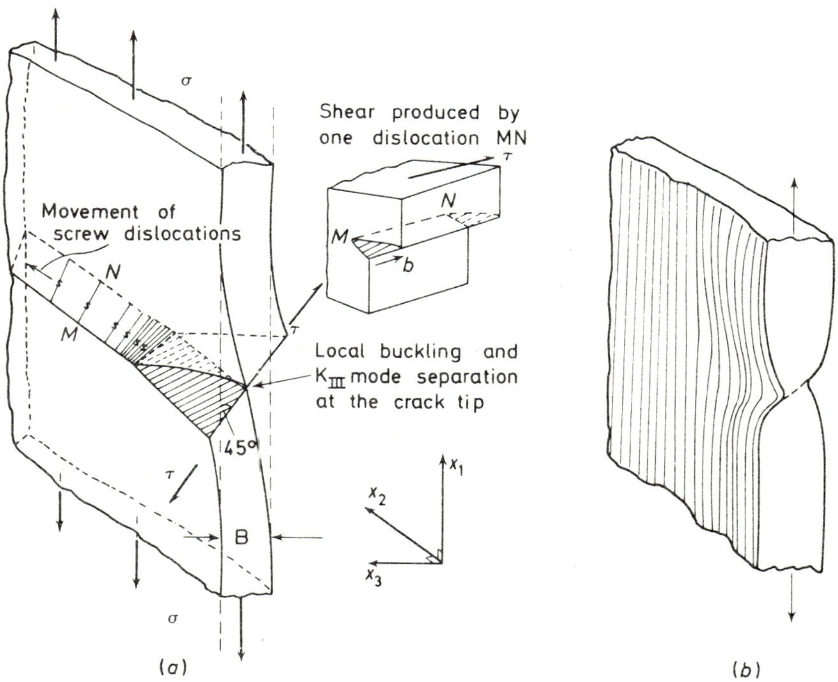

Figure 5.3 (a) K_{III} mode of separation in thin sheet. In practice, some crack opening will be associated with this shear and the dislocations will be part screw, part edge. (b) Asymmetric neck formation ahead of crack tip

If one compares equation 5.4.2 with the conventional relationship between failure stress, crack length and toughness (equation 5.2.1) it becomes possible to write the critical strain energy release rate as

$$G_{\text{crit}} = \frac{2\sqrt{2}\,\sigma_Y B}{(1+\nu)} \cong 2\sigma_Y B \tag{5.4.3}$$

i.e. the crack tip stress field, which reduces to σ_Y in the yielded region, does work by moving through a constant displacement. This definition of G_{crit} predicts a very steep linear increase with thickness, which is generally held adequately to represent the experimental results in region A of *Figure 5.2.* If these were plotted in terms of K_{crit}, rather than G_{crit}, the relationship would, of course, be parabolic in this region.

The model may not be correct in detail, because the antiplane separation may be preceded by necking and thinning of the sheet near the crack tip (*Figure 5.3b*)[7]. It is not clear whether or not the crack tip actually advances by any fracture mode before the antiplane mode can occur. The important point is that fully unstable fracture can occur by a wholly ductile fracture mechanism, once a critical crack tip displacement has been attained, because the plastic zone size at the point of instability needed to extend the crack is very much smaller than the width of the sheet. The material itself is fully ductile, but the particular assemblage of thin sheet, a long pre-existing crack and a soft loading system can produce a fracture that is brittle in engineering terms (see Section 1.6).

5.5 Plane Strain Fracture

We consider next the fracture behaviour of very thick specimens, as in region C of *Figure 5.2.* Here, total instability occurs at loads corresponding to a virtually constant toughness value and the fracture appearance is almost completely square, with very small proportions of slant ('shear lips') at the edges. We assume that the central region of a thick test piece deforms under approximately plane strain conditions. Then, the strain, e_{33}, is zero and, when yielding occurs around a crack tip, high constraints are set up, and a triaxial stress state is developed (Section 2.11).

This stress state enhances the initiation of fracture, whether stress-controlled or strain-controlled. It is easy to see that the high value of the maximum tensile stress below the notch, σ_{11}, may promote any cracking mechanisms to which the material is prone. Alternatively, if crack extension occurs when the strain in the region immediately ahead of the crack tip achieves a critical value, an effect of triaxial stress can arise through changes that it may make on the strain gradient in this region. For a given crack opening displacement, the plastic zone size in plane stress (specimen edges) is much larger than in plane strain (specimen mid-thickness) because yielding spreads under a shear stress component which incorporates the full value of the local tensile stress [σ_{11} rather than

$(\sigma_{11} - \sigma_{22})$: see Section 2.11 and *Figure 5.4*]. The strain distribution within this larger zone is correspondingly more diffuse, because the change in plastic displacement from the crack tip opening to zero at the plastic/elastic interface is accommodated in a much larger area. In plane strain, the strain gradient immediately ahead of the crack tip is steep; strain is more concentrated in that region; and crack extension is easier to achieve.

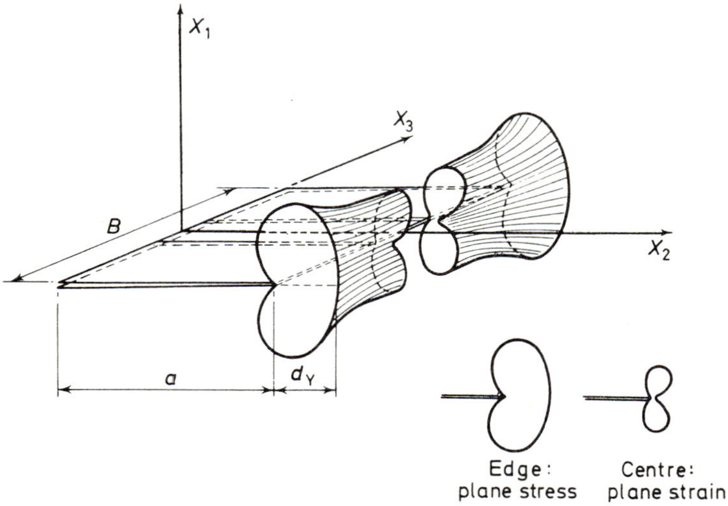

Figure 5.4 Plastic zone ahead of crack in a plate of finite thickness. At the edges of the plate $\left(x_3 \rightarrow \pm \frac{B}{2}\right)$ the stress state is close to plane stress. In the centre of a sufficiently thick plate ($x_3 \rightarrow 0$) the stress state approximates to plane strain

It has been postulated that crack advancement by simple plane strain plastic flow, as indicated in *Figure 5.5*, does not lead to a fracture that is unstable or brittle in the engineering sense[5]. The plastic flow required to produce an increment in growth is modelled as two inverse dislocation pile-ups at 45° to the tensile axis. A further increment demands the formation of two more pile-ups of identical length. These can be created only by raising the applied stress. Even if the rate of decrease of cross-section is sufficient to produce the necessary increase in stress, the yield rapidly becomes far-reaching and the piece yields generally before it breaks in half. Certainly, the fractures in region C (*Figure 5.2*) occur so far below general yield that a displacement criterion of this sort appears to be inapplicable. However, other models for ductile fracture extension, based on the assumption of visco-elastic behaviour, predict plane-strain instability. Such instability is, of course, also produced by cracking mechanisms. Detailed discussion of plane strain fracture modes will be left until Chapters 7 and 8. At this stage, it is immaterial whether the mode is stress- or strain-controlled provided that it propagates in an unstable manner at a low applied stress.

In region C, the specimens are thick, so that virtually all the load-bearing cross-section deforms in plane strain. Fracture propagates in the centre under constant critical crack tip conditions and any differences in the behaviour of the edges of the specimen are insignificant in determining failure conditions

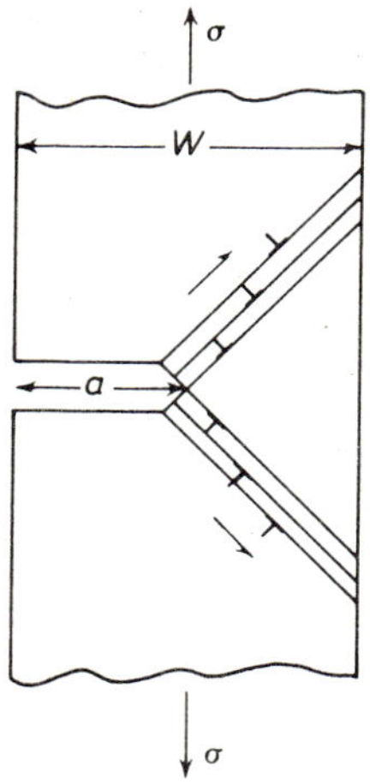

Figure 5.5 Non-cumulative ductile fracture by simple plastic extension. The fracture cannot become unstable before general yield

for the specimen as a whole. The load is observed to rise in a linear manner until a critical (low) value is reached, when catastrophic failure occurs.

5.6 The Intermediate Range[3]

In region B, the fracture behaviour is more complicated. The specimen is neither so thin that failure occurs by the sliding-off mechanism observed in region A, nor so thick that it fails by an almost completely 'plane-strain' square fracture. Instead, its thickness is such that the central and edge regions are of comparable size. The sequence of events is then as follows, as indicated by the load-displacement trace in *Figure 5.2b.* The load on a cracked specimen is raised to a value, P_p (corresponding to the stress σ_p in *Figure 5.2b*), at which some square fracture can form in the centre of the testpiece's thickness. In a very thick testpiece, such fracture would spread catastrophically because it would occupy a large proportion of the thickness, but, in the intermediate range, so much of the load is borne by the side-ligaments of the cross-section that total instability does not occur at the load P_p. The load–displacement curve may show a sudden extension for constant, or even decreasing, load if the square fracture tunnels ahead rapidly. This is known as a 'pop-in'. If the square fracture does not advance so rapidly, its presence will be detected only by a change in the compliance of the testpiece. The crack is longer; so the load–deflection curve exhibits a decreased slope (see *Figure 4.5*). Both effects are indicated in *Figure 5.2b.*

As the load is raised above P_p, the central square fracture tunnels into the centre of the testpiece and perhaps also spreads slightly in the through-thickness, X_3, direction. The side ligaments can be sheared apart when sufficient displacement at the crack tip is attained and the total crack advances in a composite fashion: the square fracture tunnelling ahead and dragging the slant (shear lips) with it.

It is critically important to realise that, as the crack grows under increasing load, the plastic zone ahead of the crack tip grows bigger and is therefore more easily able to relax the through-thickness stress σ_{33}. Less of the thickness therefore deforms in plane strain and the proportion of square fracture decreases. The criteria for final instability in this intermediate range have not been fully established, but the behaviour must run something as follows. At the thin end of the range, the initial square fracture occupies only a small proportion of the thickness cross-section; as it tunnels forward, the plastic zone size becomes large with respect to plate thickness; the through-thickness stress, σ_{33}, is relaxed; and final instability is achieved at a load, P_F, which is sufficiently large to operate a sliding mechanism for separation, analogous to that observed in region A. The sequence of events is illustrated in *Figure 5.6* and is quite consistent, both with

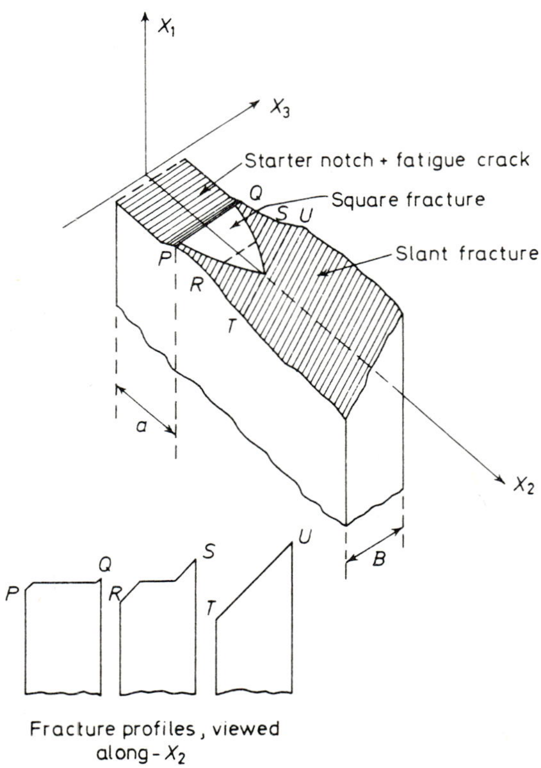

Figure 5.6 Development of slant fracture in region B

the fracture appearance of broken specimens and with the observed load–displacement records. The fracture load is lower than that for a thinner specimen, because the testpiece, at instability, contains a longer crack.

At the thick end of the intermediate range, the side ligaments bear a much smaller proportion of the total load applied to the specimen and so final instability follows the initiation 'pop-in' more rapidly, provided that the test is carried out under load control (a 'soft' system). The fracture profile at instability is now a mixture of slant and square. Under displacement control, a rapid 'pop-in' can cause the load to relax.

The behaviour observed in this intermediate range gives rise to doubts concerning the application of the Griffith–Irwin relationship as such to characterise the failure criterion. If final instability is produced by a sliding mechanism, the extent of the square fracture preceding instability depends in a rather complicated manner on both the crack length and on the thickness of the piece. Krafft, Sullivan and Boyle[2] observed that the increment of square growth depended only on thickness and was sensibly independent of initial crack length. The failure stresses that they observed for cracked aluminium panels of different sizes could not be reconciled using Griffith–Irwin methods because a growth increment of given magnitude perturbed the net stress field in a small specimen more strongly than in a large specimen. Their treatment of this situation is discussed in Section 5.8. We now summarise the effects of thickness and discuss their relevance to toughness testing in general.

5.7 Conclusions on Thickness Effects

In regions B and C, it can be seen that the fracture behaviour of a testpiece is best regarded as that of a laminate. The central part deforms under conditions close to those of plane strain deformation: the side faces can support no stress normal to the free surface and deform in plane stress. High tensile stresses and a concentrated strain gradient are present ahead of the crack tip in the centre: on the sides the stresses are much lower and the strain is spread over a larger plastic zone. 'Plane strain' 'pop-ins' are produced in the centre at a critical value of stress intensity: whether the fracture of the testpiece as a whole proceeds catastrophically at this stress or at a higher value depends on the proportion of the cross-section occupied by the shear lips.

The peak of the toughness curve occurs at a thickness of approx. 2 mm. For thicker specimens, an increasing proportion of square fracture is produced, and the total fracture toughness drops. To provide an upper bound for the curve which shows how the proportion of square fracture varies with thickness, we assume that the shear lips are of constant size (i.e. 2 x 1 mm) in specimens of all thicknesses. Strictly, this situation applies only to the initiation of the central square fracture at a constant value of G_{crit}. The curve shown in *Figure 5.10* (p. 129) can then be constructed and is found to agree remarkably well with that found experimentally, particularly for thick specimens. Some differences arise because the thinner specimens break at substantially higher loads so that the size of the plastic zone, and hence of the shear lips, at fracture is not constant.

The next assumption to be examined is that the 'pop-ins' are occurring under plane strain conditions. If this were fully substantiated for a particular alloy, it would be possible to make measurements of the stress intensity needed to produce square fracture in specimens much thinner than are needed to produce the limiting value of G_{crit}. The assumption rests first on the constancy of the widths of the shear lips at the initiation of square fracture, implying that the increase of σ_{33} from zero at the free side surfaces to the plane strain value in the centre of the piece occurs over a constant distance, and secondly on behaviour of the 'laminate', such that both 'plane strain' and 'plane stress' fractures occur under uniquely defined conditions. There is, unfortunately, no guarantee that a square fracture is characteristic of plane strain over all its thickness. In mild steel, macroscopically square fractures have been shown to occur at loads which decrease with increasing thickness until constancy is achieved, when plane strain conditions are met (see Section 7.5). The fear is that, in the aluminium alloys, a square 'pop-in' may occur in relatively thin sheet at a critical stress intensity greater than the limiting value, so that the toughness of the material for true plane strain conditions is overestimated[6].

It is interesting to try to predict the extent of relaxation of the stress σ_{33} and to compare this with the observed sizes of shear lips. At the plane strain G_{crit} value of 20 kJ m^{-2} (see *Figure 5.2a*) the radius of the *plane stress* plastic zone would be:

$$r_Y = \frac{EG}{2\pi\sigma_Y^2} \tag{5.7.1}$$

$$= \frac{7 \times 10^{10} \times 20 \times 10^3}{2\pi \times 25 \times 10^{16}} \text{ m}$$

taking appropriate values for E (70 GN m^{-2}) and σ_Y (500 MN m^{-2}) for the alloy. We obtain:

$$r_Y = \frac{140}{50\pi} \times 10^{-3} \text{ m} \cong 1 \text{ mm} \tag{5.7.2}$$

If the σ_{33} stress is assumed to relax over a distance equal to the radius of the plastic zone, the agreement between the calculated size (1 mm) of a single shear lip at the plane strain fracture toughness level and the critical specimen thickness (2 mm) at which square fracture is first observed is excellent. The 'pop-in' procedure for plane strain toughness determination is thus apparently well substantiated.

However, the only indication of plane strain is the occurrence of square fracture, and the argument is fallacious if such fracture can occur under non-plane-strain conditions. The 'pop-in' value may then be too high. Suppose, for example, that σ_{33} was relaxed over a distance of $2r_Y$ (the total length of the plastic zone) from the side face. Then, square, but non-plane-strain, fracture would occur in specimens whose thickness lay between 2 mm and 4 mm, and the 'pop-ins' would occur at progressively decreasing nominal stresses with increasing thickness.

There is a case to be made for the non-plane-strain square fracture in terms of an assumed three-dimensional stress distribution. If we examine the

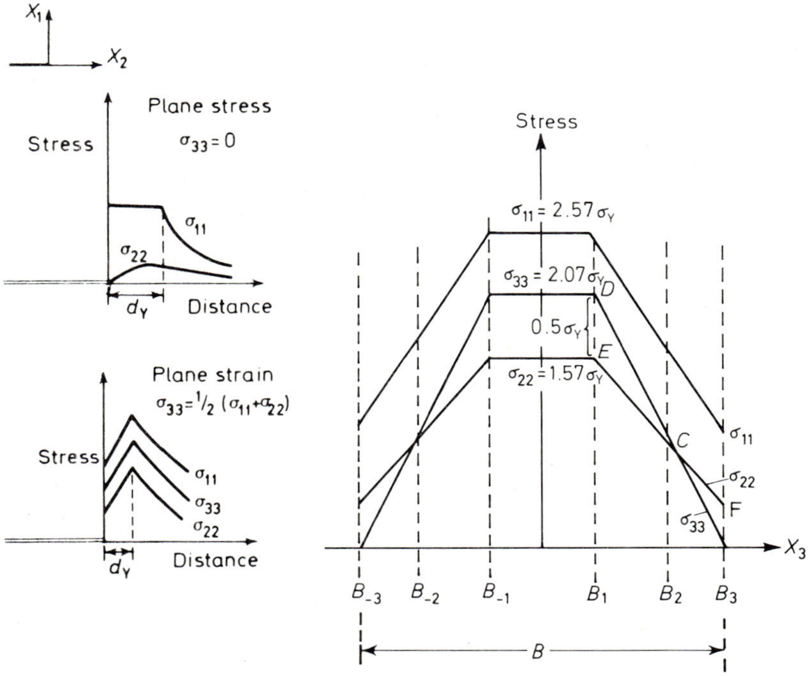

Figure 5.7 Schematic variation of stresses in plastic zone with thickness. σ_{22} falls to a value which is supposed to lie in the range 0.2–0.6 σ_Y on the free surface. Triangles DEC and B_3FC are similar: hence

$$\frac{B_3C}{CD} = \frac{B_3F}{DE}$$

Triangles B_1DB_3 and B_2CB_3 are similar; hence

$$\frac{B_2B_3}{B_1B_2} = \frac{B_3C}{CD} = \frac{B_3F}{DE}$$

variation of stresses in the plastic zone as a function of thickness, we could obtain something of the form of *Figure 5.7*.

It can be seen that there is a region, between B_1 and B_2, in which the hydrostatic component, and hence σ_{11}, is less than in plane strain, yet σ_{22} is still the smallest principal stress and so, on Tresca's criterion, yielding still occurs in the X_1X_2 plane on slip lines at 45° to the X_1 and X_2 axes as in plane strain. The reduction of σ_{11} means, however, that processes such as brittle cracking (see Chapter 7) are less favoured and a greater load must be applied to fracture such a region. A toughness calculated from the onset of square fracture in specimens of thicknesses lying between $(B_{-3}B_{-2} + B_2B_3)$ and $(B_{-3}B_{-1} + B_1B_3)$ can therefore be unduly optimistic.

We attempt roughly to estimate the magnitude of B_1B_2 as follows, for the situation given in *Figure 5.7* where we have given σ_{22} a typical plane strain value of 1.5 σ_Y and σ_{11} a value of 2.5 σ_Y. From the geometry of the similar triangles, we see that the the ratio $B_2B_3 : B_1B_2$ is equal to the ratio $B_3F : DE$, i.e. to the ratio of the value of σ_{22} on the side faces to $(\sigma_{33} - \sigma_{22})$ in the centre. This latter value is equal to 0.5 σ_Y. An estimate of the appropriate figure for σ_{22} on the side faces is not easy to obtain, but upper and lower limits, taken as 0.2 σ_Y and 0.6 σ_Y, give a range for the ratio $B_2B_3 : B_1B_2$ of 0.4–1.2.

Within the obvious errors of our estimate, we see that the thickness over which we might expect to obtain square, but non-plane-strain, fracture is comparable with that in which through-thickness yielding ought to predominate. As a rough guide, we suppose that the thickness at which plane strain fracture first occurs in a specimen is twice that at which some square fracture is first observed. This seems to correlate with a total thickness equal to twice the total extent of the plane stress plastic zone at fracture.

No direct evidence of the point appears to exist, but it is perhaps instructive to compare the critical thickness for the onset of fully plane strain fracture, 15 mm, and associated shear lip thickness, 2 mm, giving $13/15 \triangleq 85\%$ square fracture, with the figure of 70% cleavage facets needed to sustain fully brittle fracture in mild steel[8]. Cottrell[5] has explained this latter figure in terms of constrained Orowan yielding (Section 2.12) between cleavage microcracks. If σ_{33} were relaxed over a distance of twice the shear lip thickness, a figure of $11/15 \triangleq 73\%$ would be obtained for the proportion of plane strain fracture, in closer agreement with the steel results.

In thick specimens, the total toughness is controlled by that of the square fracture. So much load is transferred to the shear lip regions when the central fracture starts that they fracture catastrophically and final instability is coincident with 'pop-in'. This is the situation that gives constant toughness for all large thicknesses and is that sought in the techniques of plane strain fracture toughness testing which form the British and American standards, as described in subsequent sections.

The total fracture toughness of a specimen in the intermediate range is composed of significant contributions from both the slant and square components. Estimation of the subsequent toughness is not simple, because, as the crack tip advances under constant load, the strain energy release rate increases, so that more driving force for accelerating fracture is available, yet the plastic zone size, and hence the area that will fracture by high energy shearing, also increases, so that more work needs to be done to cause failure. The precise balance between these two factors determines the onset of total instability.

5.8 The R-curve Analysis

The problem of fractures in the intermediate range will be discussed before the standard plane strain methods are described, because due allowance for both square and slant modes lies at the heart of the proper understanding of the fracture of cracked pieces. The approach followed in this section is a version of that given by Krafft, Sullivan and Boyle[2] who attempted to make quantitative estimates of the increase in resistance to fracture as a crack grows, by means of 'resistance-curves' or 'R-curves' derived as follows.

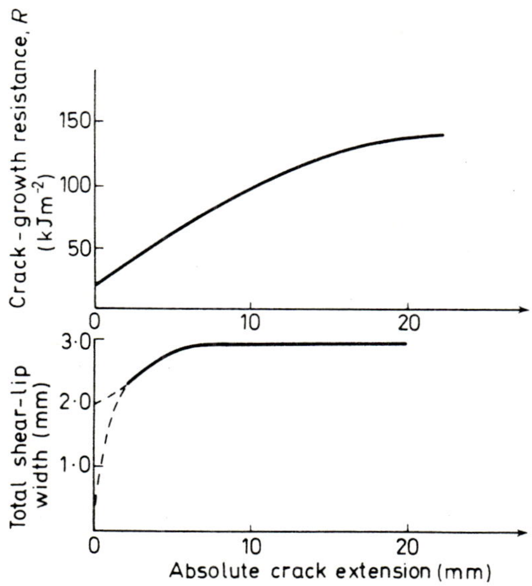

Figure 5.8 Schematic diagram, showing increase of crack-growth resistance and shear lip width with absolute crack extension in 3.1 mm *thick 7075 sheet. The extrapolation of shear-lip width to zero for very small crack extension is incompatible with the results quoted in Figure 5.2, which suggest an extrapolation to* 2 mm *(after Krafft, Sullivan and Boyle*[2]*)*

Figure 5.8 is taken from Krafft, Sullivan and Boyle, and shows both the increasing resistance to crack growth and the increasing thicknesses of shear lips as cracks propagate from initial sharp notches in 3.1 mm (0.125 in) thick plates of the aluminium alloy on which previous discussion has been based.

In the early stages, the resistance curve bears a close relationship to the increasing depth of shear lip. Krafft, Sullivan and Boyle suggest that the continuing increase in resistance once full thickness shear has been attained is associated with change in shear angle, but it has more probably arisen from testing machine relaxations in these particular experiments, since *Figure 5.2* shows that instability can be produced at a $G_{crit} = R$ value of about 100 kJ m^{-2} in direct loading tests carried out on the same thickness.

Krafft's analysis proceeds simply by assuming that the shear lips are at 45° and that the plastic zone bounded by these two 45° lines and those symmetrically disposed at 45° across the crack plane is subjected to a uniform 'plastic work density' $(\mathrm{d}W_p/\mathrm{d}V)$ which is supposed to be constant (*Figure 5.9*). Then, if the two shear lips together occupy a fraction SB of the cross-section, the thickness of a shear lip's 'plastic volume' is $BS/2$. For a given increase in crack length da, the total plastic volume of both shear lips is

$$2 \times \left(\frac{BS}{2} \times \frac{BS}{2}\right) \mathrm{d}a = \frac{B^2 S^2}{2} \mathrm{d}a \tag{5.8.1}$$

and the plastic work done is thus $(\mathrm{d}W_\mathrm{p}/\mathrm{d}V)\dfrac{B^2S^2}{2}\mathrm{d}a$. The work done in producing unit area of square fracture is assumed also to be constant and is denoted by $(\mathrm{d}W_\mathrm{s}/\mathrm{d}A)$.

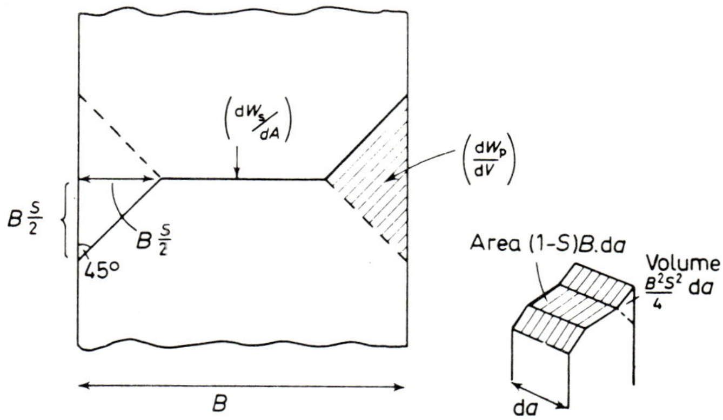

Figure 5.9 Krafft's model for the calculation of crack-growth resistance (after Krafft, Sullivan and Boyle[2])

Then for a given increase in length da, we have for the work done:

$$\mathrm{d}W = \left(\frac{\mathrm{d}W_\mathrm{s}}{\mathrm{d}A}\right) B\,\mathrm{d}a + \left(\frac{\mathrm{d}W_\mathrm{p}}{\mathrm{d}V}\right) \frac{B^2 S^2}{2} \mathrm{d}a \tag{5.8.2}$$

in the original Krafft derivation, or, treating the shear lip fracture as part of the 'plastic work density':

$$\mathrm{d}W = \left(\frac{\mathrm{d}W_\mathrm{s}}{\mathrm{d}A}\right) B(1-S)\,\mathrm{d}a + \left(\frac{\mathrm{d}W_\mathrm{p}}{\mathrm{d}V}\right) \frac{B^2 S^2}{2} \mathrm{d}a \tag{5.8.3}$$

Dividing through by the incremental area, $B\mathrm{d}a$, we have (for unit thickness):

$$R = G = \frac{\mathrm{d}W}{\mathrm{d}a} = \left(\frac{\mathrm{d}W_\mathrm{s}}{\mathrm{d}A}\right)(1-S) + \left(\frac{\mathrm{d}W_\mathrm{p}}{\mathrm{d}V}\right)\frac{BS^2}{2} \tag{5.8.4}$$

Krafft *et al.* fitted their experimental data to this expression [ignoring the $(1-S)$ factor in the first term] to obtain values for $(\mathrm{d}W_\mathrm{s}/\mathrm{d}A)$ and $(\mathrm{d}W_\mathrm{p}/\mathrm{d}V)$ and showed agreement for both 3 mm (0.125 in) and 4.5 mm (0.1875 in) thick sheet. The value of $(\mathrm{d}W_\mathrm{s}/\mathrm{d}A)$ was found to be about 13.5 kJ m^{-2} and $(\mathrm{d}W_\mathrm{p}/\mathrm{d}V)$ about 145 kJ m^{-3}.

The critical point in the fitting of such a relationship appears to rest simply on the use of the square of the fraction of shear lips. Assuming that the initial shear lips have a *constant* thickness of 2 mm, we take a value of 20 kJ m^{-2} for (dW/dA) for the square fracture and 200 kJ m^{-2} for an 'effective work' for shear lip fracture (the lower limiting value for thick specimens and the peak G_{crit} value for 2 mm specimens respectively). Then the toughness values in the intermediate range can be shown to be well represented by the expression:

$$G_{crit} = 200S^2 + 20(1 - S) \quad (5.8.5)$$

where S is the *initial* fractional proportion of shear lips, as calculated by assuming that the shear lips are 2 mm thick. Values of the parameters are given in *Table 5.1* and the fit is indicated in *Figure 5.10.*

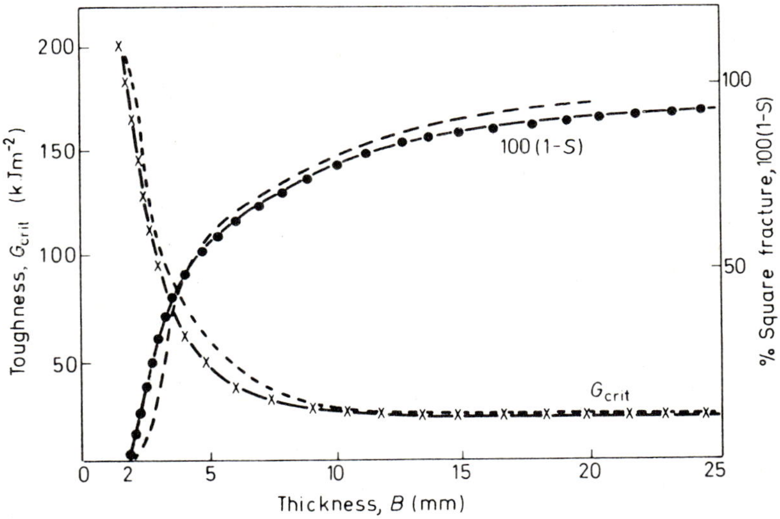

Figure 5.10 Calculations of percentage square fracture and toughness (see Table 5.1) —·—·- *% square fracture* $100(1 - S)$ *calculated assuming constant thickness shear lips, i.e.* $1 - S = \frac{B-2}{B}$ *where B is in* mm. –x–x–x G_{crit} *calculated from the formula* $G_{crit} = 200S^2 + 20(1 - S)$ *where S is the proportion of slant fracture* $= \frac{2}{B}$ *and* G_{crit} *is in* kJ m^{-2}. *The figures* 200 *and* 20 *represent the upper and lower limiting toughnesses respectively.* – – – – *experimental results from Figure 5.2*

It will be noted that this expression makes use of the experimental value for the work done in fracturing a specimen of thickness equal to two shear lips, so that the $B/2$ factor in equation 5.8.4 is automatically included.

Despite the fact that this is only a lower estimate, because the *initial* fraction of slant fracture has been used, agreement with experiment is obviously good.

Krafft, Sullivan and Boyle also examined the way in which the shear lip fraction increased as the crack grew (e.g. *Figure 5.8*). It was found that S depended principally on the absolute crack extension and it was to some extent possible to relate the increase in load during crack extension with the increasing proportion of shear lips. In general, however, an accepted theory for explaining fractures in the mixed mode range (region B, *Figure 5.2*) does not exist and '*R*-curves' have to be determined experimentally for a given specimen geometry. Standard testing is aimed at producing consistent 'plane strain' toughness values.

Table 5.1

Specimen thickness (mm)	S	$1-S$	S^2	$200S^2$	$20(1-S)$	G_{crit}
2.0	1	0	1	200	0	200
2.5	0.8	0.2	0.64	128	4	132
3.0	2/3	1/3	4/9	88	7	95
4.0	1/2	1/2	1/4	50	10	60
6.0	1/3	2/3	1/9	22	13.3	35.3
10.0	1/5	4/5	1/25	8	16	24
12.5	0.16	0.84	0.0256	5	17	22
25.0	0.08	0.92	0.0064	1.3	18.4	19.7
50.0	0.04	0.96	0.0016	0.3	19.2	19.5

5.9 Plane Strain Fracture Toughness (K_{IC}) Testing

The aim of plane strain fracture toughness testing is to obtain reproducible values for the lower limiting critical toughness of a material, tested in sufficiently thick section for the shear lip contribution to the toughness to be neglected. Simultaneously, the fracture must proceed under quasi-elastic conditions, i.e. the size of the plastic zone at fracture must be very much less than the cross-section of the specimen. A third requirement is that the region of interest at the crack tip must be very much smaller than the crack length, so that the critical fracture event may be characterised by a single-valued K parameter (see Section 3.10). The critical toughness value in mode I opening is then denoted by K_{IC}. The recommended procedures for plane strain toughness testing attempt to meet these three requirements in specimens of a size which is economical on usage of material and which can be easily tested in the laboratory.

The centrally cracked wide plate demands large specimen sizes, with attendant problems regarding the capacities of testing machines. The first attempts to reduce specimen size led to the selection of a single-edge-notched (SEN) geometry as indicated in *Figure 5.11*. In relatively thin plate, these were used

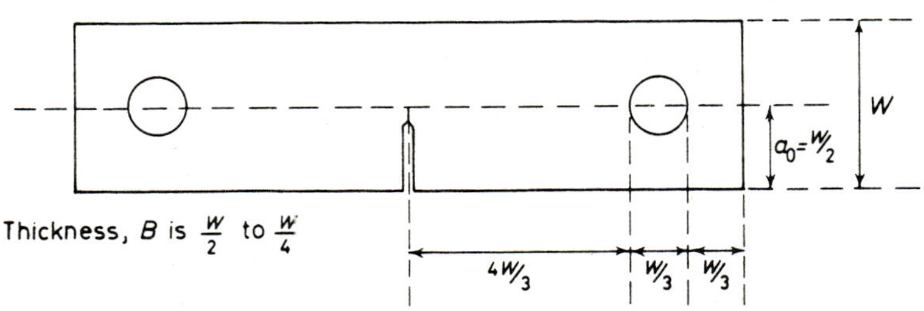

Figure 5.11 Single-edge-notched tension testpiece

with some success whilst the 'pop-in' procedure appeared tenable, but the still fairly large sizes required to produce instability at the lower critical value led to a search for specimen designs in which a given K-value could be produced in smaller pieces, and hence with lower loads. The term '*K-capacity*' is used to characterise this property. Additionally, the SEN testpiece was not wholly satisfactory in that its asymmetrical geometry under nominally uniform tension tended to produce an additional bending component which could not be allowed for easily.

The designs currently recommended in ASTM or British Standards are as shown in *Figure 5.12a* and *b*. Both are SEN specimens subjected to a bending moment. The bend specimen is deformed under three-point loading: the 'compact tension specimen' (CTS) is subjected to point loading through pins

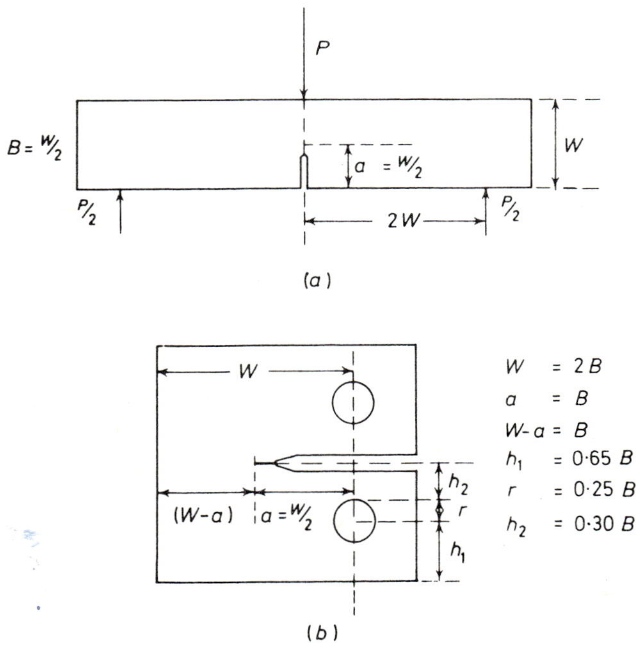

Figure 5.12 Standard toughness testpieces. (a) Single-edge cracked bend testpiece. (b) Compact tension testpiece (CTS)

above and below the crack faces. Variations of the pin-loading system, using screw-threads or screw-thread/pin combinations (WOL specimen, *Figure 9.6*) have also been tried.

The K-calibration of specimens of complex geometry is not easy. Modified stress-functions can be developed, so that the boundaries of the specimen remain stress-free, but the modifications must, by definition, be concerned with conditions remote from the crack tip. Much careful investigation of the functions has to be carried out to ensure that the modifications made produce sensible results in the critical K singularity field near the crack tip. The stress obviously tends to infinity as the tip is approached, but we may expect to detect differences in the region just ahead of the tip.

The alternative method of producing K-calibrations is to make use of the compliance technique (Section 4.4) either by measuring the compliance experimentally or by calculating the overall elastic response in various regions of the specimen.

Table 5.2 COMPLIANCE COEFFICIENTS FOR STANDARD TESTPIECE GEOMETRIES (*Figures 5.12a* and *5.12b*)

For the bend testpiece

$$K = \frac{3PL}{BW^{\frac{3}{2}}}\left[1.93\left(\frac{a}{W}\right)^{\frac{1}{2}} - 3.07\left(\frac{a}{W}\right)^{\frac{3}{2}} + 14.53\left(\frac{a}{W}\right)^{\frac{5}{2}} - 25.11\left(\frac{a}{W}\right)^{\frac{7}{2}} + 25.80\left(\frac{a}{W}\right)^{\frac{9}{2}}\right]$$

where $L = 2W$. Hence, $K = \frac{P}{BW^{\frac{1}{2}}} Y_1$ where $Y_1 = f_1\left(\frac{a}{W}\right)$

Value of Y_1

$\frac{a}{W}$	Y_1	$\frac{a}{W}$	Y_1	$\frac{a}{W}$	Y_1	$\frac{a}{W}$	Y_1
0.450	9.1	0.455	9.23	0.510	10.96	0.515	11.14
0.460	9.37	0.465	9.52	0.520	11.33	0.525	11.52
0.470	9.66	0.475	9.81	0.530	11.71	0.535	11.91
0.480	9.96	0.485	10.12	0.540	12.12	0.545	12.33
0.490	10.28	0.495	10.44	0.550	12.55		–
0.500	10.61	0.505	10.78				

For the CTS testpiece

$$K = \frac{P}{BW^{\frac{1}{2}}}\left[29.6\left(\frac{a}{W}\right)^{\frac{1}{2}} - 185.5\left(\frac{a}{W}\right)^{\frac{3}{2}} + 655.7\left(\frac{a}{W}\right)^{\frac{5}{2}} - 1017\left(\frac{a}{W}\right)^{\frac{7}{2}} + 638.9\left(\frac{a}{W}\right)^{\frac{9}{2}}\right]$$

Hence $K = \frac{P}{BW^{\frac{1}{2}}} Y_2$ where $Y_2 = f_2\left(\frac{a}{W}\right)$

Value of Y_2

$\frac{a}{W}$	Y_2	$\frac{a}{W}$	Y_2	$\frac{a}{W}$	Y_2	$\frac{a}{W}$	Y_2
0.450	8.34	0.455	8.45	0.510	9.90	0.515	10.05
0.460	8.57	0.465	8.69	0.520	10.21	0.525	10.37
0.470	8.81	0.475	8.93	0.530	10.54	0.535	10.71
0.480	9.06	0.485	9.19	0.540	10.89	0.545	11.07
0.490	9.32	0.495	9.46	0.550	11.26		–
0.500	9.60	0.505	9.75				

Values of Y for intermediate a/W values may be found fairly accurately by linear interpolation.

The standard specimens have been calibrated by several methods, which have been refined until complete agreement has been achieved. Information is presented to the user of the standard procedures in terms of a compliance coefficient (the Y-function) which enables the load on a specimen to be converted directly to the K-value. The main variable in testpiece preparation is the ratio of crack length (a) to testpiece width (W) (see later), and *Table 5.2*, which is taken from the British Standard, shows Y as a function of (a/W) for standard bend and CTS testpieces. The appropriate polynomial forms for K are also given. We now proceed to examine the restrictions placed on specimen dimensions.

5.10 Specimen Dimensions: Crack Length

The restriction on crack length is treated with reference to the distribution of the tensile stress σ_{11} ahead of a crack (from equation 3.10.2) as given by

$$\sigma_{11} = \frac{K_1}{\sqrt{(2\pi r)}} + \ldots \text{ series} \tag{5.10.1}$$

where r is the distance ahead of the tip. K_1 is the leading term in a series of coefficients and dominates only for very small values of r. For different testpiece geometries, we determine stress distributions and K-coefficients as described above. K is calculated as a single-valued parameter only as r tends to zero.

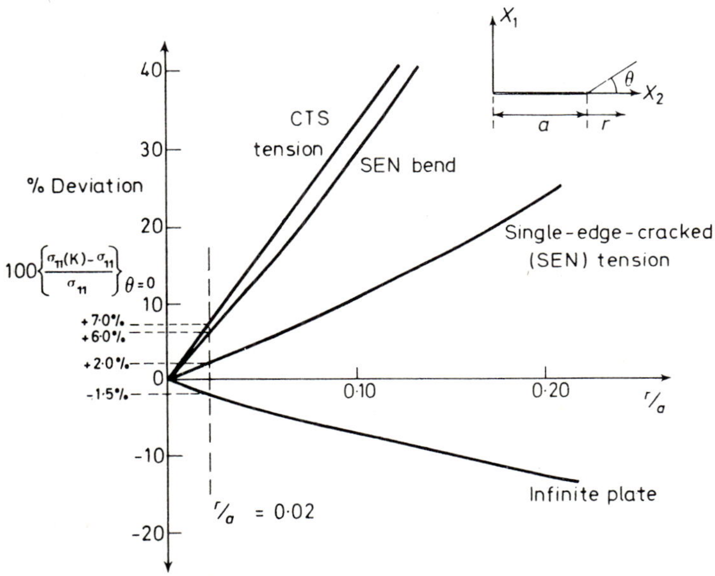

Figure 5.13 Variation between crack tip stress as calculated from a single-valued stress intensity factor, $\sigma_{11}(K)$, and that calculated from the full series, σ_{11}, expressed as the percentage $100\left[\frac{\sigma_{11}(K) - \sigma_{11}}{\sigma_{11}}\right]_{\theta=0}$ and plotted v. distance ahead of the crack tip (after Wilson[9])

Now, in *Figure 5.13* we compare stress distributions ahead of a crack tip for different testpieces loaded to produce the same crack tip ($r \to 0$) stress intensity. As r increases from zero, we expect that the stress will be described less and less accurately by a single value of K, because the extra terms in equation 5.10.1 begin to contribute. It is critically important also to realise, however, that the stress distributions are different in different testpieces, so that the *deviations* from a single valued K parameter description also differ. If we are to obtain agreement between experimental fracture stresses in different specimens through a K approach, we must ensure that the fracture events are limited to a region ahead of the crack which is so small, compared with the other specimen dimensions, that the single K-parameter describes the stress field in the critical region with almost complete accuracy. The specification of the size of such a region is one of many value judgments made in fracture toughness testing. It is usually taken to be a size equal to one-fiftieth of the crack length. In terms of fracture criteria, this is interpreted as being equivalent to saying that the radius of the plane strain plastic zone at fracture must also be less than $0.02a$. If the radius of the plane stress plastic zone is r_Y, the radius of the plane strain plastic zone, r_{IY}, is taken to be one-third of r_Y, because the plane strain yield stress is raised by constraint to a maximum value of $3\sigma_Y$ (see Section 2.12). Thus:

$$r_{IY} = \tfrac{1}{3} K_{IC}^2/2\pi\sigma_Y^2 \leqslant 0.02a \qquad (5.10.2)$$

where K_{IC} is the critical value of stress intensity at fracture (see page 130). Reference to *Figure 5.13* shows that this criterion produces, in the *elastic* stress analysis of the crack tip stress fields: about 7% error for a CTS specimen; about 6% error for SEN bend; about 2% error for SEN tension; and about 1.5% error for the infinite plate solution[9].

We can estimate this last error by recalling the Westergaard solution for the stress ahead of a crack. From equation 3.7.1 (with change of co-ordinates) we have:

$$\sigma_{11} = \sigma\left(1 - \frac{a^2}{x_2^2}\right)^{-\frac{1}{2}} \qquad (5.10.3)$$

Substituting $r = (x_2 - a)$ and expanding, we obtain

$$\sigma_{11} = \frac{\sigma(a+r)}{\sqrt{(2ar)}}\left(1 + \frac{r}{2a}\right)^{-\frac{1}{2}}$$

$$\sigma_{11} = \frac{\sigma(a+r)}{\sqrt{(2ar)}}\left(1 - \frac{r}{4a} + \frac{3}{32}\frac{r^2}{a^2} - \ldots\right) \qquad (5.10.4)$$

Putting $r = 0.02a$ and considering terms in r up to power 1, equation 5.10.4 becomes

$$\sigma_{11} = \frac{\sigma(1.02a)(0.995)}{\sqrt{(2ar)}} = 1.015\,\sigma\sqrt{\left(\frac{\pi a}{2\pi r}\right)} \qquad (5.10.5)$$

Comparison of this more accurate form with the simple expression

$$\sigma_{11} = \sigma\sqrt{\left(\frac{\pi a}{2\pi r}\right)} = \frac{K}{\sqrt{(2\pi r)}} \qquad (3.10.2)$$

used previously to give the σ_{11} stress shows that the simple expression underestimates the value by about 1.5%, as indicated in *Figure 5.13*. Calculations of errors for the other testpiece geometries follow the same procedure, but involve more complicated expressions for the σ_{11} stress because the polynomial stress functions used to determine crack tip stresses contain more significant terms.

Since the principle which underlies the crack length criterion depends on the deviation of the real stress distribution from the assumed single-valued K-parameter distribution, it is not clear why the criterion is not expressed on the percentage deviation, even though this would imply different r_{IY}/a ratios for different testpiece geometries. Nevertheless, the *ratio* is at present standardised as 0.02, implying the different deviations shown in *Figure 5.13*.

5.11 Specimen Dimensions: Ligament

The second restriction is on the width of the specimen ligament, i.e. $(W-a)$. If $(W-a)$ is too small, compared with the size of the plastic zone at fracture, the quasi-elastic solution for stress intensity breaks down, because the closeness of the stress-free boundaries significantly affects the crack tip stress field.

To provide a basis for the $(W-a)$ restriction, we examine the solution[10] for the spread of plasticity in an infinite series of cracks whose centres are spaced $2W$ apart (see *Figure 5.14*), replacing shear stresses by tensile stresses.

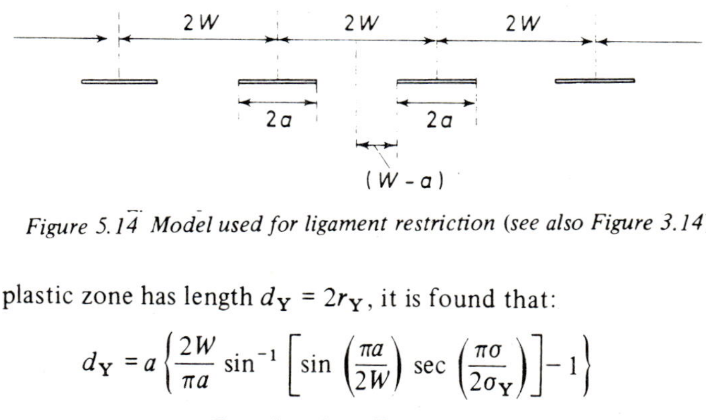

Figure 5.14 Model used for ligament restriction (see also Figure 3.14)

If the plastic zone has length $d_Y = 2r_Y$, it is found that:

$$d_Y = a\left\{\frac{2W}{\pi a}\sin^{-1}\left[\sin\left(\frac{\pi a}{2W}\right)\sec\left(\frac{\pi\sigma}{2\sigma_Y}\right)\right] - 1\right\} \qquad (5.11.1)$$

i.e. as $(a/W) \to 0$, so $d_Y \to a\left[\sec\left(\frac{\pi\sigma}{2\sigma_Y}\right) - 1\right]$ as for the isolated crack in an infinite body (equations 3.13.7 and 3.13.9). Taking $d_Y = 0.04a$ (d_Y is the *total* extent of the plastic zone), we find that the stress in an infinite body needed to produce this amount of plasticity is given by:

$$\sec\left(\frac{\pi\sigma}{2\sigma_Y}\right) = 1.04 \qquad (5.11.2)$$

At this stress level, we examine the effect of $(W-a)$ on the spread of the plastic zone, as shown in *Table 5.3*, derived from equation 5.11.1 by substituting different values of (a/W). Values for $d_Y = 0.02a$ are included for comparison.

The usual value taken for toughness testpieces is $(W - a) = a$, i.e. $W/a = 2$. We see from *Table 5.3* that this means that the size of the plastic zone can be as much as 25% greater in the testpiece than would be predicted for an isolated

Table 5.3. ESTIMATE OF LIGAMENT RESTRICTION [assuming $(d_Y/a) = 0.04$ or 0.02 for the infinite plate]

Hence $\sec\left(\frac{\pi\sigma}{2\sigma_Y}\right) = 1.04$ or 1.02

W/a	(d_Y/a) (from equation 5.11.1)	
	$(d_Y/a)_\infty = 0.04$	$(d_Y/a)_\infty = 0.02$
4/3	0.1	0.045
3/2	0.07	0.035
2	0.05	0.025
3	0.045	0.022

crack in an infinite body at the same stress intensity. In terms of the equivalent elastic crack, the effect on K, given by equation 4.6.6, may be written as:

$$\begin{aligned}\delta K/K &= \sqrt{(1 + r_Y/a)} - \sqrt{(1 + r'_Y/a)}\\ &= (1 + 0.025)^{\frac{1}{2}} - (1 + 0.02)^{\frac{1}{2}}\end{aligned} \tag{5.11.3}$$

and is found to be less than 0.5%. This conclusion holds only for the very low stress intensities associated with the plastic zone size $r_{IY} = 0.02a$. For higher stresses, the errors become much larger, and eventually markedly affect any substitutions into an equivalent elastic crack formula (see Section 4.6), based on plastic zone size, for any testpiece of finite width.

5.12 Specimen Dimensions – Thickness

The limitations on specimen thickness derive from the results on aluminium alloys described in Section 5.5 and from similar experience obtained on maraging steels. The thickness required to give 'plane strain' behaviour (total instability at the lower critical toughness) is taken to be:

$$B \geqslant 2.5\left(\frac{K_{IC}}{\sigma_Y}\right)^2 \tag{5.12.1}$$

where K_{IC} is the plane strain fracture toughness. Substituting values for the aluminium alloy from *Figure 5.2*, equation 5.12.1 gives the condition $B \geqslant 15.5$ mm which may be seen fairly to represent the experimental observations. In terms of the radius of the plastic zone, we have:

$$B \geqslant 5\pi r_Y \tag{5.12.2}$$

using equation 5.7.1, where r_Y is the radius of the plastic zone in plane stress. In plane strain, the radius r_{IY} is taken to be one-third that in plane stress, as before. Thus, approximately:

$$B \geqslant 15\pi r_{IY} \triangleq 47 r_{IY} \tag{5.12.3}$$

This value is extremely close to that decided upon previously for the crack length restriction ($a = 50r_{\text{IY}}$, Section 5.10) and the two dimensions are planned to be identical in the standard testpieces. The final limits become:

$$a = (W - a) = B \geqslant 2.5 \left(\frac{K_{\text{IC}}}{\sigma_{\text{Y}}}\right)^2 \tag{5.12.4}$$

and these must all have been satisfied before a K_{IC} result can be classed as 'valid'.

5.13 Test Procedure

Once specimens have been made to the recommended design, it is necessary to introduce a sharp crack at the tip of the stress concentrator. *Figure 5.15* shows

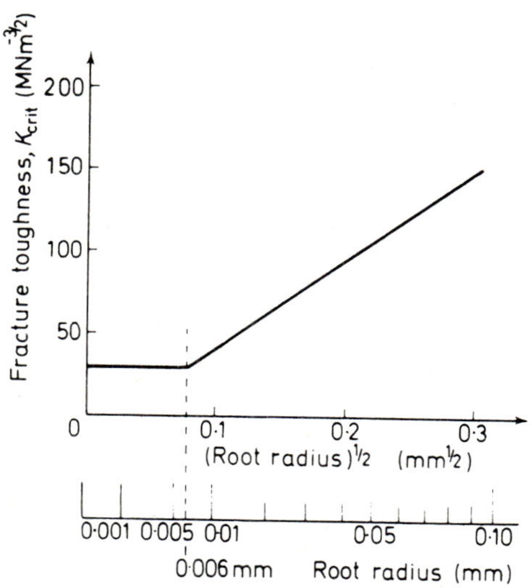

Figure 5.15 Schematic variation of toughness with notch root radius. Data taken as for that of a high-strength steel

the effect of using concentrators which are not infinitely sharp: it can be seen that the final toughness decreases rapidly with root radius until a limiting 'cut-off' radius is obtained. To introduce sharp cracks, fatigue cracking is usually employed. The peak stresses achieved during fatiguing must, however, be very much less than those producing fracture, or it is found that the toughness value depends on the fatigue loads used, e.g. *Figure 5.16*. If the maximum stress intensity during fatigue is denoted by K_f, a crack may initially be grown at values of $K_f \leqslant 0.75\, K_{\text{IC}}$: the last 1.25 mm or 2.5% W, whichever is greater, must, however, be grown by intensities $K_f \ngtr 0.67\, K_{\text{IC}}$, to preclude effects of the fatigue on the subsequent toughness. The crack front must be kept as straight as possible: at the end of pre-cracking, the central and surface portions of the crack should not differ in length by more than 5% and the crack plane should not twist or tilt by more than 5°. The final a/W value must lie between 0.45 and 0.55 (see Sections 5.11 and 5.12).

The standard testpieces have been calibrated to provide a value of K for a given value of load, using standard tables as *Table 5.2*, provided that the total

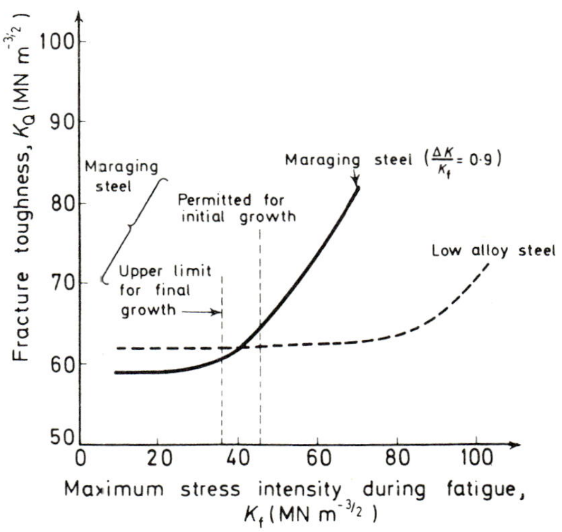

Figure 5.16 Effect of fatigue stress level on fracture toughness. (The low alloy steel results indicate that higher values of K can be tolerated in a fatigue test than in a subsequent fracture toughness test carried out under slower rates of loading, and suggest that the restrictions on fatique stress level may be unduly conservative) (after May[16])

behaviour of the piece is such that fracture occurs under what are nominally elastic conditions. The test procedure provides a method of estimating how closely this condition is approached by the examination of load–displacement curves, as follows.

Load is measured by a conventional load-cell, which must be accurate to better than ±1%. Displacement is measured by means of a 'clip-gauge' mounted across the open mouth of the stress concentrator as shown in *Figure 5.17*. The gauge is composed of two cantilever beams separated by a spacer block. Resistance wire strain gauges are affixed to the tension and compression faces of each beam and all four gauges are used as the arms of a Wheatstone's Bridge. The deviation from the true displacement has to be within 0.003 mm for displacements up to 0.5 mm and within 1% of higher values. Load and displacement are then plotted as the ordinate and abscissa respectively on an $X–Y$ recorder.

Typical load–displacement curves for different materials and different thicknesses are shown in *Figure 5.18*. Trace (*a*) represents ideal behaviour. Total fracture or, at least, substantial crack advance is obtained at a characteristic load and the trace remains linear until this load is reached. The load can obviously be used directly, via the calibration tables, to calculate K_{IC}.

Conversely, in specimens which are not sufficiently thick, giving traces such as (*c*) or (*d*), where total fracture or limited cracking is preceded by large amounts of plasticity, it is apparent that K_{IC} values, as such, cannot be obtained, and, if values of toughness are needed, *R*-curve analyses (Section 5.8)

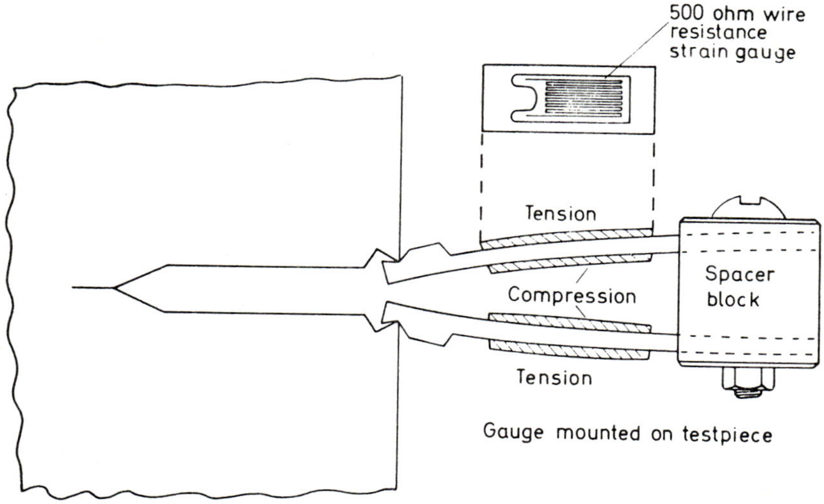

Figure 5.17 Design and assembly of clip gauge

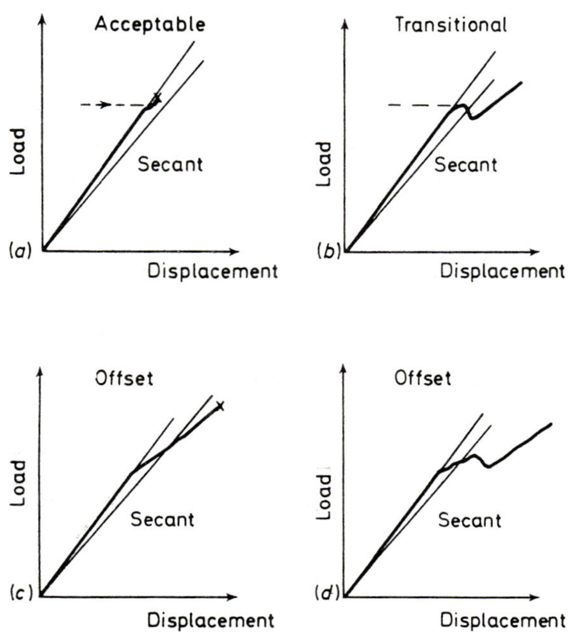

Figure 5.18 Schematic load–displacement curves. (For the secant construction see Figure 5.19 and text)

should really be adopted. Trace *5.18(b)* represents transitional behaviour, which will be discussed in the next section, together with the procedure for deriving a toughness value from a rising-load curve.

5.14 The Offset Procedure

A criterion to decide whether or not the plasticity preceding pop-in is large or small needs to be applied in situations such as that shown in *Figure 5.19* (or *5.18b*) where there is some curvature to the trace, preceding an obvious

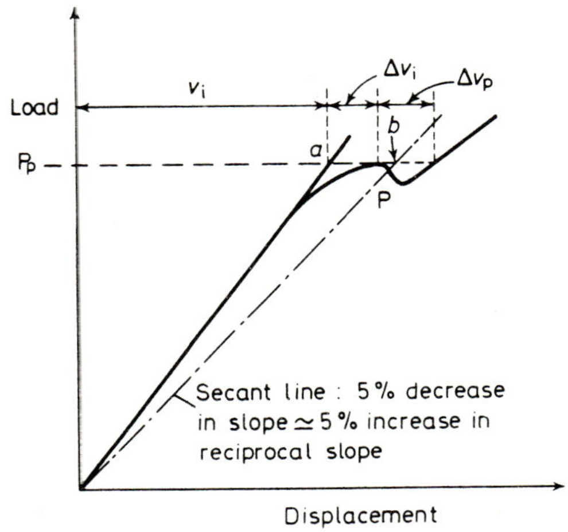

Figure 5.19 The secant construction for a trace showing an obvious pop-in. (An 'obvious' pop-in is one for which Δv_p, the extra displacement due to pop-in, is greater than the distance ab between the elastic modulus and the secant at the load P_p – i.e. the pop-in is $> 0.02\ a_0$). On this trace, $P_Q = P_p$. Had the secant line intersected, at P, to the left of P_p, P_Q would have been taken as P (see text)

sudden pop-in which produces an increased displacement under decreasing load. The standard criterion is derived as follows. Suppose that P_p is the maximum load preceding pop-in and that v_i is the displacement associated with the elastic loading of a testpiece containing a crack of length a_0 to the load P_p. Then Δv_i represents the non-elastic displacement prior to pop-in. This may arise from plasticity effects or from a change in specimen compliance if there has been any slow crack growth prior to pop-in. It is this second situation that is analysed to provide the criterion.

If Δv_i is due to an increase in crack length of magnitude Δa, it is assumed that Δa shall not exceed the value previously associated with the region dominated by the K singularity (equation 5.10.2), i.e.

$$\Delta a/a_0 \leqslant 0.02 \qquad (5.14.1)$$

To convert this back to a displacement criterion, use must be made of experimentally determined compliance curves which relate displacement (v) to load

(P) as a function of relative crack length (a/W). Expressions take the form:

$$v \cdot \frac{EB}{P} = \mathrm{f}\left(\frac{a}{W}\right) \tag{5.14.2}$$

where E is Young's modulus and B is thickness. Then, at constant load:

$$\frac{\Delta v}{v} = \frac{\Delta v\, EB/P}{v\, EB/P} = \frac{\mathrm{f}\left(\frac{a}{W} + \frac{\Delta a}{W}\right) - \mathrm{f}\left(\frac{a}{W}\right)}{\mathrm{f}\left(\frac{a}{W}\right)} \tag{5.14.3}$$

Since $\Delta a \ll a_0$ (the condition was $\Delta a = 0.02a_0$), we may write:

$$\frac{\Delta v_i}{v_i} = \frac{1}{\mathrm{f}} \cdot \frac{\mathrm{df}}{\mathrm{d}(a_0/W)} \cdot \left(\frac{\Delta a_0}{W}\right) \tag{5.14.4}$$

where f stands for $\mathrm{f}\left(\frac{a_0}{W}\right)$. Thus

$$\frac{\Delta v_i}{v_i} = \left[\frac{a_0}{W} \cdot \frac{1}{\mathrm{f}} \cdot \frac{\mathrm{df}}{\mathrm{d}(a_0/W)}\right] \frac{\Delta a_0}{a_0} \tag{5.14.5}$$

Or, since we have decided that $\Delta a_0 \ngtr 0.02a_0$

$$\frac{\Delta v_i}{v_i} \ngtr 0.02 \left[\frac{a_0}{W} \cdot \frac{1}{\mathrm{f}} \cdot \frac{\mathrm{df}}{\mathrm{d}(a_0/W)}\right] \tag{5.14.6}$$

$$\ngtr 0.02\, H \tag{5.14.7}$$

where H is a calibration factor derived experimentally from load–displacement curves as a function of (a_0/W). The limitation on deviation from linearity may be expressed in terms of the reciprocal slope of the ('secant') line joining the origin to the maximum load, P_p, on the trace (see *Figure 5.19*). Since the displacement at this point P_p is $v_i + \Delta v_i$ we have:

$$\frac{v_i + \Delta v_i}{P_p} \leqslant \frac{v_i}{P_p}\,[1 + 0.02\, H] \tag{5.14.8}$$

For the recommended range of values of (a_0/W), lying between 0.45 and 0.55, it is found that 0.02 H has an average value of 0.05 for SEN bend specimens (*Figure 5.20*). This leads to a requirement that the deviation from linearity before pop-in should represent a change in reciprocal slope of not more than 5% on the load–displacement trace, where load is the ordinate and displacement is the abscissa.

The standards put this requirement rather differently. A straight line is drawn through the initial part of the trace, corresponding to elastic loading, and a secant line is drawn with a 5% decrease in slope (see *Figure 5.19*). For traces more or less at 45° to the axes (the standard specifies inclinations between 40° and 65°) this corresponds reasonably enough to the increase in reciprocal slope. The point P represents the lowest load at the intersection of, or tangency to, the load–displacement curve. Then for a trace of the type shown, the load, P_Q, used to determine toughness is equal to P or to any higher load preceding P. The toughness value is denoted by K_Q.

For traces of a type which do not show a definite pop-in (e.g. *Figure 5.21*), a different procedure is adopted. Again, the 5% secant line is drawn and the point $P \equiv P_Q$ located. The difference in displacement between the trace and the elastic loading line at the load P_Q is measured. Let this be denoted by v. Then a similar displacement, v', is measured, from the elastic loading line to the actual

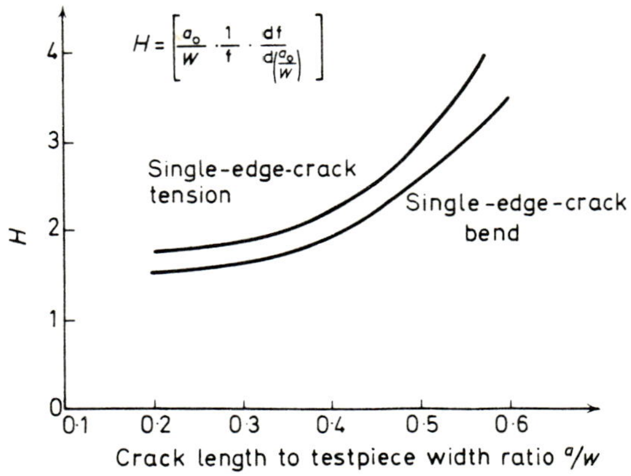

Figure 5.20 Variation of calibration factor, H, with crack length

trace, at a load of 0.8 P_Q. If v' is greater than 0.25 v, it is considered that gross plasticity or instrumental errors may be occurring, leading to unrepresentative values of K_{IC}.

The 5% secant line represents the change in compliance due to crack extension equal to the radius of the plane strain plastic zone: $\Delta a = r_{IY} = 0.02a_0$. The procedure is designed to test whether or not P_Q can be associated with crack extension. If v' is very small, then the dramatic increase in displacement

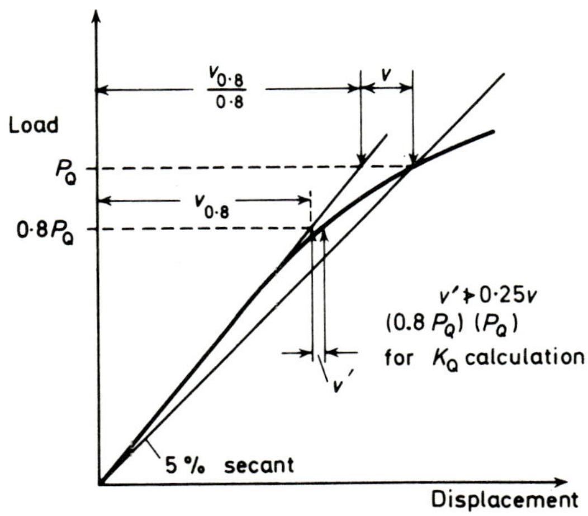

Figure 5.21 Offset procedure used when the trace does not show an obvious 'pop-in'

between 0.8 P_Q and P_Q can be attributed to cracking rather than to plasticity alone. The criterion itself is empirical, but may be rationalised as follows. If the displacement at 0.8 P_Q due to plasticity is v' and the total elastic displacement is $v_{0.8}$, we have, from equation 5.14.8, if we treat the plastic zone as increasing the effective crack length (4.6.6):

$$\frac{v'}{v_{0.8}} = \frac{r_{IY}}{a_0} H \tag{5.14.9}$$

where r_{IY} is the radius of the plane strain plastic zone at 0.8 P_Q. Since r_{IY} is proportional to P^2 (i.e. to σ^2 or K^2), the radius of the zone at P_Q will be $r_{IY}/(0.8)^2$. The elastic displacement at P_Q is $v_{0.8}/0.8$ because the elastic slope is linear, if the crack has not extended. Then we may write for the *plastic* displacement, v, at a load P_Q:

$$0.8 \frac{v}{v_{0.8}} = \frac{r_{IY}}{(0.8)^2 a_0} H \tag{5.14.10}$$

By elimination between equations 5.14.9 and 5.14.10 we obtain

$$v' = (0.8)^3 \, v \cong 0.5 \, v \tag{5.14.11}$$

In other words, the displacement at P_Q produced simply by effects due to plasticity would be about twice that produced at 0.8 P_Q. By laying down the stipulation that v must be more than four times v', the criterion determines that cracking of magnitude comparable with that of the plastic zone radius must have occurred between 0.8 P_Q and P_Q.

There are several unsatisfactory features with this procedure. First, it is very difficult practically to measure v' with any degree of accuracy. Secondly, if cracking begins at P_Q, the procedure can be more restrictive than that used when there is definite evidence of a fast 'pop-in'. This is because the total change in displacement at P_Q, due both to plastic zone and to crack extension, must lie within the 5% secant line (equivalent to a change in crack length of $0.02\,a_0$). For constant- or decreasing-load 'pop-ins', only the amount of plasticity preceding fracture has to fall within the 5% secant line. The amount of displacement produced by the cracking subsequent to initiation may be very large. Thirdly, there is no clear indication as to the load, between 0.8 P_Q and P_Q, at which cracking started. This must seriously affect the reproducibilities of fracture toughnesses measured on a rising load curve. From the trace alone, some indication of the point of initiation could presumably be obtained by constructing the line between 0.8 P_Q and P_Q which shows the effect of plasticity on displacement (for a load hP_Q, $v_h = v'/h^3$ where $0.8 < h < 1$) and comparing this with the actual trace to detect any sharp discontinuity in slope. This procedure would, however, be very tedious and prone to error, because v' is difficult to measure.

The main problem with the offset construction is that it is trying to do two things at the same time. On the one hand, it is trying to detect crack initiation: on the other it is trying to ensure that the radius of the plane strain plastic zone prior to such initiation is less than $0.02a$. These two aims are extremely difficult to satisfy simultaneously in a way that is consistent with the previous procedures. It would be much more satisfactory to detect initiation directly, using additional instrumentation such as a piezo-electric crystal (barium titanate or PTZ: lead titano-zirconate) probe which could respond to the acoustic emission produced when cracking occurred. The load–displacement trace could then be analysed to assess non-linear effects.

The very nature of a rising load curve implies, however, that some stable crack growth occurs after initiation and account should be taken of both the increase in crack length and (shear lip) resistance to crack extension before calculating the K_{IC} value. Preferably such tests should be treated like those of specimens of intermediate thickness and subjected to a full R-curve analysis (see Section 5.8).

In all cases, the plastic zone size associated with a K_Q value must be compared with specimen dimensions before the K_Q value can be said to represent a valid K_{IC} determination.

5.15 Toughness Results

A full category of results produced by fracture toughness testing is well beyond the scope of this book. The effects of major variables in testpiece dimensions, such as thickness (Section 5.3) and notch radius (*Figure 5.15*), have already been described and will be referred to again in Chapter 7, where notched-bar fracture behaviour and microscopic fracture mechanisms are discussed. Three further major variables which produce marked effects on toughness are the temperature of testing, the applied strain-rate and, obviously, the material tested. The bare results will be presented in this section and such discussion as can be related to micro-mechanisms of fracture will be included in Chapters 7 and 8.

5.16 Effects of Temperature and Strain Rate

Effects of temperature on K_{IC} are shown most strongly in relatively low-strength structural steels. Non-ferrous alloys and very high-strength steels, such as the maraging steels, show rather small variation of toughness with temperature, up to 100°C. In an analogous manner, the low-strength steels show a sharp energy transition in the conventional notched-bar impact test, whilst the other alloys do not.

The main problem in testing a steel of low yield strength is that, if its toughness is at all high, the size of testpiece required to satisfy the valid K_{IC} criteria, i.e.

$$B \cong a \cong (W-a) > 2.5\left(\frac{K_{IC}}{\sigma_Y}\right)^2 \qquad (5.12.4)$$

will be very large indeed. Critical experiments have been carried out by Wessel[11], using CTS testpieces of different sizes, up to 300 mm thick, as shown in *Figure 5.22a*. *Figure 5.22b* shows the form of the results obtained for a steel, designated A533 Grade B Class I (0.23% C, 1.35% Mn, 0.5% Ni, 0.5% Mo),

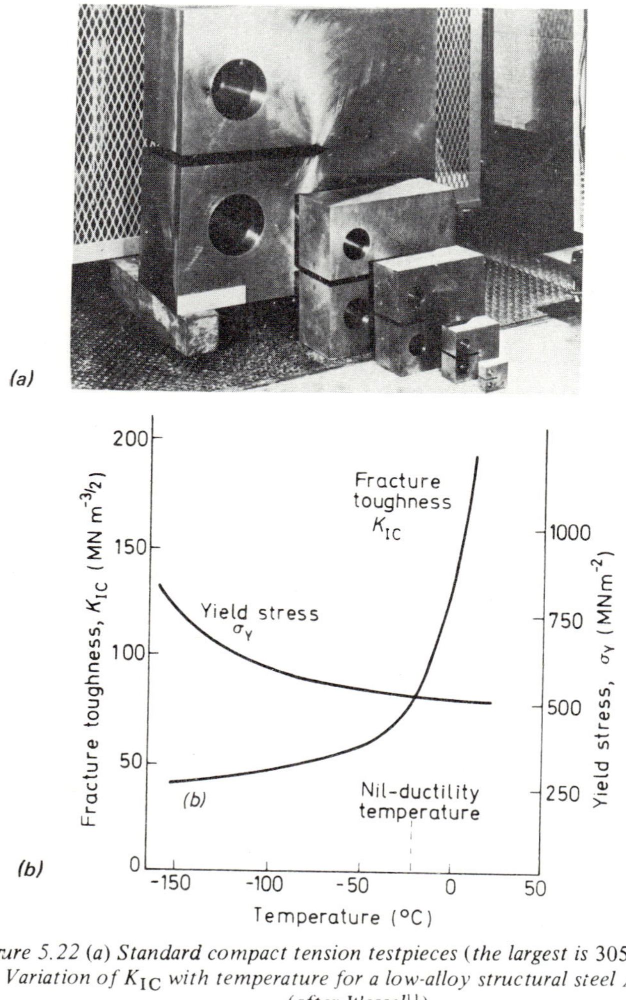

Figure 5.22 (a) Standard compact tension testpieces (the largest is 305 mm thick) (b) Variation of K_{IC} with temperature for a low-alloy structural steel A533B (after Wessel[11])

heat-treated to a yield stress of 500 MN m^{-2} at room temperature. It can be seen that there is a very steep increase in the K_{IC} value above a temperature of about −20° C. At −100° C, the values of σ_Y and K_{IC} from the figure, when substituted into equation 5.12.4 show that the K_{IC} result is valid in testpieces above about 17 mm thick. At room temperature, where K_{IC} is about 150 MN m$^{-3/2}$ and σ_Y is about 500 MN m^{-2}, the critical dimensions can be calculated as at least 225 mm. It is of fundamental importance to realise that this transition in K_{IC} has occurred under linear elastic plane strain conditions.

In other words, it is a transition characteristic of the material and is not accentuated by the gross relaxations which may accompany an initially modest increase of toughness with temperature in smaller testpieces. This point will be discussed further in Chapters 7 and 8.

Moderate increases in strain rate produce rather small effects on K_{IC}. Results[12] for a semi-killed steel are shown in *Figure 5.23*. It can be seen that

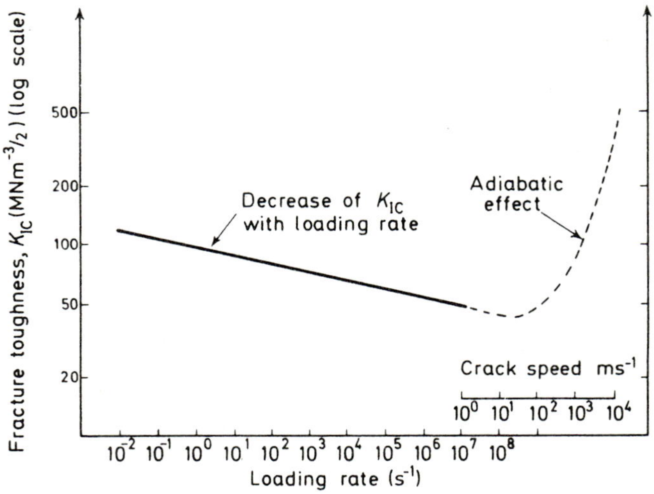

Figure 5.23 Variation of fracture toughness with loading rate in a semi-killed steel (after Radon and Turner[12])

about an order of magnitude increase in loading rate is required to decrease K_{IC} by 10%. To some extent, the effect is small because very large testpieces are not usually able to be tested. The materials studied therefore tend to be those of high strength and low toughness, in which the effects of strain rate on flow characteristics are much less than for a mild steel. At very high loading rates (ahead of a running crack, for example) it is possible for the deformation to occur so quickly that the heat generated by plastic flow cannot be fully dissipated into the bulk of the specimen. The temperature ahead of the propagating crack may then be raised substantially. Under these *adiabatic conditions*, the toughness of the region just ahead of the crack tip may be greatly increased. If it is desired to stop running cracks in a structure, it is wise to design on the minimum value of dynamic fracture toughness, K_{ID} (Knott[6]).

5.17 Material Properties

This section will be kept short deliberately to draw attention to one or two general points that can be made concerning the relationship of fracture toughness to other material properties. It has been indicated in the previous section that only relatively low-strength steels have a very marked dependency of toughness on temperature, at low temperatures.

The main variable affecting the toughness of a particular type of material is its yield strength. *Figure 5.24* shows the way in which toughness varies with

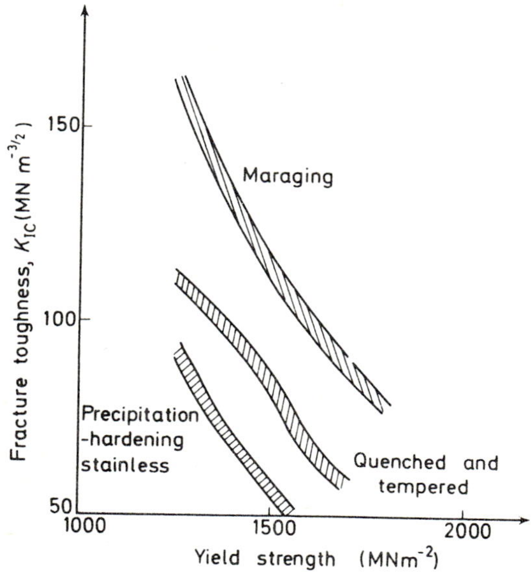

Figure 5.24 Variation of fracture toughness with strength level, for different classes of steel (after Pellini et al.[13]*)*

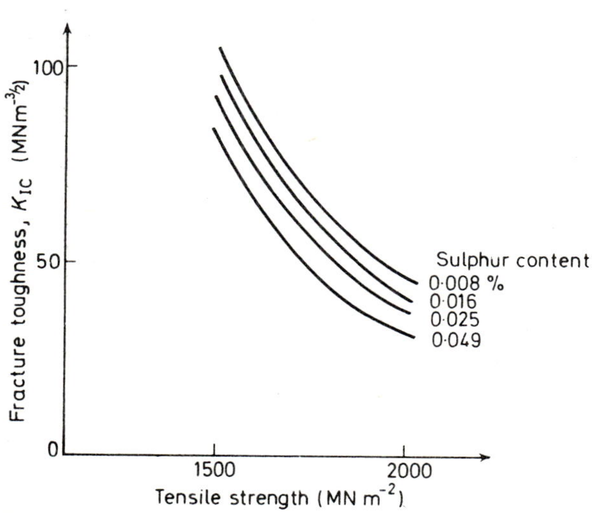

Figure 5.25 Effect of sulphur content on the fracture toughness of a quenched and tempered low-allow steel (after Feige and Murphy[15]*)*

yield strength for conventionally quenched and tempered steels, for maraging steels and for precipitation-hardened stainless steels[13]. In all cases, it is observed that plane strain toughness decreases markedly with increase in yield strength. The improvements gained by using maraged steels at very high strength levels in high-duty applications are obvious. A similar effect of strength level on K_{IC} is found for aluminium alloys[14].

There is also an effect of non-metallic inclusion content on toughness, as shown in *Figure 5.25*, where K_{IC} is plotted *v* tensile strength for a quenched and tempered low alloy steel containing different sulphur levels[15]. It is clear that a high inclusion content produces low toughness. The effect is perhaps even greater in aluminium alloys where brittle intermetallic particles (containing iron and silicon) can provide sites for crack nucleation. Trace impurity elements also have a marked effect on the toughness of steels if they produce grain-boundary embrittlement.

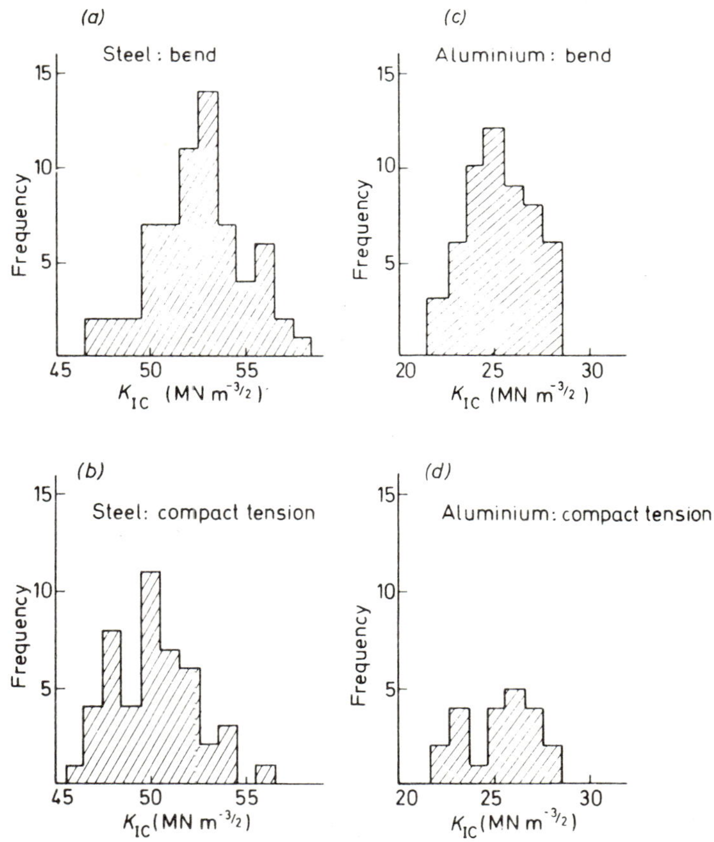

Figure 5.26 Preliminary results of collaborative determinations of K_{IC} (Courtesy of M. J. May E. F. Walker)

5.18 Conclusions

Established procedures are available for measuring reproducible values of plane strain fracture toughness, K_{IC}, provided that due care and attention are given to specimen size requirements and to the analysis of the load–displacement traces. The minimum specimen dimensions imply that all dimensions are about fifty times greater than the radius of the plane strain plastic zone at fracture. With respect to crack length, a similar criterion in terms of testpiece compliance is reflected in the requirement that the decrease in slope of the load–displacement curve prior to instability or definite 'pop-in' shall not exceed 5%. The recommended procedure for determining K_{IC} in testpieces which show a rising-load trace is less satisfactory.

The worth of the standard procedure is finally demonstrated by *Figure 5.26 (a–d)* which shows K_{IC} results determined on a steel and an aluminium alloy, using both SEN bend and CTS testpieces, by some twenty different British laboratories. It is apparent that satisfactory agreement between the different laboratories can be obtained through careful use of the recommended procedures and that the inclusion of K_{IC} values in standard material specifications is quite feasible. In practice, values are likely to be specified mainly for rather high-strength alloys, because it is only for these that the required specimen sizes and testing facilities are not unduly large.

The question of how to measure the toughness of low-strength, tough materials in relatively small laboratory test specimens is one which has still not been fully answered. The following chapter describes the problems involved and the methods which have been developed to solve them.

REFERENCES

1. Irwin, G. R., Kies, J. A. and Smith, H. L., *Proc. Am. Soc. Test. Mat.*, **58**, 640 (1958)
2. Krafft, J. M., Sullivan, A. M. and Boyle, R. W., *Proc. Symp. Crack Propagation*, Cranfield, 8 (1961)
3. Brown, W. F. and Srawley, J. E., *Am. Soc. Test. Mat.*, Spec. Tech. Pub. No. 381 133 (1965)
4. Tetelman, A. S. and McEvily, A. J., *Fracture of Structural Materials*, Wiley, New York (1967)
5. Cottrell, A. H., *Proc. R. Soc.*, **A276**, 1 (1963)
6. Knott, J. F., *Mater. Sci. Engng*, 7, 1 (1971)
7. Bradshaw, J., *Euromech. Conference* No. 39, Ronneby (Sweden), June (1972)
8. Hodgson, J. and Boyd, G. M., *Trans. R. Inst. nav. Archit.*, **103** (1961)
9. Wilson, W. K., *Am. Soc. Test. Mat.*, Spec. Tech. Pub., No. 410, 75 (1966)
10. Smith, E., *Proc. R. Soc.*, **A285**, 46 (1965)
11. Wessel, E. T., *Practical Fracture Mechanics for Structural Steel*, Paper H, UKAEA/ Chapman and Hall (1969)
12. Radon, J. C. and Turner, C. E., *J. Iron Steel Inst.*, **204**, 842 (1966)
13. Pellini, W. S. *et al.*, *Naval Res. Lab. Rep.*, No. 6300, Washington, D.C. (1965)
14. Kaufman, J. G. and Hunsicker, H. Y., *Am. Soc. Test. Mat.*, Spec. Tech. Pub. No. 381, 290 (1965)
15. Feige, N. G. and Murphy, T., *Metals. Eng. Q.*, 7, 53 (1967)
16. May, M. J., *Amer. Soc. Test. Mat.*, Spec. Tech. Pub., No. 463, 42 (1970)

Six

GENERAL YIELDING FRACTURE MECHANICS

6.1 Introduction

In Chapter 5 it was shown that the use of standard procedures leads to the measurement of consistent values of plane strain fracture toughness, K_{IC}. To obtain 'valid' results in relatively tough materials, it is, however, necessary to use testpiece dimensions so large that the testpieces may not be representative of the behaviour of the sections actually used in service. In addition, the material user generally requires that a quality control test be carried out on each batch of material to ensure that it meets the properties specified. It is obvious that a standard fracture toughness test would be unsuitable for such a purpose in any but high-strength, brittle alloys, because it would require far too much material for the testpiece. There is, consequently, much interest in the possibility of making alternative measurements of a material's resistance to fast fracture, using small testpieces, which do not require a large amount of material and which may be tested easily in the laboratory. To take full advantage of linear elastic fracture mechanics stress analyses, the toughness parameter measured should be able to be related directly and quantitatively to the material's fracture toughness.

Two such parameters have been proposed. The first is the critical value of the crack-tip opening displacement or COD (Section 3.13) at fracture: the second is the critical value of a quasi- 'strain-energy release rate' (J_{IC}), derived assuming non-linear elastic behaviour[1].

6.2 The Crack Opening Displacement – Principles

It is generally accepted that the crack-tip stress or strain field cannot be characterised by a single-valued parameter when yielding in the crack-tip region becomes extensive. In particular, when a specimen has yielded generally, the applied stress cannot be used to calculate either the local stresses or strains at the crack tip. Independently, however, Cottrell[2] and Wells[3] focused attention on the amount of crack opening prior to crack extension as a parameter which might be treated as a characteristic of the crack-tip region, for a given material tested under a given set of conditions.

Cottrell employed the concept to explain why small testpieces, cut from a large ship-plate which had fractured well before general yield, with a fracture surface exhibiting more than 90% bright 'crystalline' cleavage facets, broke well after general yield, with a completely 'fibrous' fracture appearance, when tested at the same temperature. The argument runs as follows. A given value of COD needs to be 'accommodated' by a specific size of plastic zone. Well below general yield, the relationship for plane stress tension is given (equation 3.13.14 et seq.) by:

$$\delta = 2\pi \frac{\sigma_Y}{E} r_Y \tag{6.2.1}$$

where r_Y is the plastic zone radius. Then, if we take a fixed length of crack and a fixed COD, we can see that the question of whether a piece breaks before or after general yield is resolved simply by the position of the far boundary surface of the piece, opposite the crack. If it is so close that the plastic zone traverses the net cross-section before the critical COD value (δ_{crit}) is attained at the crack-tip, the piece is ductile: if the boundary is so remote that δ_{crit} is attained first, the piece is brittle. Cottrell estimated the size of testpiece needed to exhibit brittle behaviour. Using the expression (see equation 3.14.4) to give the antiplane sliding displacement, S, at *a* crack-tip:

$$S = \frac{4\tau_Y}{\pi\mu} a \ln \left[\sec\left(\frac{\pi q}{2\tau_Y}\right)\right]$$

and taking $\tau_Y = \mu/1000$, the values of $10^4 S/a$ and c/a can be calculated for different ratios of (q/τ_Y) as shown in *Table 6.1*. Cottrell then took a value for

Table 6.1

q/τ_Y	0.1	0.5	0.75	0.9	0.95	0.99
$10^4 S/a$	0.16	4.4	12.1	23.6	32.4	53
c/a	1.01	1.41	2.6	6.4	12.8	64
a	4.5 m	178 mm	63 mm	33 mm	23 mm	15 mm
c	4.55 m	254 mm	165 mm	212 mm	292 mm	978 mm

the critical displacement of 0.075 mm and derived the individual figures for *a* and *c* given in the fourth and fifth rows of the table.

It is clear that fracture can occur at about half the yield stress if the piece is sufficiently large (some 250 mm) and contains a sufficiently long crack (180 mm). Conversely, it is clear that failures will not occur at normal service stresses ($0.5\,\tau_Y$ – see Section 1.5) unless the defects present in a structure are so large as to be obviously detectable. This argument rests, of course, on the assumption that the critical displacement remains at the value of 0.075 mm.

The effect is also shown clearly in *Figure 6.1*[4]. Here, three testpieces containing identical stress concentrators have been fractured at the same temperature of −80°C. The deformation preceding fracture has been revealed by ageing and etching in Fry's reagent. It is obvious that the testpiece with the smallest ligament has yielded generally before fracture: those of large size have broken before general yield, but after comparable amounts of local (accommodation) yielding.

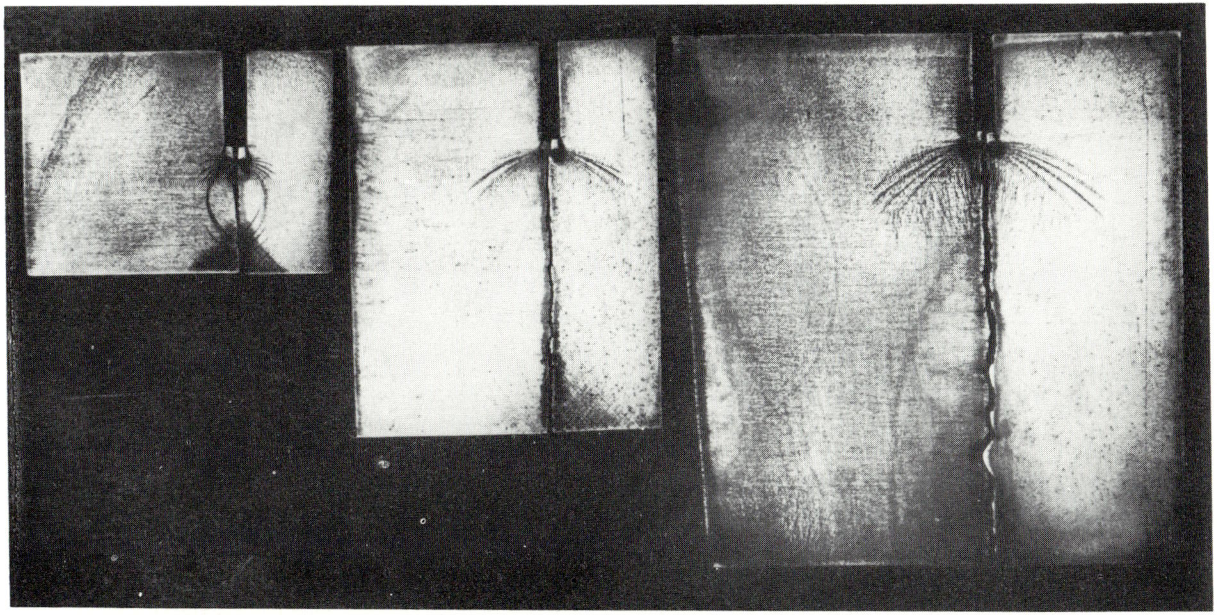

Figure 6.1 Sections of mild steel specimens fractured at 200 K, *the notch geometry being the same in all specimens. The plastic deformation preceding fracture has been etched, using Fry's reagent* (× 0.7)

In each case, the value of COD at fracture, i.e. δ_{crit}, was sensibly constant, as postulated.

The difference in fracture appearance between the large plate and the small testpiece in terms of the proportions of 'crystallinity', as described above, is most probably a consequence of the acceleration of a small amount of fibrous fracture in service leading to the initiation of cleavage cracks in regions which are subjected locally to high strain rates. This point will be taken up again in Chapter 8.

The Cottrell/Wells COD concept introduces a parameter which appears to characterise fracture, under fairly closely defined conditions, both before and after general yield. The sliding displacement used by Cottrell (3.14.4) may equally be replaced by an opening displacement δ_{crit}, for tensile loading (3.13.11). The possibility arises of measuring the value of δ_{crit} for a material on a small testpiece which breaks well after general yield and using this same value to predict the failure stress of a large structure which breaks before general yield, using the equation (equation 3.13.11):

$$\delta_{crit} = \frac{8\sigma_Y}{\pi E} a \ln \left[\sec \left(\frac{\pi \sigma_F}{2\sigma_Y} \right) \right] \tag{6.2.2}$$

where σ_F is the failure stress. For a large structure, which breaks well before general yield, the term (σ_F/σ_Y) is small and we may write

$$\sigma_F = \sqrt{\left(\frac{E \sigma_Y \delta_{crit}}{\pi a} \right)} \tag{6.2.3}$$

interpreting the critical strain energy release rate as the incremental work done by the (constant) crack-tip stress (σ_Y) in the yield zone, when moving through a (constant) displacement, δ_{crit}:

$$G_{crit} = \sigma_Y \delta_{crit} \tag{6.2.4}$$

The finite element results, described in Section 3.18, suggest that, for a crack in plane strain, equation 6.2.4 would take the form: $G_{crit} \cong 2\sigma_Y \delta_{crit}$.

A test of the quantitative relationships between G_{crit} and δ_{crit}, for quasi-elastic fractures and of the extension of the δ_{crit} concept to post-yield fractures demands that experiments are carried out on testpieces of different sizes, so that, at constant temperature, some break before, and some after, general yield. The testpieces indicated in *Figure 6.1* all fractured at stresses too high for valid linear elastic values of G_{crit} to be obtained. Results have been obtained which indicate that, for some steels, equation 6.2.4 gives an appropriate relationship between the two toughness parameters[5]. Similarly, the toughness results given in *Figure 5.22* imply critical values of COD at temperatures greater than −20°C, where fractures extend initially by fibrous mechanisms, which are close to those which would be predicted by a model for fibrous fracture discussed in Chapter 8. A full experimental verification of the proposed relationships between G_{crit} and δ_{crit} is however at present lacking, although, very recently, further experimental results which support the form of equation 6.2.4, have been presented[6].

Even if the precise numerical relationship is not known, there is still value in making measurements of δ_{crit} in materials which break after general yield, to provide a rating of their relative toughnesses at a given temperature. The test then has merit for the purposes of quality control, because measurements may be made on small testpieces.

It should be made clear that the critical value of COD, as defined by, e.g. equation 6.2.2 applies only to the initiation of further cracking. It does not characterise the point of total instability, derived from energetic considerations, as would a K_{IC} value. There may be a substantial difference between the value of δ_{crit} at initiation and that at total instability.

6.3 Crack Opening Displacement – Practical Measurement

The aim is to measure a value of δ_{crit}, characteristic of the mode of fracturing of a material at the tip of a sharp crack, whether this is situated in a large piece subjected to an applied stress much less than the general yield stress or in a small testpiece which is fully yielded and work-hardened. The main difficulties encountered in practical measurements have been to obtain values associated with the crack tip region and to detect the onset of crack extension. Variables

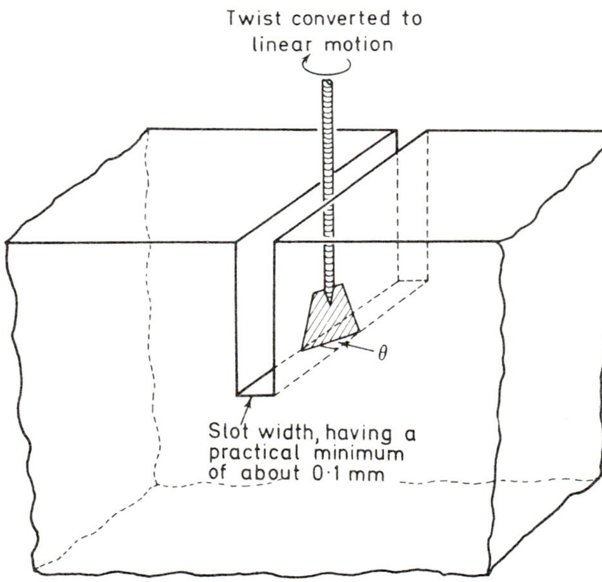

Figure 6.2 Principle of a 'paddle' COD-meter. Spring loading forces the paddle onto the bottom of the slot and a torsion spring presses the extremities of the paddle against the slot sides. As the slot opens, the angle, θ, increases and the rotation of the rod is converted into a linear movement

such as testpiece geometry, thickness and loading system have also been investigated. The development of techniques is best followed chronologically.

Early attempts to determine the ductility of material at stress concentrators involved the measurement of strains at the roots of notches of finite root radius, although these were made using reference marks which, in effect, gave displacements directly. Notch root displacements have also been measured using reference grids inscribed around the notch. Here, composite specimens were used in an attempt to determine plane strain values.

The first direct attempts to measure COD as such for a crack-like geometry were made[7] using the 'paddle' device illustrated in *Figure 6.2*. Specimens were machined with slots some 0.10–0.15 mm wide (using a fine jeweller's saw or a rubber-bonded abrasive cutting wheel), which were not extended by fatigue or any other pre-cracking process. The paddle was located diagonally across the bottom of the slot, and was mounted so that, as the slot opened, the paddle could rotate by an amount proportional to the opening. The rotation was converted to a linear movement, which was detected by a linear transducer. The device had the advantage that it operated at the very end of the slot, but the disadvantage that it could not be used directly with precracked testpieces.

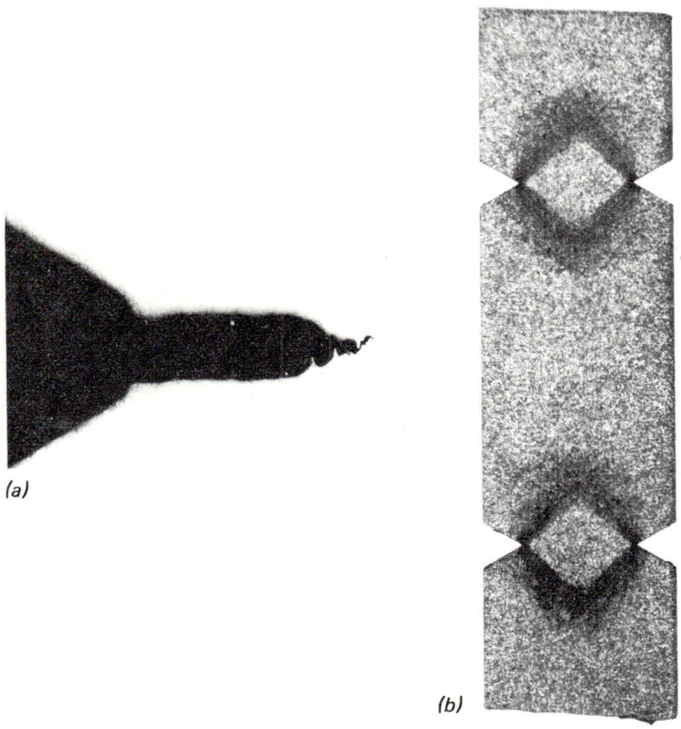

Figure 6.3 (a) Opening of fatigue crack in the unfractured half of a double-cracked specimen (× 180). *(b) Etching of double-notched specimen to demonstrate that identical deformation occurs on both notched cross-sections* (× 3)

A second problem in these early tests was the detection of crack initiation. At very low temperatures, fracture occurred by a brittle cleavage mechanism and the final fracture load was the maximum attained during the test. Here there was no doubt that the initiation of cleavage led immediately to final fracture. At higher temperatures, however, more characteristic of those found in service application for the structural steels studied, the clear-cut fracture event was replaced by a less well-defined sequence in which the trace of load *v*. COD rose to a maximum value and then decreased before much in the way of macroscopic fracture could be detected. The COD at maximum load was then taken as the value, δ_{crit}, pertaining to crack initiation.

An alternative method of measuring COD attempted to provide data for the pre-cracked situation by employing double-notched specimens[8]. Here, by careful fatigue cracking, it was possible to produce cracks of roughly equal length at each notch. The (ductile) specimen was then fractured, and the COD at the unfractured pre-cracked notch was obtained by metallographic sectioning and direct measurement on the micrograph. This COD then represents the value just prior to final instability. *Figure 6.3a* shows the opening of such a crack and *Figure 6.3b* indicates how similar the deformation across the two notched cross-sections remains. In *Figure 6.3a* it can be seen that a small amount of further fracture has occurred ahead of the fatigue crack: this is regarded as the amount of fibrous growth just prior to total instability. In alloy steels, the development of brittle nitride layers around the notch root was sought as an alternative to fatigue for the production of sharp cracks. The COD results obtained with these nitride layers were uniformly lower than those for fatigue cracks, presumably because the sudden fracture of a nitride layer produced an effect of dynamic loading which affected the steel's ductile/brittle transition temperature at the crack-tip[9]. *Figure 3.13* shows the surface of a cracked nitride specimen, which demonstrates the classical slip behaviour described in Section 3.15.

For fatigue-cracked specimens, values of δ_{crit} at room temperature and +80°C were generally identical for the steel studied, but decreased with increasing yield stress and differed for the hardest condition, where +80°C was above the transition temperature and room temperature was below[8]. The values of δ_{crit}, even for ductile fibrous initiation, were substantially less than those obtained from slotted specimens.

Results of this sort and knowledge of the effects of root radius on apparent K_{IC} values (*Figure 5.15*) in linear elastic fracture mechanics led to an investigation of the effect of stress concentrator width on critical values of COD. Concurrently, it was necessary also to identify the first stages of crack extension from the original stress concentrator.

It was realised that, if COD were to be measured in testpieces which contained sharp cracks, the 'paddle' type of device could no longer be used. There were obvious practical advantages to be gained if COD instrumentation could be identical to that used in linear elastic fracture toughness tests and so a technique was developed in which the opening at standard knife-edges mounted on a specimen was measured using a clip-gauge (*Figure 5.17*) and related directly to crack-tip COD by prior calibration[10]. The calibration was determined both experimentally and theoretically[11]. In an experimental calibration, a specimen is unloaded from a particular knife-edge opening (V_g) and is then sectioned and prepared metallographically so that optical measurements of the crack opening (δ) can be made. For a series of specimens, it is then possible to draw a calibration graph of V_g *v.* δ. An example of such a graph for a bend specimen[12] is shown in *Figure 6.4a*. Here, there is a marked variation in n for low displacements, but

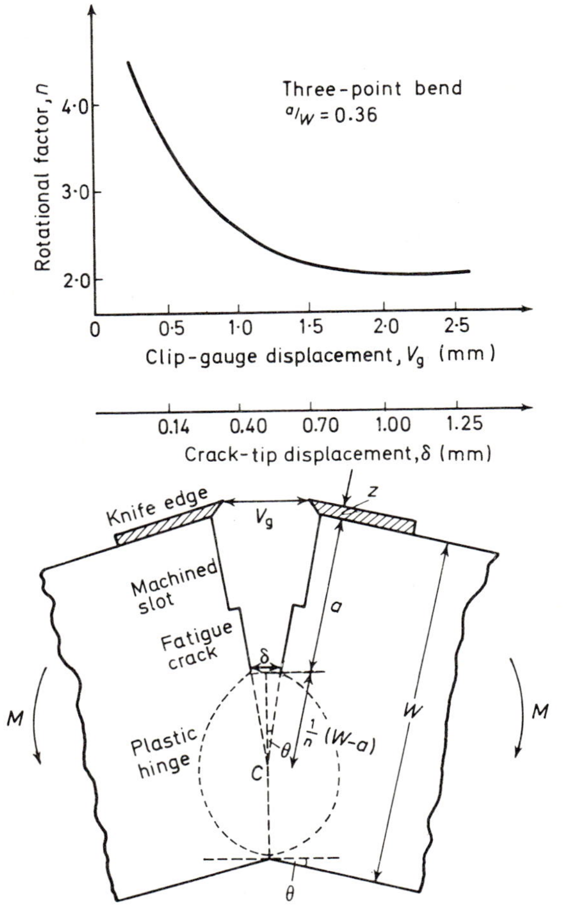

Figure 6.4 (a) Variation of rotational factor with displacement. (b) Diagram showing relationship between crack-tip displacement, δ, and knife-edge displacement, V_g. For a rigid rotation about C:

$$\frac{\delta n}{(W-a)} = \frac{V_g}{\left[z + a + \frac{1}{n}(W-a)\right]}, \quad \text{i.e. } \delta = \frac{V_g}{1 + \left[\frac{n(a+z)}{(W-a)}\right]}$$

If n = 2, C is at the neutral axis, as predicted by slip-line field theory

a value, $n = 2$, is approached as V_g increases (i.e. $V_g = 2\delta$). This is at the lower end of a range of values ($2 < n < 3$) found experimentally, but agrees with a theoretical calculation. The meaning of a particular value of n may be readily understood by reference to *Figure 6.4b* which shows the geometry of the deforming bend specimen and the form of the plastic deformation associated with the bending. The slip-lines form two plastic 'hinges' (see Section 2.12) along which the rigid ends of the specimen slide. If $n = 2$, the centre of rotation for deeply notched specimens lies on the neutral axis predicted by slip-line field theory. For four-point bend specimens, the slope n is approximately constant and equal to 2.7, whilst for the compact tension geometry it is about 2.5.

These results are confirmed by an analysis of the compliances of notched bars of different (a/W) values in materials of different ratios of yield stress to modulus[13]. A proposed standard for COD testing includes these theoretical calibration factors for general application, but, at the present stage, it is wise to carry out independent experimental calibrations for any new material.

6.4 Crack Opening Displacement – Results

Results on effects of slot width on COD have been produced by Smith[12,14]. Using testpieces made in a free-machining mild steel which had been fully heat-treated after fatigue-cracking or slotting, he investigated first the effect, at room temperature, of slot width on COD at maximum load, δ_{max}. The results, shown in *Figure 6.5*, indicate that δ_{max} increases in virtually a linear manner with slot

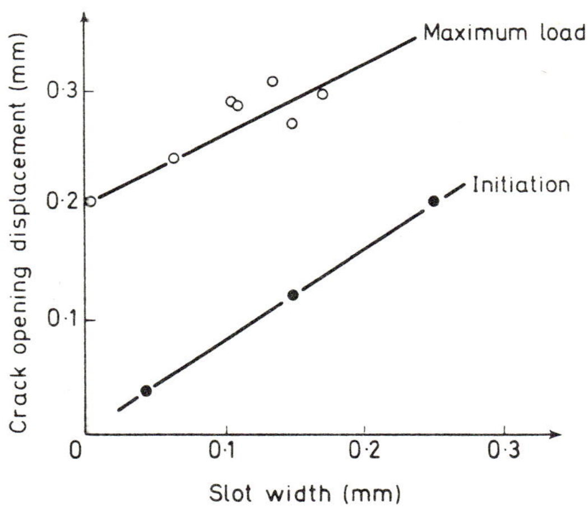

Figure 6.5 Variation of crack opening displacement with slot width (after Smith and Knott[14])

width. Widths intermediate between 0.15 mm and the assumed zero value for a fatigue crack were obtained by fatiguing, straining a little to open the crack and then heat-treating fully as before. It is not clear from these results alone, however, whether the observed effect is due to the way in which slot width affects initiation or whether it arises from an influence of slot width on the maximum in the load–COD trace. Clarification of this point demanded that the initiation of crack extension be detected.

Detailed studies were made on specimens containing either fatigue cracks, 0.15 mm or 0.25 mm wide slots. Specimens were unloaded from different points along the load–COD curve and were then broken open in liquid nitrogen. Any fibrous crack growth at room temperature could be detected as an obvious 'thumbnail' showing in the crystalline cleavage facets typical of the 77 K fracture. By patient experiments of this type, it was possible to draw graphs of COD *v.* fibrous crack length as shown in *Figure 6.6.* Extrapolation of these

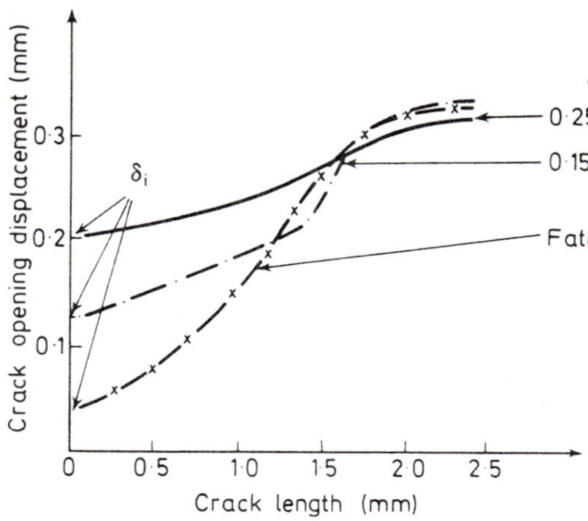

Figure 6.6 Effect of width of stress concentration on variation of crack opening displacement with crack growth (after Smith and Knott[14]*)*

graphs to zero crack growth is then a comparatively straightforward matter and defines an unambiguous measure of the COD at initiation, δ_i.

These values of δ_i are shown as a function of slot width in *Figure 6.5*. It is apparent that it is the variation of δ_i with width which forms the main variable and that, for these specimens, δ_{max} can be represented approximately by the relationship:

$$\delta_{max} = \delta_i + 0.17 \text{ mm} \tag{6.3.1}$$

From the displacements measured on the slotted specimens, it is possible to calculate a value of fracture strain, ϵ_f, if it is assumed that this strain occupies a gauge length equal to the width of the slot. Thus for the 0.25 mm slot,

$$\delta_i = 0.203 \text{ mm}; \qquad \epsilon_f = \frac{0.203}{0.25} = 81\% \tag{6.3.2}$$

for the 0.15 mm slot:

$$\delta_i = 0.120 \text{ mm}; \qquad \epsilon_f = \frac{0.120}{0.15} = 80\% \tag{6.3.3}$$

If it is assumed that this same strain causes fracture in the fatigue-cracked specimen it is possible to calculate an effective gauge length, l_f, for the fatigue crack by putting a value of δ_i into the relationship

$$\epsilon_f = 80\% = \frac{\delta_i}{\text{gauge length}} = \frac{\delta_i}{l_f} \quad (6.3.4)$$

For the fatigue cracks in the mild steel, l_f was calculated as 0.046 mm, indicating that any slot narrower than this figure would give a value of δ_i equal to that for the fatigue crack. This limiting slot width is, however, very much the property of a particular material. It is, for example, much larger than the critical root radius required for the limiting K_{IC} determinations shown in *Figure 5.15* (0.006 mm).

Attempts to define δ_i by less tedious methods have generally made use of the electrical potential method. This will be described in more detail in Chapter 9, but briefly consists of passing a constant d.c. current through a specimen, so that any crack extension, which alters the electrical resistance of the piece, can be detected by observing the change in potential between a pair of probes mounted across the open mouth of the crack. Potential methods have been used to detect initiation with success in aluminium alloys and some steels, but, in other steels, the crack extension is so gradual a process that no sharp change at δ_i is detected. Some sort of offset potential may then be defined as that corresponding to initiation. Attempts to detect initiation have also been made using piezo-crystal accelerometers to detect any stress-wave emission when a crack forms: again, the definition produced depends on how suddenly the fracture extension occurs. Detection by the observation of change in compliance due to crack growth is an insensitive technique for a fully yielded specimen. Such further results as have been obtained, however, generally indicate that fracture in ductile, structural materials can initiate substantially before the maximum in the load-COD trace.

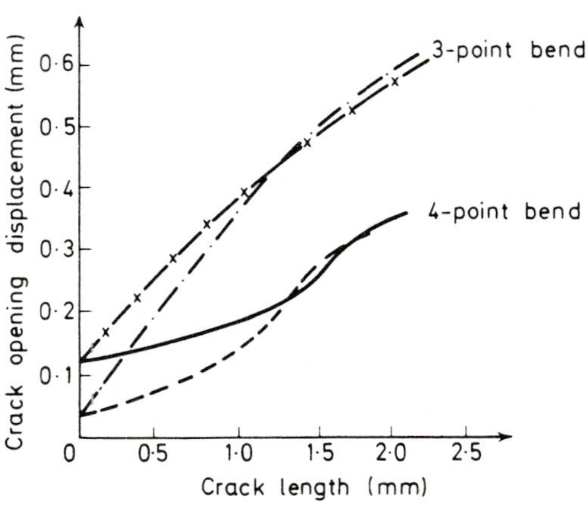

Figure 6.7 Variation of crack opening displacement with crack growth in three and four-point bend speciments. ——— 0.15 mm *slot 4-point bend;* – – – *fatigue crack 4-point bend;* – x – 0.15 mm *slot 3-point bend;* –·– *fatigue crack 3-point bend* (*after Chipperfield, Knott and Smith*[15])

The value of δ_i does appear to be a material characteristic, independent of specimen geometry. *Figure 6.7* shows a comparison between four- and three-point bend specimens and similar results are obtained for four-point bend and compact tension specimens[15]. Despite differences in the rate of increase of COD with fibrous crack growth and in the point of maximum load, very good agreement between the δ_i values is found. The effect of thickness is indicated in *Figure 6.8*. Here it can be seen that δ_i is independent of thickness for thicknesses

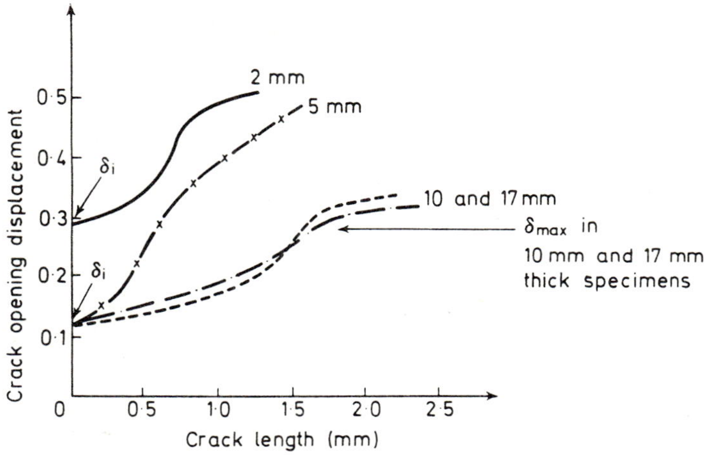

Figure 6.8 Effect of specimen thickness on crack opening displacement. ——— 2 mm *thick;* – x – 5 mm *thick;* –·– 10 mm *thick;* – – – 17 mm *thick*

of 5 mm or greater, (δ_i was also found to have the same value for a 63 mm thick specimen), but is substantially increased for a 2 mm thick specimen. It is anticipated that the critical minimum thickness will vary with the critical value of δ_i (by analogy with thickness effects in the linear elastic range, see Section 5.7). It is interesting to note that, if we take the plastic zone radius in a quasi-elastic case corresponding to the δ_i value, for a fatigue crack, of 0.037 mm (see equation 6.2.1)

$$r_Y = \frac{E}{\sigma_Y}\frac{\delta_i}{2\pi} \tag{6.3.5}$$

and substitute for σ_Y a value of, say, 700 MN m^{-2}, as representative of the work-hardened flow stress in the notch tip region, we obtain:

$$r_Y = \frac{200 \times 10^3}{700} \times \frac{0.037}{2 \times 3.14}$$

$$\cong 1.7 \text{ mm}$$

In other words, for an analogous LEFM situation, we would expect that a specimen would need to be greater than about 3.4 mm to develop conditions conducive to square fracture in its centre (see Section 5.7). This value lies between the 2 mm and 5 mm limits observed in the present case and may serve to indicate that the fact that general yield has occurred has not relaxed through-thickness stresses in the centre of the specimen. This view is supported by the behaviour of thicker specimens. At initiation, a thumbnail initiates across all the central part of the notch, leaving unfractured ligaments approximately 2 mm thick on each side. Confirmation of this plane strain deformation has recently been obtained in precracked 'V'-notched specimens[6].

The δ_i value for ductile fracture therefore appears to bear many resemblances to the plane strain fracture toughness, K_{IC}. To gain reproducible values, there is a thickness requirement and a need to use sharp cracks or very narrow slots (< 0.046 mm wide in the free-machining mild steel). There is also an a/W requirement, which will be described in conjunction with notch depth effects in Chapter 7.

The aim with COD testing has been to produce a method which facilitates the measurement of a given value of δ from knife-edge displacements. Only recently has emphasis been placed on the need to detect initiation and to record δ_i values. Since δ_i values are obviously more pessimistic with regard to crack-tip ductility than are δ_{max} values, there is still discussion as to which figure is more relevant to service application. The problem is perhaps put into perspective if it is realised that there may be as much as an order of magnitude difference between δ_i values obtained on fatigue-cracked specimens and the δ_{max} figures measured previously on slotted specimens and apparently used successfully to predict service behaviour. (These predictions were, however, based on experimental calibrations using large slotted plates which were often subject to the same errors as the small slotted testpieces[7].) To compare δ_i and δ_{max} scientifically, it is necessary to study the growth of a fibrous crack until maximum load is reached. Some similarities with the behaviour of fracture toughness specimens of 'intermediate' thickness (Section 5.6) will be discerned.

For a specimen of free-cutting mild steel, 5 mm thick or greater, the initial thumbnail occupies the total thickness, apart from the two exterior ligaments, each approximately 2 mm thick, as described previously. In a conventional screw-driven testing machine, the specimen is loaded further by displacement of its ends at a constant rate. The central thumbnail deepens with increase in load and spreads slightly in the through-thickness direction. The exterior ligaments are thus forced to bear even more load and strain. As maximum load is approached, it appears that the ligaments are really behaving as independent 'tensile' specimens and the maximum is reached more or less as these ligaments neck and fracture in a manner comparable with uniaxial behaviour. The value of δ_{max} is then characteristic of the extension and through-thickness contraction of these ligaments. In this respect, it is apparent that the behaviour of a 2 mm thick specimen ought to be generally similar to that of a single ligament in a thick specimen. Striking confirmation of this is given if the *initiation* of fracture in a 2 mm specimen is compared with the *onset of maximum load* in thicker pieces. The through-thickness necking (see *Figure 6.9*) and extension phenomena are closely similar and the δ_i value for the 2 mm thickness corresponds almost perfectly with the δ_{max} values in 10 or 17 mm thicknesses (*Figure 6.8*).

The δ_{max} values are therefore based on a necking phenomenon in the ligaments. This is unlikely to be the same in service as in a bend specimen and must be understood for every loading configuration likely to be encountered. Moreover, the emphasis on the displacement in the ligaments providing the criterion for maximum load pertains only to displacement-controlled tests. Under load control, the rapid transference of load to relatively narrow ligaments at initiation

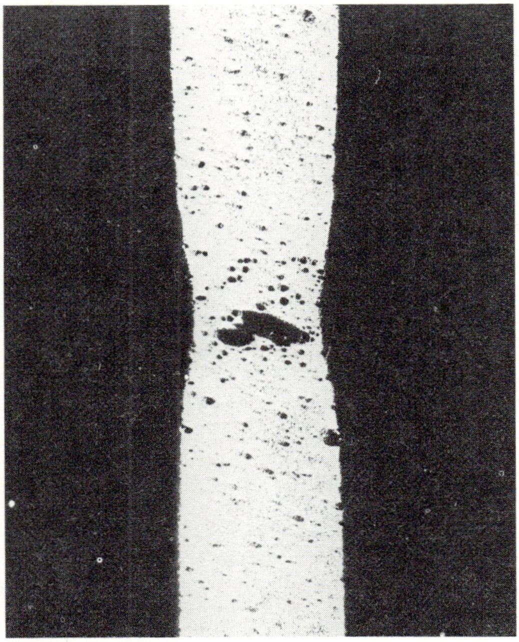

Figure 6.9 Plan section, just below slot tip, in 2 mm *thick specimen loaded to produce a small amount of fibrous fracture (note similarity to a necked tensile specimen)* (× 10) *(courtesy of R. F. Smith)*

in a thick specimen means that the difference between the load at initiation and that at instability will be minimal in thick specimens. This is exactly analogous to the K_{IC} situation (Section 5.5).

The most pertinent point of all, however, is that the ligament flow and necking process is time-dependent[20]. If a specimen is held under a constant load greater than that at initiation, the COD is observed to increase until the specimen finally breaks. The δ_{max} values are comparable to those found previously, but the stress necessary to produce them has been decreased by the time available for thermally-activated flow processes.

The proposed standard on COD testing describes only the experimental methods for measuring critical values of δ. Exactly which values are to be taken as critical and how they are to be used in practical application is not discussed. For the various reasons given above, it is considered advisable only to quote δ_i values in material specification and to employ only δ_i values in engineering design. The initial collaborative tests which were made did not take into account sufficiently the need to measure δ_i or to control loading rate or displacement rate. In addition, tests were often carried out in the transition range where the amounts of fibrous growth preceding fracture varied greatly, leading to a wide scatter in COD. The point is made by comparing *Figure 6.10a*, showing data points from a number of laboratories, with *Figure 6.10b* where the COD results are plotted *v.* the amount of fibrous fracture[16]. As in *Figure 6.7* (3-point bend), it is seen that δ increases markedly with the fibrous component.

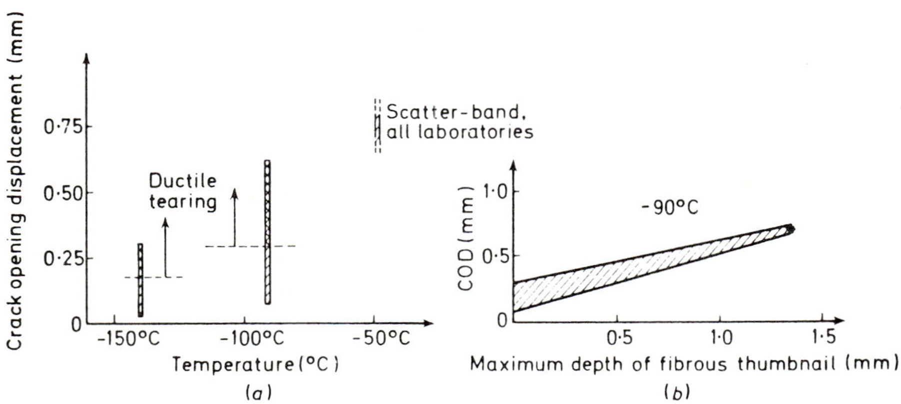

Figure 6.10 (*a*) *Variation of* COD *with temperature. Results from seven laboratories on a Swedish mild steel.* (*b*) *Variation of* COD *with depth of fibrous thumbnail at* −90°C (*after Nichols et al.*[16])

On the other hand, the extremely consistent results obtained by Smith provide a logical and scientific basis for COD measurement[12,14]. The results depend strongly on the micromechanisms of fibrous fracture at a stress concentrator and detailed discussion of these will be deferred until Chapter 8. The following section describes a proposed alternative method for measuring fracture toughness, using small specimens.

6.5 The J Integral: Principles

An alternative method for assessing the toughness of a cracked body, which cannot be regarded as linear elastic, relies on the determination of an energy term, which expresses the change in potential energy, when a crack is extended by an amount da, in a manner analogous to that of the strain energy release rate, G, in the linear elastic condition. The theory is developed for a non-linear elastic body[17].

We may define for such an elastic body, a single-valued strain energy density function (as in equation 2.6.4)

$$W = W(\epsilon_{mn}) = \int_0^{\epsilon_{mn}} \sigma_{ij}\, d\epsilon_{ij} \tag{6.5.1}$$

For a simple linear elastic body, subjected to a single tensile stress σ, giving a strain ϵ, this expression is equivalent to

$$W = \frac{\text{strain energy}}{\text{unit volume}} = \int_0^{\epsilon} \sigma_{11}\, d\epsilon_{11} = \tfrac{1}{2}\sigma . \epsilon \qquad (6.5.2)$$

We then consider the change in potential energy when a body containing an elliptical hole and subjected to boundary tractions T_i^0, along part of its surface S_T, and displacements u_i^0 along a different part of its surface S_u, has the hole increased in size, by a volume increase ΔV, equivalent to a surface increase ΔS (see *Figure 6.11*). The body is initially loaded with tractions T_i^0 on S_T and displacements u_i^0 on S_u. The initial stress in the body is σ_{ij}^0; the initial strain ϵ_{ij}^0.

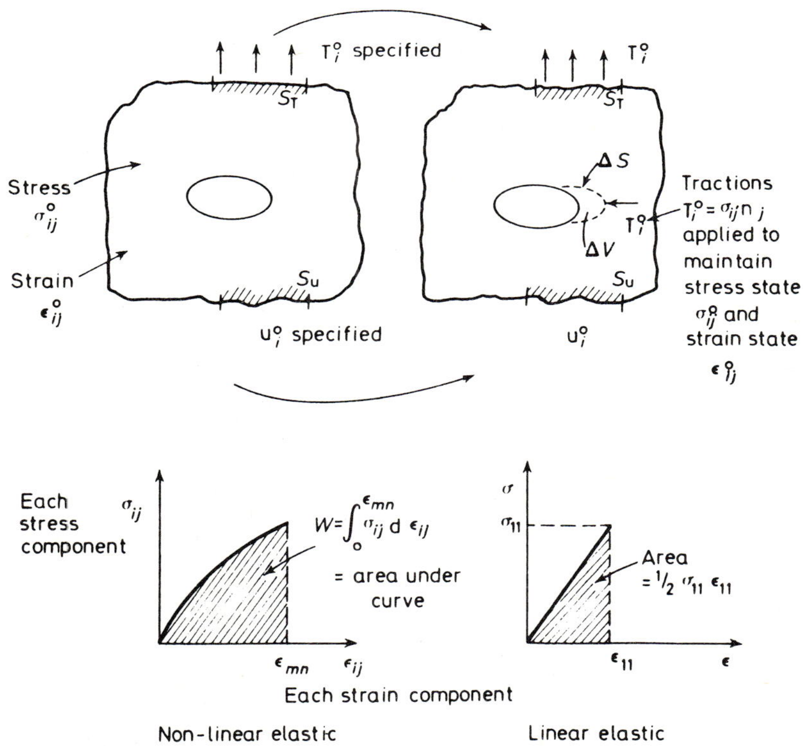

Figure 6.11 Change in potential energy for expansion of hole by ΔV (see text)

The hole is then expanded, but tractions $T_i^0 = \sigma_{ij}^0 n_j$ are applied to the new surface ΔS to maintain the initial deformation state σ_{ij}^0, ϵ_{ij}^0 in the body. Then, holding the *loadings* on S_T and S_u fixed, the tractions on ΔS are reduced to zero to produce the new deformation state $\sigma_{ij}^0 + \Delta\sigma_{ij}$, $\epsilon_{ij}^0 + \Delta\epsilon_{ij}$. If T_i^* and u_i^* are the loadings and displacements on ΔS during this process, the total change in potential energy may be written as:

$$-\Delta U = \int_{\Delta V} W(\epsilon_{mn}^0)\,dV - \int_{\Delta S\; u_i^*=u_i^0,\; T_i^*=T_i^0}^{u_i^*=u_i^0+\Delta u_i,\; T_i^*=0} \{T_i^*\,du_i^*\}\,dS \qquad (6.5.3)$$

i.e. the net energy reduction is the strain energy of the material removed minus the (negative) work done when the tractions on the new crack surface reduce to zero.

For the case of linear elasticity, this equation reduces simply to:

$$-\Delta U = \frac{1}{2}\int_{\Delta V} \sigma_{ij}^0 \epsilon_{ij}^0\,dV - \frac{1}{2}\int_{\Delta S} T_i^0\,\Delta u_i\,dS \qquad (6.5.4)$$

since $\Delta T_i = -T_i^0$ on the void surface. For a sharp crack tip, the change in volume on extension is zero, so that only the surface integral remains. For an increase in crack length da, the expression then becomes equivalent to that derived earlier for the change in potential energy or 'strain energy release', $G\,da$ (equation 4.5.2). Writing $\Delta S = 2\,da$, we have:

$$G\,da = \int_0^{da} \sigma\sqrt{\left(\frac{a}{2r}\right)} \cdot \frac{2(1-\nu^2)}{E}\sigma\sqrt{(2a)}\sqrt{(da-r)}\,.\,dr \qquad (6.5.5)$$

where the general traction T_i^0 has been replaced by the single component

$$\sigma_{11} = \frac{K_1}{\sqrt{(2\pi r)}} = \sigma\sqrt{\left(\frac{a}{2r}\right)} \qquad (6.5.6)$$

and Δu_i has been replaced by the opening displacement

$$u_{11} = \frac{2(1-\nu^2)}{E}\sigma\sqrt{(2a)}\sqrt{(da-r)} \qquad (6.5.7)$$

In this case G is defined per unit thickness.

To return to the general form for change in potential energy, involving reduction in strain energy and work done by applied tractions, it is possible to set up the inequality:

$$0 \leqslant (-\Delta U) - \int_{\Delta V} W(\epsilon_{mn}^0)\,dV \leqslant \int_{\Delta S} \Delta T_i \Delta u_i\,dS \qquad (6.5.8)$$

where ΔT_i and Δu_i are the changes in traction and displacement respectively. Now, if we extend a notch tip normal to itself by an infinitesimal amount dn, and note that $\Delta T_i\,(\equiv \Delta\sigma_{ij} n_j)$ and Δu_i are both first order quantities, so that their product – the upper bound of the inequality – approximates to zero compared with the volume integral, which is first order, we may write:

$$-dU = \int_{\Delta V} W(\epsilon_{mn}^0)\,dV \qquad (6.5.9)$$

Now consider a flat-surfaced notch in a two-dimensional stress field (i.e. all stresses depend only on x_1 and x_2) (see *Figure 6.12*). The notch has upper and lower surfaces parallel to x_2 and a smooth curved tip denoted by Γ_t. Then for an extension da, maintaining the crack tip shape constant, we note that $\Delta V = \mathrm{d}a.\mathrm{d}x_3.\mathrm{d}x_1$ and using equation 6.5.9 obtain for $U' = U$/unit thickness (in the x_3 direction):

$$-\mathrm{d}U' = \int_{\Delta A} W(\epsilon^0_{mn})\,\mathrm{d}x_1\,\mathrm{d}a \tag{6.5.10}$$

or

$$-\frac{\mathrm{d}U'}{\mathrm{d}a} = \int_{\Gamma_t} W(\epsilon^0_{mn})\,\mathrm{d}x_1 \tag{6.5.11}$$

The change in potential energy is therefore replaced by a path independent line integral around the notch tip.

6.6 The J-Integral – Definition

More generally, it is possible to define[1] a line integral, J, on any curve Γ surrounding the notch tip, starting from the lower surface and ending on the upper surface of the notch such that J is given by:

$$J = \int_{\Gamma} [W\,\mathrm{d}x_1 - T\frac{\partial u}{\partial x_2}\,\mathrm{d}s] \tag{6.6.1}$$

where the curve is traversed in the anticlockwise direction, s is arc length and $T \equiv \sigma_{ij}n_j$ is the traction vector on Γ according to an outward unit vector $\boldsymbol{n}$ (with direction cosines n_j) normal to the curve Γ. When Γ is identical to Γ_t, $T = 0$ because the free surface of the notch can withstand no stress normal to it, and so $J = -\mathrm{d}U'/\mathrm{d}a$ (J would be identical to the strain-energy release rate G for linear elastic fixed grip conditions). It may be shown that the difference $(J_1 - J_2)$ for two curves Γ_1 and Γ_2 is zero and the conclusion drawn is that the J integral is path independent and formally equivalent to the change in potential energy when the notch is extended by an amount da. The same conclusion is reached if the change in potential energy is calculated for a sharp crack.

We therefore recognise a parameter, J, derived from non-linear elastic behaviour, which bears the same function as does G in linear elastic theory. Its use has been extrapolated to deal with elastic/plastic and even fully yielded situations. Discussion of these applications follows, but there is one feature of the path-independent nature of J which gives useful information on the forms of stresses and strains at a crack tip for linear elastic loading. We write equation 6.6.1 as

$$J = \int_{\Gamma} \left(\tfrac{1}{2}\sigma_{ij}\epsilon_{ij}\,\mathrm{d}x_1 - T_i\frac{\partial u_i}{\partial x_2}\,\mathrm{d}s\right) \tag{6.6.2}$$

for linear behaviour. Then, if we take a circle of radius r, where $r \to 0$, as our curve Γ_t, it is clear that only the singular crack tip stresses or strains (Section 3.7) are important. Now, in polar co-ordinates, $x_1 = r \sin \theta$, for the definition of x_1 given in *Figure 6.12* Hence, we have $dx_1 = r \cos \theta \, d\theta$ and $ds = r \, d\theta$. Since, by path independence, J is independent of r, the coefficients of dx_1 and ds must be of the order of $1/r$. For linear elastic behaviour, stress is proportional to

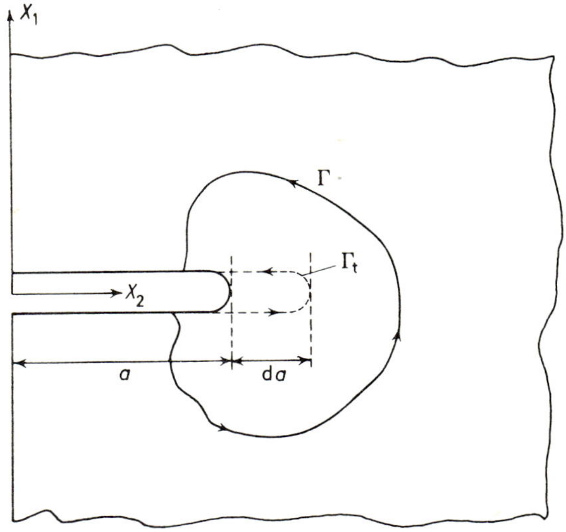

Figure 6.12 Relationship between change in potential energy and line integral

strain; since the dimensions of the product have order $1/r$, the stress singularity is shown to have order $r^{-\frac{1}{2}}$, as evidenced by the results derived from stress functions (e.g. equation 3.7.5).

Problems arise in determining, and even in defining, the value of J in a real situation, which is likely to be elastic/plastic rather than non-linear elastic. We treat first the situation where a yielded region, very small with respect to crack length and specimen width and small compared with the region governed by the crack tip singularity, exists at a crack tip. This is similar to the situation described in Section 4.6 and is referred to as small-scale yielding. Here, the J-integral is evaluated by using a Γ-contour which passes through elastic material encompassing the plastic zone. If Γ is taken as a circle of radius r, we may, in an infinite body, allow r to tend to infinity, so that the J-integral becomes identical to the strain-energy release rate, G, for a linear elastic body:

$$J = G = \frac{K^2}{E}(1 - \nu^2) \text{ in plane strain} \tag{6.6.3}$$

Here a plastic-zone crack-length 'correction' (Section 4.6) ought to be used for K, because we have assumed that a small plastic zone is located around the crack tip, i.e.

$$K = \sigma \sqrt{[\pi(a + r_{IY})]} = \sigma \sqrt{\left[\pi\left(a + \frac{K^2}{6\pi\sigma_Y^2}\right)\right]} \tag{6.6.4}$$

The most appealing feature of the J-integral is that it can be evaluated by choosing Γ-contours which deal with the crack-tip fields or with boundary conditions as appropriate. For example, for the configuration of a crack + yield zone in plane-strain tension (Section 3.18), J has been evaluated from the shear strain energy density in the centred fans of plastic slip-lines immediately above and below the crack tip (the local Prandtl field) as the contour shrinks to zero.

6.7 The J-Integral in Cracked Testpieces

The most straightforward experimental method of evaluating J derives from its definition as the rate of change of potential energy with crack length[18] (equation 6.6.1 et seq). *Figure 6.13* shows schematic non-linear loading curves corresponding to specimens with crack lengths a and $(a + \delta a)$ under displacement

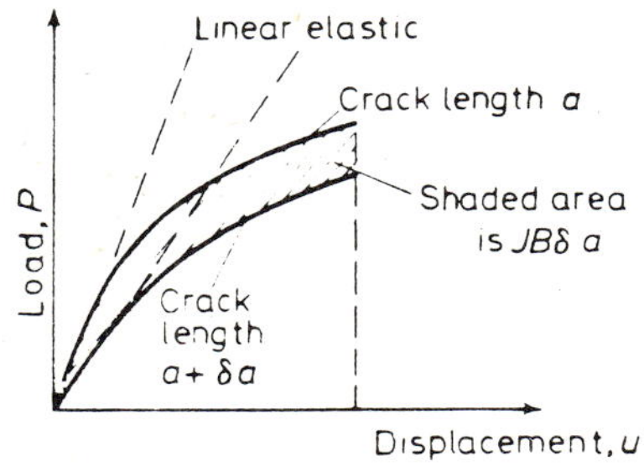

Figure 6.13 The J-integral (Displacement control)

control. The shaded area represents an energy change $JB\delta a$. The displacement must be measured at the loading points (cf. Section 4.4).

Experimental values of J may be obtained readily from these curves. At a given displacement, the potential energy, U, may be found directly by using a planimeter to determine the area under a loading curve for a given crack length, a_1. Then, at the same displacement, energies may be determined for crack lengths $a_2, a_3, \ldots$, etc., in a similar fashion. For a given standard crack length, a_0, the value of J at the displacement u is found from the tangent of the curve $U\ v.\ a$, taken at a_0. Schematic results[18], obtained by this procedure, are shown in *Figure 6.14a* and *6.14b*. The experiments are not, however, easy to perform and it is of interest to try to predict the form of curve given in *Figure 6.14b* from first principles.

For linear elastic behaviour (at low loads or displacements), we have: $u = CP$ where P is the load and C is the elastic compliance of the system for a given crack length. The potential energy per unit thickness for this condition is then given by:

$$U = \frac{1}{2}\frac{Pu}{B} = \frac{1}{2}\frac{CP^2}{B} \tag{6.7.1}$$

omitting the prime on U. Then, by definition, we have:

$$J = \frac{\mathrm{d}U}{\mathrm{d}a} = \frac{1}{2}\frac{\mathrm{d}C}{\mathrm{d}a}\frac{P^2}{B} \qquad (6.7.2)$$

Substituting back for u as CP, we obtain:

$$J = \frac{1}{2}\frac{\mathrm{d}C}{\mathrm{d}a}\frac{u^2}{C^2 B} \qquad (6.7.3)$$

Hence, for linear elastic behaviour, the dependence of J on displacement is parabolic, for a given crack length.

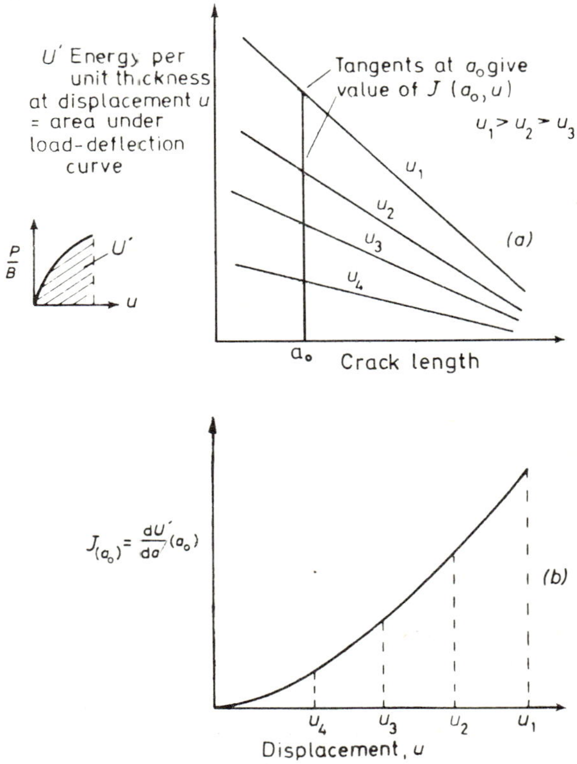

Figure 6.14 (*a*) *Variation of energy with crack length and displacement* (*Data representative of that for a low-alloy steel*). (*b*) *Variation of J-integral with displacement* (*for the steel, almost independent of crack length*)

For rigid/plastic material, the displacement u is unlimited at the limit load (general yield load), $P = P_{GY}$, whilst for $P < P_{GY}$, $u = 0$. The work done in extension is therefore $P_{GY} . u$. Hence, we derive J as

$$J = -\frac{u}{B}\frac{\partial P_{GY}}{\partial a} \qquad (6.7.4)$$

where $(\partial P_{GY}/\partial a)$ is evaluated for the crack length of interest. We therefore obtain a linear relationship between J and u for rigid/plastic behaviour.

In real specimens, a transition occurs from linear elastic behaviour at low loads to fully plastic behaviour beyond the general yield load (limit load) P_{GY}. *Figure 6.15a* shows schematic load–displacement curves for crack lengths a and $(a + \delta a)$ and *Figure 6.15b* shows how J, calculated for these specimens, would

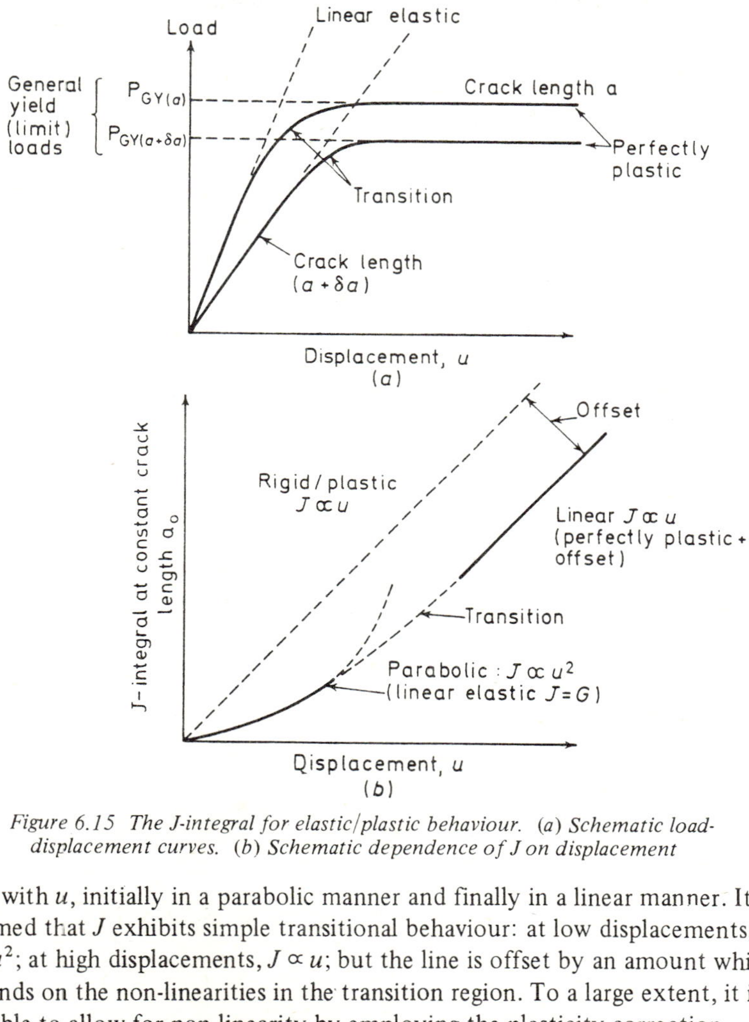

Figure 6.15 The J-integral for elastic/plastic behaviour. (a) Schematic load-displacement curves. (b) Schematic dependence of J on displacement

vary with u, initially in a parabolic manner and finally in a linear manner. It is assumed that J exhibits simple transitional behaviour: at low displacements, $J \propto u^2$; at high displacements, $J \propto u$; but the line is offset by an amount which depends on the non-linearities in the transition region. To a large extent, it is possible to allow for non-linearity by employing the plasticity correction (Section 6.6) up to limit load P_{GY}; i.e. by taking the effective crack length as:

$$a^* = (a + r_Y) = \left(a + \frac{K^2}{2\pi\sigma_Y^2}\right) \quad \text{in plane stress} \tag{6.7.5}$$

$$a^* = (a + r_{IY}) = \left(a + \frac{K^2}{6\pi\sigma_Y^2}\right) \quad \text{in plane strain} \tag{6.7.6}$$

In the following section, we consider how detailed 'J-calibration' tables may be constructed.

6.8 Calculation of J in Bending

Consider a cracked specimen subjected to a bending moment, M, which we will take initially as pure bending. Then the total energy stored in the piece, U_{tot}, is given by the sum of that which would exist if no crack were present, U_1, and that due to the presence of the crack, U_2:

$$U_{tot} = U_1 + U_2 \tag{6.8.1}$$

In pure bending, for a beam of depth W, thickness B and length between loading points L, U_1 is given by:

$$U_1 = \frac{6M^2L}{EBW^3} \tag{6.8.2}$$

where E is Young's modulus. For three-point bending an additional contribution U_1 arises from the work done by the shear stresses. By definition, $\partial U_2/\partial a$ is the strain energy release rate per unit thickness; hence:

$$U_2 = \int_0^a GB\,\mathrm{d}a = \int_0^a \frac{K^2}{E}(1 - \nu^2)B\,\mathrm{d}a \tag{6.8.3}$$

in plane strain (see Section 4.5), where a is the crack length. However, from elastic compliance calibrations (e.g. *Table 5.2*), we may express K as a function of (a/W), e.g. for three-point bending:

$$K = \frac{3PL}{BW^{\frac{3}{2}}}\,\mathrm{f}\left(\frac{a}{W}\right) \tag{6.8.4}$$

The displacement at the loading point, u, is composed of a part arising from the bending of the beam and a part from the presence of the crack, and may be found from the differential of total energy with respect to load:

$$u = \frac{\partial U_{tot}}{\partial P} = \frac{\partial U_1}{\partial P} + \frac{\partial U_2}{\partial P} \tag{6.8.5}$$

The first term may be calculated by standard strength-of-materials calculations for beam deflections. The second term may be calculated from equations 6.8.3 and 6.8.4 as a polynomial, i.e.

$$u = P\,\mathrm{f}_1\left(\frac{a}{W}, W, L, B, E\right) \tag{6.8.6}$$

If plasticity has occurred at the crack tip, the crack length a^* (from equation 6.7.6 for plane strain) is substituted:

$$u = P\,\mathrm{f}_1\left(\frac{a^*}{W}, W, L, B, E\right)$$

$$= P\,\mathrm{f}_1\left[\frac{(a + K^2/6\pi\sigma_Y^2)}{W}, W, L, B, E\right] \tag{6.8.7}$$

Expressions such as these, derived for three-point bend and CTS specimens, have been used to compute the dependence of J on displacement in the transition region[18] (*Figure 6.15b*), and the predictions agreed well with the results of experimental compliance calibrations.

The J-integral can thus be evaluated for a given specimen, either experimentally or by calculation. The next section describes such evidence as exists for postulating that a critical value of J, J_{IC}, represents a characteristic failure criterion in situations where fracture is preceded by substantial amounts of yielding.

6.9 The J_{IC} Fracture Criterion

It was stated, in Section 6.2, that no single parameter could be used to characterise fracture in a yielding situation. The COD experiments show that a critical value of the crack tip displacement can be used to define the initiation of fibrous fracture for a limited set of conditions, but that thickness, for example, can drastically alter the critical value, if no plane-strain thumbnail can form.

Experiments in which the value of J at fracture has been measured in two specimen geometries, designed to give markedly different stress states, appear to show that a constant J_{IC} figure characterises fracture in both geometries[19]. It is argued that the crack blunting occasioned by the large displacements preceding fracture reduces the effects of plastic constraint on the fracture criterion.

The two geometries studied were a thick, centre-cracked tension specimen and a deeply cracked bend specimen. The slip-line fields at limit load in these specimens are indicated in *Figures 2.17* and *2.15*. In the tension specimen, there is no stress elevation ahead of the crack tip and the limit load, P_{GY}, is given by:

$$P_{GY} = 2\tau_Y . 2(W - a)B \qquad (2.12.14)$$

where τ_Y is the shear yield stress. In three-point bending, the limit moment, M_{GY}, is given by (preceding 2.12.12)

$$M_{GY} = 0.608\tau_Y B(W - a)^2$$

which is 1.216 times that of an uncracked bar of width $(W - a)$. Experimental measurements of M_{GY} were found to agree well with this expression. The material tested was a medium strength (approx 900 MN m^{-2} yield stress) alloy steel and J values were determined experimentally, as described in Section 6.7.

It was necessary accurately to determine the (gross) displacement corresponding to fracture initiation, because any crack extension alters the compliance of a specimens are indicated in *Figures 2.17* and *2.15*. In the tension specimen, there fracture initiation was found fortuitously to coincide with maximum load, in contrast to the mild steel behaviour in COD tests reported in Section 6.3.

The values of J_{IC} were measured for the displacements corresponding to fracture initation and are tabulated in *Table 6.1* together with further results for rather similar specimen geometries.

These values appear to be reasonably constant and tend to indicate that, in terms of energy release, the higher limit loads associated with constrained test-pieces are more or less compensated for by higher displacements in the less constrained specimens.

Table 6.1 CRITICAL VALUES OF J_{IC}. (after Landes and Begley[19])

Specimen type	*Dimensions* (mm)	*Temp.* (K)	J_{IC} (kJ m^{-2})
	Ni–Cr–Mo–V steel		
Centre cracked	25 × 25 × 57	394	172
Double thickness bend bars	12 × 20 × 40	394	187
Single thickness bend bars	12 × 10 × 40	366	167
Compact tension (8T)	406 × 203 × 488	394	175
Bend bars double size	24 × 20 × 96	366	179
	A 533B Class II steel		
Compact tension (2T)	102 × 51 × 122	298	165
Compact tension (1T)	51 × 25 × 61	298	180

There is, then, evidence that a critical parameter, J_{IC}, adequately represents fracture in specimens of different geometries. The values of J_{IC} also appear to extrapolate well from those of G_{IC} at lower temperatures. It is, of course, vitally important that J_{IC} should be measured at the point of initiation because any crack growth prior to final instability will affect the compliance of the specimen. Measurement of J_{IC} is a new testing technique and much work remains to be done to establish fully the viability of the approach and the detailed dependence of the critical values on variables such as temperature, strain rate, stress state and so forth. In steels, it is possible that low triaxiality induces fibrous fracture, whilst high triaxiality induces cleavage (see Chapter 7). In such circumstances it seems unreasonable that identical J_{IC} values should be obtained in widely differing specimen geometries.

6.10 Conclusions

Both COD and J_{IC} attempt to characterise ductile fracture events by single parameters which may be related to the critical value of energy release rate when a large piece breaks well before general yield. The COD focuses attention on the crack tip region and is able to be related directly to the micro-mechanisms of fracture in a region of area less than some 0.01 mm^2: J relates to macroscopic work terms or to crack tip conditions, depending on the Γ-contour chosen.

In a generally yielded specimen, J is directly proportional to the displacement of the loading points, u. In bend, u is proportional to the angle of bend, θ, and so J is proportional to θ. In notched bars, it has been shown, that, for a given specimen geometry, θ is proportional to notch root displacement. In other words, it is highly probable that J is directly proportional to COD for a given specimen geometry. This relationship remains to be explored. The J parameter is, in principle, easier to measure, but, in practice, is probably subject to at least as many errors as COD. In the transition region, J has no obvious advantage over the LEFM approach, with plastic zone correction. The main defect, as with K_{IC}, is that the critical value of J_{IC} has no physical basis. The next two chapters discuss the extent to which a knowledge of crack tip fracture mechanisms can help to provide such a basis.

REFERENCES

1. Rice, J. R., *J. appl. Mech.* (*Trans. ASME*), 379 (June, 1968)
2. Cottrell, A. H., *Iron and Steel Institute Spec. Rep.*, 69, 281 (1961)
3. Wells, A. A., *Crack Propagation Symposium Proceedings*, Cranfield College of Aeronautics **1**, 210 (1961)
4. Knott, J. F., *Mater. Sci. Engng*, 7, 1 (1971)
5. Thornton, D. *Engng. Fract. Mech.*, **2**, 125 (1970)
6. Robinson, J. N. and Tetelman, A. S., *Third International Congress on Fracture*, Munich (April, 1973)
7. Burdekin, F. M. and Stone, D. E. W., *Jnl Strain Anal.*, **1**, 145 (1966)
8. Knott, J. F., *J. Iron Steel Inst.*, **204**, 1014 (1966)
9. Knott, J. F., *Metallurgia*, 77, 93 (1968)
10. Elliott, D., Walker, E. F. and May, M. J., *Conf. Practical Application of Fracture Mechanics to Pressure-vessel Technology*, Institute of Mechanical Engineers, 217 (1971)
11. Ingham, T., Egan, G. R., Elliott, D. and Harrison, T. C., *Conf. Practical Application of Fracture Mechanics to Pressure, Vessel Technology*, Institute of Mechanical Engineers, 200 (1971)
12. Smith, R. F., *Ph.D. Thesis*, University of Cambridge (1972)
13. Wells, A. A., *The Mechanics of the Fracture Transition in Yielding Materials*, NDACSS, CODA panel report, 46 (1970)
14. Smith, R. F. and Knott, J. F., *Conf. Practical Application of Fracture Mechanics to Pressure-vessel Technology*, Institute of Mechanical Engineers, 65 (1971)
15. Chipperfield, C. G., Knott, J. F. and Smith, R. F., *Third International Congress on Fracture*, Munich (April, 1973)
16. Nichols, R. W. *et al.*, *Practical Fracture Mechanics for Structural Steels*, Paper F, UKAEA/Chapman and Hall (1969)
17. Rice, J. R., *Fracture – An Advanced Treatise* (edited by H. Liebowitz), Academic Press, New York, 191 (1968)
18. Bucci, R. J., Paris, P. C., Landes, J. D. and Rice, J. R., *Am. Soc. Test. Mat., Spec. Tech. Pub.* No. 514, 40 (1972)
19. Begley, J. A. and Landes, J. D., *Am. Soc. Test. Mat.* Spec. Tech. Pub. No. 514, 1 (1972)
20. Green, G., Smith, R. F. and Knott, J. F., British Steel Corporation Conference on *Mechanics and Mechanisms of Crack Growth*, Churchill College, Cambridge (April, 1973)

Seven

NOTCHED BAR FRACTURE MECHANICS AND THE MICRO-MECHANISMS OF CLEAVAGE FRACTURE

7.1 Introduction

In previous chapters, we have considered macroscopic methods of measuring toughness parameters, such as critical stress intensity, J-integral or crack opening displacement. We shall now pay attention to the local fracture events ahead of a stress concentrator to try to gain a physical understanding of how material properties control the limiting values of toughness. The aim of this is twofold. First, by knowing how mechanical factors affect the micro-mechanisms of fracture near a stress concentrator, we can comment on how generally applicable any particular toughness value may be. Secondly, by identifying the microstructural features which give poor toughness, we may be able to eliminate them in the design of materials with improved resistance to fast fracture.

The majority of the chapter concerns the behaviour of medium- and low-strength steels in specimens containing notches, rather than sharp cracks. The reason for dealing with these materials is that the microscopic fracture events are more clearly defined than those in high strength alloys. The reason for discussing behaviour in notched specimens is that more is known about elastic/plastic stress distributions, as related to fracture mechanisms, in these, than in pre-cracked pieces. The analysis of notched-bar behaviour may be used to distinguish mechanical and microstructural effects and also to estimate the value of results obtained from more traditional methods of toughness testing.

7.2 Notched-bar Impact Testing

Usually, notched-bar toughness tests are carried out under impact conditions (see Section 1.7). Specimens are broken in a machine which allows a pendulum hammer to descend on the specimen from a fixed height. By recording the height of the swing when the hammer has smashed through the specimen, the amount of energy absorbed in fracture is determined. This fracture energy is usually measured as a function of test temperature and the results are presented in the form of an 'impact transition curve'. A curve, typical of those obtained for a low-strength steel, is drawn in *Figure 1.3.* Several 'transition temperatures' may be defined. We shall refer later to two temperatures: that at which the fracture appearance is 50% fibrous, 50% crystalline (the fracture appearance transition temperature or FATT) and that corresponding to the temperature at which the curve changes from the almost horizontal line at low temperatures to the steep slope in the transition region. This is the 'nil-ductility temperature' or NDT. We shall also be interested in the 'upper shelf' energy level when the curve flattens off at high temperatures. These various parameters are indicated in *Figure 1.3.*

It is difficult to relate the energy measurements directly to the local fracture mechanisms in the notched impact specimen because the work done involves contributions from that needed to initiate fracture and that needed to propagate fracture through the specimen, in addition to the plastic work associated with the indentation of the specimen by the hammer. Instrumentation may be attached to the hammer or to the specimen but it is easier to consider first the results which have been obtained in slow notch-bend, where loads and displacements can be measured directly.

7.3 Slow Notch-bend Testing

The general fracture behaviour of mild steel notched specimens deformed in four-point bending is shown as a function of temperature in *Figure 7.1a* (see Knott[1]). The general yield load curve gives the loads P_{GY} needed to spread yielding completely across the net section of the specimen, and may be related to the uniaxial yield stress by the expression (following equation 2.12.10)

$$M_{GY} = 0.63\tau_Y(W-a)^2B \tag{7.3.1}$$

where M_{GY} is the bending moment ($\frac{1}{2}P_{GY}$ x bending arm) and τ_Y is the shear yield stress ($\tau_Y = \sigma_Y/2$ on the Tresca yield criterion or $\sigma_Y/\sqrt{3}$ on the Mises criterion). The temperature, T_{GY}, is that below which the fractures are brittle in the sense of our definition in Section 1.6. The low temperature fractures are generally produced by cleavage mechanisms. The sequence of photographs in *Figure 7.1b* shows how the amount of yielding preceding fracture increases with temperature (see Knott and Cottrell[2]). Above T_{GY}, the specimens deform generally before they break. Consequently, the amount of work required to fracture them increases rapidly with temperature. To a first approximation the transition point, T_{GY}, may be regarded as analogous to the nil-ductility-temperature (NDT) in impact testing (see Section 7.2). In several steels, however, it is observed that the fracture load curve runs nearly parallel to the general yield load curve over a substantial temperature range above T_{GY}, and that the macroscopic deformation preceding fracture remains fairly small until a temperature T_W is reached. This temperature is associated with the yielding of the gross section of a specimen before fracture by the spread of plastic 'wings' from the top face[3,4] (see *Figure 7.1*). In such steels, and perhaps generally, T_W rather than T_{GY} is to be identified with the NDT.

The behaviour may be related to the increasing amounts of plastic deformation which precede fracture as the temperature is raised. At low temperatures, the plastic zone is small, compared with the cross-section and the specimen is macroscopically brittle. At T_{GY}, the zone is just sufficient to give general yield. At T_W it can be made bigger only by raising the applied load, because the net section has been work-hardened, and the increase in load is sufficient to yield the gross section. Above T_W, the loads and displacements rise rapidly, because some of the effect of the notch on triaxiality is removed by the gross deformation. The onset of fibrous fracture may occur above or below T_W, depending on the magnitude of the notch-tip fracture strain required to initiate fracture and on the ratio of notch to specimen depth. This strain is usually a function of a material's inclusion content. In a very pure iron, a specimen with a fairly shallow notch may be bent double, without initiating fracture. For a similar matrix, but containing a high volume fraction of closely spaced inclusions, fracture may be initiated at low notch strains, with correspondingly small overall angles of bend.

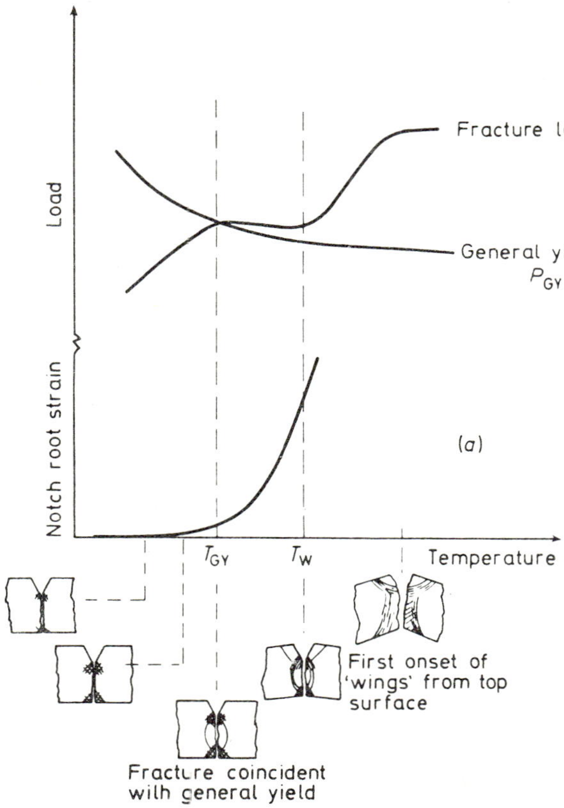

Figure 7.1a Schematic fracture behaviour of mild steel in slow notch-bend test

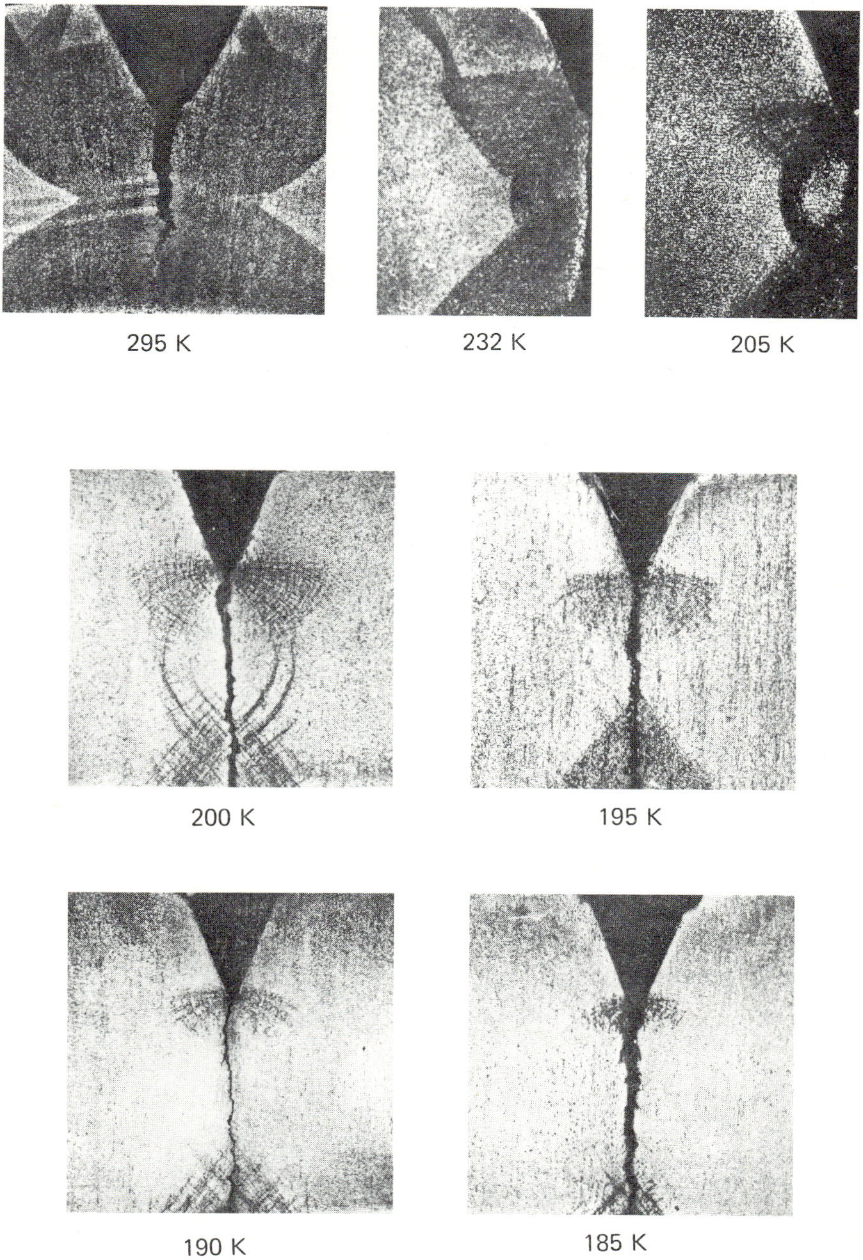

Figure 7.1b The deformation preceding fracture in notched bars, broken at low temperatures (the deformation is shown by etching in Fry's reagent) (×3)

The phenomena which need to be explained are therefore: the production of cleavage fracture as a function of temperature; the 'engineering' ductile/brittle transition at T_{GY}; the onset of gross plastic deformation at T_W and the transition from cleavage to fibrous initiation. We begin by considering the behaviour associated with cleavage fracture.

7.4 The Tensile Stress Criterion for Cleavage Fracture

The first really successful attempt to explain the effect of a notch on the cleavage fracture process in mild steel was made by Orowan[5], who pointed out that the slip line field at general yield for a specimen containing two deep external cracks (*Figure 2.13*) implied that the peak tensile stress in the specimen was given by:

$$\sigma_{11(\max)} = 2\tau_Y(1 + \pi/2) = 2.57\sigma_Y \tag{7.4.1}$$

using Tresca's yielding criterion. He supposed that cleavage fracture was controlled by the magnitude of tensile stress at the 'yield point', and that the notched bar therefore behaved in a more brittle manner than a tensile specimen because the tensile stress was higher at a given temperature. Over a range of temperature, the situation would be as indicated in *Figure 7.2*, assuming that

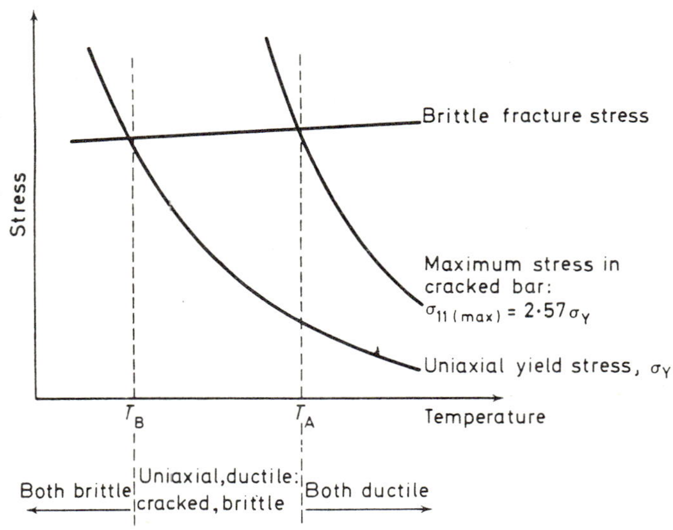

Figure 7.2 Orowan's theory of brittle fracture

the 'brittle fracture stress' – the critical value of tensile stress needed to produce cleavage fracture – was relatively independent of temperature. The tensile stresses at yield in smooth, σ_Y, and notched $2.57\sigma_Y$, bars are shown to increase markedly with decrease in temperature. Orowan's model predicts that, above T_A, both notched and smooth specimens are ductile, because they yield at a stress lower than the fracture stress. Between T_A and T_B, the notched bar is brittle because it breaks before it 'yields', but the tensile specimen remains ductile. Below T_B, both types of specimen exhibit brittle behaviour.

The main faults with Orowan's argument are that distinctions between local and general yield in the notched piece are not drawn. The model is, therefore, not sufficiently flexible to incorporate the requirement that cleavage crack nuclei must be initiated by slip-bands or by mechanical twins even when the piece is macroscopically brittle. It does, however, draw attention to the importance of tensile stress in promoting brittle behaviour. This view received support from the experiments of Hendrickson, Wood and Clark[6], but tended to be ignored by contemporary dislocation theories of fracture, which predicted that the total behaviour was controlled by the localised stresses at the head of a dislocation pile-up. A dislocation model for cleavage fracture produced by Cottrell[7] appreciated the importance of tensile stress, but, again did not discriminate properly between local and general yield conditions. This model will be discussed in more detail in Section 7.10.

By comparing *Figure 7.1* with the Orowan predictions shown in *Figure 7.2*, it is clear that the temperatures T_{GY} (*Figure 7.1*) and T_A (*Figure 7.2*) are equivalent in that they both represent the point at which fracture is coincident with general yield. Below this temperature, it is clear (from *Figure 7.1*) that fracture actually occurs after successively decreasing amounts of local plastic deformation around the notch, even though the total fracture behaviour is macroscopically brittle.

In the absence of any reliable analyses for the stresses in the locally yielded zone, Knott[3] examined the importance of tensile stress in determining the onset of cleavage fracture by testing specimens of annealed mild steel containing notches of different included angles. From Section 2.12 it is found that the maximum tensile stress in a notched bar at general yield is given as a function of notch angle by the expression:

$$\sigma_{11(\max)} = 2\tau_Y \left(1 + \frac{\pi}{2} - \frac{\theta}{2}\right) \qquad (7.4.2)$$

Then, by determining the temperatures T_{GY} at which fracture is just coincident with general yield for a range of angles, it proves possible to calculate the value of $\sigma_{11(\max)}$ at fracture as a function of temperature, using the appropriate value of uniaxial yield stress at each value of T_{GY}. Typical values of $\sigma_{11(\max)}$ at fracture are plotted in *Figure 7.3a* from which it may be deduced that the basic Orowan concept: that fracture occurs at a critical value of tensile stress, σ_F, which is virtually independent of temperature; is substantiated for these steels. Further support is gained from metallographic evidence which indicates that a cleavage crack first forms below the notch root where the tensile stress is high[8] (*Figure 7.3b,c*).

The sequence of events, for a bar of constant notch angle at temperatures below T_{GY} is then as follows. At low temperatures, the uniaxial yield stress is high and the size of plastic zone needed to elevate the tensile stress by stress intensification (Section 2.12) to the critical value is small. Fracture therefore occurs well before general yield. As the temperature increases, the uniaxial yield stress decreases and the plastic zone at fracture must therefore be correspondingly larger. Eventually, at T_{GY}, it is of a size such that general yield and fracture coincide. The fracture criterion may be stated simply as:

$$Q\sigma_Y = \sigma_F \tag{7.4.3}$$

where Q represents the stress intensification produced by the plastic zone.

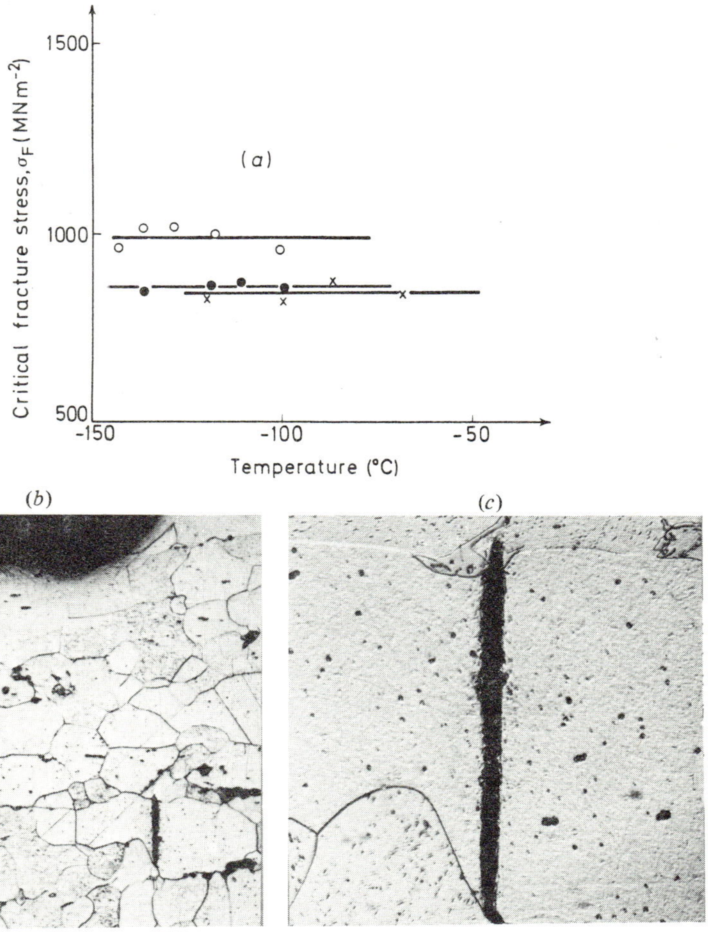

Figure 7.3 (*a*) *Experimental values of maximum local tensile stress at fracture for three mild steels.* (*b*) *Initiation of a cleavage crack in the region of high tensile stress well below the notch root* (x 70). (*c*) *Detail of cleavage crack; showing association with grain-boundary carbide* (x 400)

Knowing σ_F over a range of low temperatures, it proves possible to calculate Q for any particular specimen geometry, by measuring values of σ_Y over the temperature range. The method has been used for a piece of Charpy dimensions, tested in four-point bend. Results are given in *Figure 7.4*, and show the increase in Q with size of plastic zone (P/P_{GY}), much as indicated earlier (Section 2.13). Very recently, finite element computer programs have been developed to determine the stresses in the plastic zone (Section 3.18). These have been applied to calculate values of σ_F as a function of temperature and give fairly close agreement with the experimental results, despite the fact that the maximum value of σ_{11} is found to lie some distance behind the plastic/elastic interface, particularly for the larger plastic zones. A direct test of the tensile stress criterion in 3% silicon–iron, based on the finite element analysis, is shown in *Figure 7.5*, from which it is clear that the general concept of the temperature independent fracture stress is well supported[9].

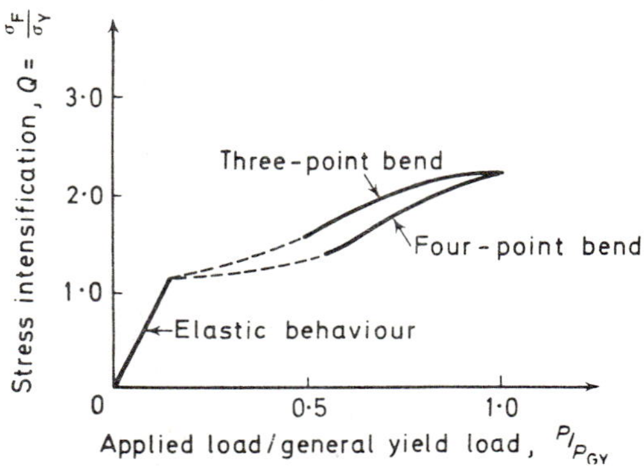

Figure 7.4 Experimental determination of stress intensification in specimens of Charpy geometry $\left(\frac{a}{W} = 0.2\right)$, *assuming general yield slip-line field solution. (In four-point bend, the geometry is insufficiently deep to give fully constrained yielding: see Figure 2.16b)*

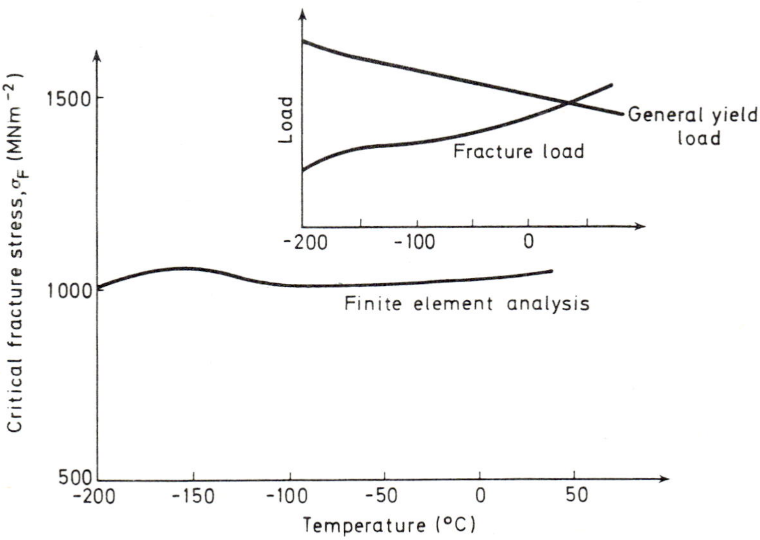

Figure 7.5 Variation of critical fracture stress with temperature in Iron–3% silicon. (Stresses calculated by finite element analysis. Owen and Griffiths[9])

The criterion may now be explored in more detail, firstly to show how it can be used to explain effects of specimen geometry on cleavage fracture and, secondly, to relate it to micromechanisms of cleavage fracture. The first geometrical variable to be considered is that of thickness.

7.5 Effects of Specimen Thickness

The main effects of thickness[3,10] on the occurrence of fracture in notched mild-steel specimens are indicated in *Figure 7.6*. Not only is the obvious transition from cleavage to fibrous fracture strongly affected, but the temperature, T_{GY}, at which cleavage fracture is coincident with general yield, also varies with thickness. Typical results are quoted in *Table 7.1*. If a particular temperature, such as T_X in *Figure 7.6* is chosen for close examination, the thickness effect is

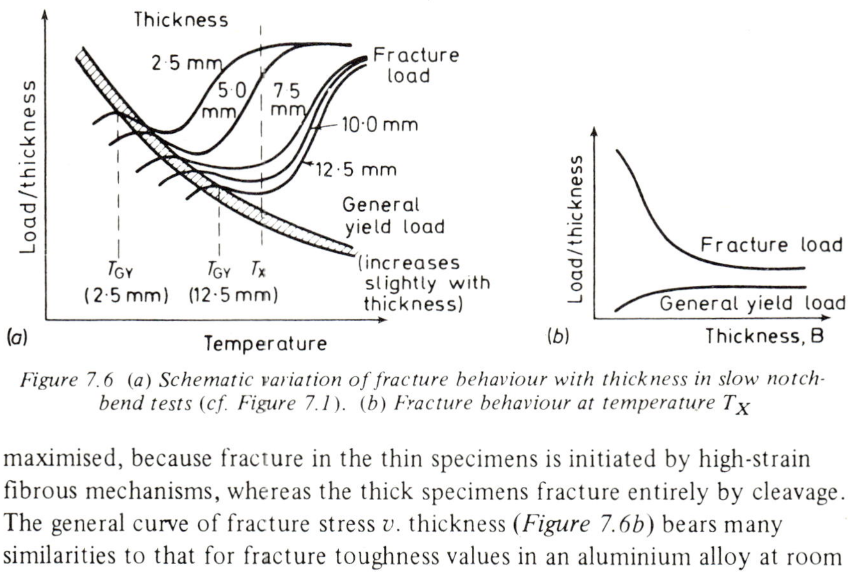

Figure 7.6 (a) Schematic variation of fracture behaviour with thickness in slow notch-bend tests (cf. Figure 7.1). (b) Fracture behaviour at temperature T_X

maximised, because fracture in the thin specimens is initiated by high-strain fibrous mechanisms, whereas the thick specimens fracture entirely by cleavage. The general curve of fracture stress *v.* thickness (*Figure 7.6b*) bears many similarities to that for fracture toughness values in an aluminium alloy at room

Table 7.1 VARIATION OF TRANSITION TEMPERATURE, T_{GY}, WITH THICKNESS

Thickness (mm)	12.5	10	7.5	5	2.5
Temperature, T_{GY}°C	−102	−108	−114	−120	−126
$Q = \sigma_F/\sigma_Y$	2.51	2.48	2.36	2.22	2.15

temperature (*Figure 5.2a*). It should be noted that, for the mild steel, 'square' fractures in moderate thickness may occur at loads substantially higher than the critical fracture load for very thick sections (see Section 5.7). We are therefore led to distrust a square fracture appearance as any definitive indication of whether or not limiting 'plane strain' conditions have been attained.

Use may be made of the experimental values of T_{GY} to calculate the maximum tensile stress, and hence the stress intensification, in a specimen at general yield as a function of thickness. If the local fracture stress is given by σ_F over the temperature range of interest (*Figure 7.3*), we have, at general yield:

$$\sigma_{11(max)} = Q_{(B)GY}\sigma_Y = \sigma_F \tag{7.5.1}$$

Hence, by measuring values of σ_Y in uniaxial tensile tests, $Q_{(B)GY}$ may be obtained as a function of thickness. Results are shown in *Figure 7.7* for a four-point bend specimen. It is apparent that $Q_{(B)GY}$ decreases rapidly below a thickness of about 2.5 mm. Even above 2.5 mm, $Q_{(B)GY}$ increases slightly with testpiece thickness.

The analogy with the aluminium alloy may be pursued. If the mild steel specimens were of much larger section, so that they broke under virtually linear

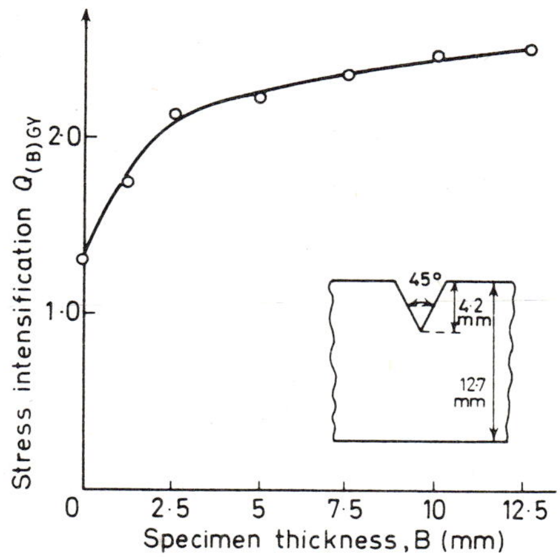

Figure 7.7 Variation of stress intensification, $Q_{(B)GY}$, with testpiece thickness

elastic conditions, one would expect that the critical thickness might correspond to a characteristic size of plastic zone at fracture. It has been shown[2] that the average plastic strain across the notch root at general yield is some 7%, i.e. the associated notch tip displacement is about 0.25 mm (the root radius) x 0.07 = 0.0175 mm. Under linear elastic conditions, the radius of plastic zone associated with such a displacement is:

$$r_Y \cong \frac{\delta}{2\pi}\frac{E}{\sigma_Y}$$

$$\cong 1.8 \text{ mm}$$

taking an average value of about 400 MN m^{-2} for the low temperature yield stress. We might therefore expect (from Section 5.7) that some effect on decreasing the fracture load would be observed in specimens greater than about 3.5 mm thick, but that any plane strain fracture requires a thickness of at least 7 mm. These estimates hold strictly for conditions well below general yield. However, they agree fairly well with the calculated variation of $Q_{(B)GY}$ with thickness.

The general picture is then that, in thin specimens, stress intensification does not build up to a very high value, because through-the-thickness relaxation of stresses occurs. The applied load must therefore be raised to a rather high value to produce fracture. There is an intermediate range of thicknesses in which some triaxiality persists at general yield, but where the peak stress is not as high as in very thick specimens. There is evidence, from macroscopic measurements of general yield loads and comparisons of these with the values predicted by plane-strain slip-line field theory, that plane-strain conditions exist in thick specimens up to, and, indeed, well beyond, general yield (see also Section 6.3). The gross transitions in fracture load and ductility occurring at temperatures T_W (see *Figure 7.1*) are usually attributable to the relaxations in stress state produced by gross-section yielding, rather than by yielding through the thickness. This point is amplified by considering effects of notch depth on yield and fracture behaviour.

7.6 Effects of Notch Depth

The fracture loads and general yield loads of thick notched specimens containing notches of different depths[11] are drawn in *Figure 7.8*. It is apparent that specimens with shallow notches possess lower transition temperatures. The values of

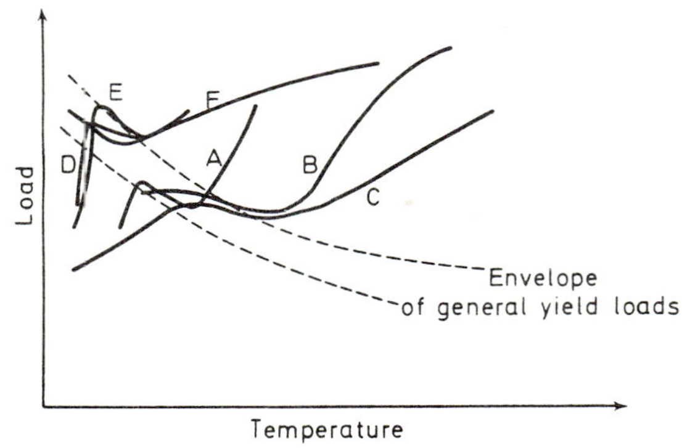

Figure 7.8 Schematic variation of fracture behaviour with notch depth in slow notch-bend. Net section depth ($W - a$) held constant in all cases at 8.75 mm.

Curve	*A*	*B*	*C*	*D*	*E*	*F*
Notch depth, a (mm)	1	2	3.75	2	3.75	1
Thickness, B (mm)	12.5	12.5	12.5	1.25	1.25	1.25

Notice that effects of notch depth on the transition do not occur in thin specimens

T_{GY} as a function of notch depth are indicated in *Table 7.2*. Over the range of shallow notch depths, the value of stress intensification at general yield may be calculated from:

$$\sigma_{11(\max)} = Q_{(a)GY}\sigma_Y = \sigma_F \tag{7.6.1}$$

where σ_Y and σ_F have been determined separately as before (Section 7.5). It is readily concluded that some form of relaxation of the triaxial stress state has occurred in thick shallow-notched bars at general yield. The reason for this relaxation may be deduced from the modes of yielding, as discussed in Section 2.12. To produce general yield in a 45° V-notched bar in bending requires a moment of:

$$M = 0.63\tau_Y(W - a)^2B \tag{7.3.1}$$

whereas the moment required to yield a rectangular bar of dimensions equal to the gross section is

$$M = 0.5\tau_Y W^2B \tag{7.6.2}$$

Table 7.2 VARIATION OF TRANSITION TEMPERATURE, T_{GY} WITH NOTCH DEPTH

Notch depth (mm)	3.75	2	1	3.75	2	1
Thickness (mm)	12.5	12.5	12.5	1.25	1.25	1.25
T_{GY} (°C)	−75	−85	−107	−128	−133	−140
$Q = \sigma_F/2\tau_Y$ exptl	2.18	2.04	1.79	1.51	1.49	1.39
Q theoretical	2.18	2.08	1.70	–	–	–

(see also *Table 2.2*)

The critical ratio of notch/section depth needed to produce constrained general net-section yielding before gross yielding has been shown to be (*Figure 2.16b*): $a/W = 0.41/1.41 \cong 0.3$. It will be noticed from *Table 7.2*, that the low values of $Q_{(a)GY}$ are obtained only for notches which are more shallow than this critical ratio. The relaxation occurs simply by the onset of gross-section yielding as indicated in *Figure 2.19*. Plastic 'wings' spread in from the top surface before the full 'hinge' pattern can develop. These events do not occur in thin specimens because the constraint factor is low.

The implications of these effects of notch depth on toughness testing procedures are important. It is clear that the notch must be deep for fully constrained net-section yielding to be developed. This shows also that even fracture toughness tests carried out well below general yield would be misleading if the pre-cracks were not sufficiently long: the specified ratio, $a/W = 0.45 - 0.55$ is, however, more than sufficient for the standard bend or CTS testpiece.

Problems may still arise in the application of fracture toughness values to design against fracture in service. A value obtained from a highly constrained testpiece may be unduly pessimistic if the material is used in structures where deep stress concentrators do not exist. Equally, a value may be optimistic if triaxial residual tensile stresses are present in service.

The problems are increased if small testpieces have to be used, to measure toughness through COD or *J*-integral procedures. Here general yielding precedes fracture and there is much more chance of gross-section yielding. Effects have been shown clearly in experiments where a comparison[12] was made between deep- and shallow-notched cylindrical (double-notched) testpieces of an alloy steel tempered to a 0.2% proof stress of 950 MN m^{-2}. For the deep notch, yielding was constrained, the fracture initiated by brittle intergranular mechanisms and the COD at the unfractured notch was low. For the shallow notch, triaxial stresses were relaxed, the mode of initiation changed to fibrous and the COD was large.

Effects of both thickness and notch depth have been explained in terms of the maximum value of tensile stress below the notch, which must reach a critical value to cause cleavage fracture. We shall now attempt to relate this critical value to the micro-mechanisms of fracture.

7.7 Micro-mechanisms of Cleavage Fracture

As discussed in Section 4.3 the theoretical fracture strengths of crystals are very much greater than those observed in practice. The discrepancies may be explained if real crystals contain atomically sharp flaws which propagate in the way suggested by Griffith[13], i.e. when the thermodynamic balance between the potential energy release rate and the surface energy required to separate bonds at the crack tip is attained. For iron, taking Young's modulus to be 200 GN m^{-2}, surface energy to be 2 J m^{-2}, and a typical fracture stress to be some 1 GN m^{-2} (cf. *Figure 7.3*), we have for the critical flaw size, a_{crit}:

$$a_{\text{crit}} = \frac{2E\gamma}{\pi(1-\nu^2)\sigma_F{}^2} \qquad (7.7.1)$$

$$\cong 0.3\ \mu\text{m}$$

Pure irons may be made which do not contain inherent, atomically sharp flaws as large as 0.3 μm and yet which cleave at low fracture stresses.

For iron tensile specimens, it has been shown that even the most apparently brittle fractures, in testpieces broken at 77 K, fail at stresses coincident with the uniaxial yield stress. The point was proved by comparing the fracture stress in tension with the yield stress in compression for a variety of different grain sizes, as indicated in *Figure 7.9*[14]. In fine grain sizes, yielding precedes fracture: in coarse grain sizes, it is clear that yielding and fracture are coincident. Similarly, even in very brittle notched bars, cleavage fracture is preceded by a small amount of local yielding around the notch root. The load required to produce fracture when the notch region is in tension is equal to that which produces slip or twinning when the notch is in compression.

These various results led to the conclusion that yielding precedes fracture, because yield is necessary to produce a cleavage crack nucleus. The yield may take the form of slip or twinning. In each case, the models proposed for nucleation involve the blocking of a slip-band or twin by an obstacle, such as a grain boundary or second-phase particle. The model for a slip-band is indicated in *Figure 7.10a* where a pile-up of dislocations in a grain of diameter d produces a highly-concentrated stress at each end of the band. We suppose for a tensile specimen that a shear stress, τ, deriving from an applied tensile stress $\sigma(\tau \cong \sigma/2)$, acts on the slip-band. Since slip proceeds by dislocation movement, the shear which the stress τ is trying to produce is opposed by a 'friction stress' τ_i, which includes contributions from the inherent resistance of the lattice to slip (the Peierls–Nabarro force), from small precipitates, from clusters of solute atoms or

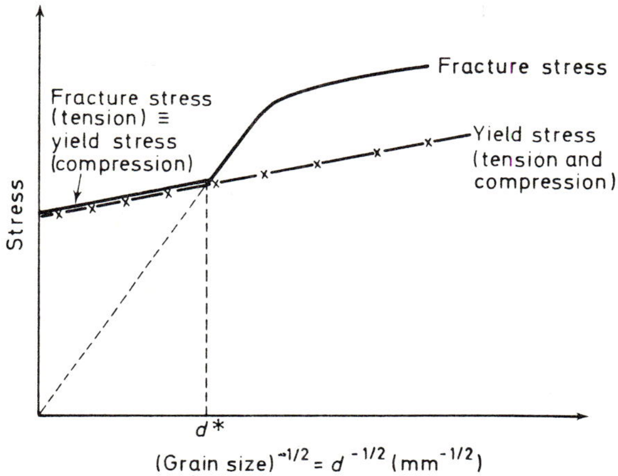

Figure 7.9 Schematic variation of yield and fracture stresses with grain size. Mild steel at 77 K (after Low[14])

point defects (if the material has, for example, been subjected to neutron irradiation), and from the annealed-in dislocation density. For a twin, τ_i is very small and is often taken as zero.

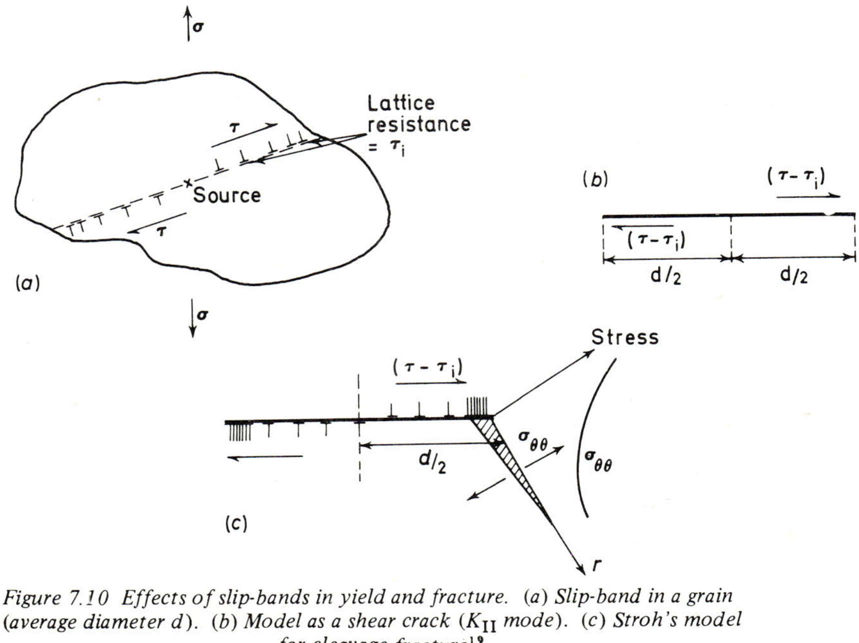

Figure 7.10 Effects of slip-bands in yield and fracture. (a) Slip-band in a grain (average diameter d). (b) Model as a shear crack (K_{II} mode). (c) Stroh's model for cleavage fracture[19]

The stress distribution around the slip-band subjected to an applied shear stress, τ, is equated to that for a freely-slipping crack, of length equal to the grain diameter, under an applied stress $(\tau - \tau_i)$ (*Figure 7.10b*). The mode of deformation is that of Mode II shear (Section 3.8), and the local stresses ahead of the slip-band therefore take the form:

$$\sigma_{12} = \frac{K_{II}}{\sqrt{(2\pi r)}} f_1(\theta) + \ldots \tag{7.7.2}$$

or

$$\sigma_{\theta\theta} = \frac{K_{II}}{\sqrt{(2\pi r)}} f_2(\theta) + \ldots \tag{7.7.3}$$

where we write, for K_{II}:

$$K_{II} = (\tau - \tau_i)\sqrt{(\pi d/2)} \tag{7.7.4}$$

This result has been used in two very important ways. We consider first its use as a model of the yielding behaviour of a polycrystalline specimen.

7.8 Yielding: the Petch Relationship[15,16]

Considering the expression for the concentrated shear stress, σ_{12}, and taking $f_1(\theta) = 1$, we may write:

$$\sigma_{12} = (\tau - \tau_i)\left(\frac{d}{4r}\right)^{\frac{1}{2}} \tag{7.8.1}$$

We are now in a position to model the propagation of a Lüders band along the gauge length of a smooth tensile specimen at the lower yield stress, by postulating that the grain ahead of the slip-band will yield when σ_{12} attains a critical value, τ^*. Then from equation 7.8.1, the applied shear stress, τ_Y, required to spread yield from grain to grain is given by:

$$\tau_Y = \tau_i + (4r)^{\frac{1}{2}}\tau^*.d^{-\frac{1}{2}} \tag{7.8.2}$$

On the average, the distance, r, to the nearest dislocation source in the unyielded grain and the value of the stress, τ^*, needed to operate that source may be taken as constant and we write

$$\tau_Y = \tau_i + k_Y^S d^{-\frac{1}{2}} \tag{7.8.3}$$

where k_Y^S is a constant pertaining to the shear stress state.

The shear stresses may be written in terms of the applied tensile stress through an orientation factor: $\sigma = m\tau$, where m may be taken as approximately 2.2 for randomly-oriented body-centred-cubic polycrystals. The final expression for the tensile yield stress is then:

$$\sigma_Y = \sigma_i + k_Y d^{-\frac{1}{2}} \tag{7.8.4}$$

where k_Y is a constant. This equation was first derived by Hall[15] and Petch[16] and is generally called the Petch equation.

The value of k_Y relates to the process by which mobile dislocations are produced in the unyielded grain. For thermally activated unpinning of dislocations from carbon or nitrogen atmospheres, k_Y is dependent on temperature, but it has been shown that, in normalised or annealed mild steel, mobile dislocations are created in, or close to, the common grain boundary at local stresses which are so high that the process is not affected by the thermal energies associated with low temperatures[17].

The second important use of equations 7.7.3 and 7.7.4 is to model fracture behaviour.

7.9 Cleavage Fracture: Stroh's Theory[18]

An early dislocation model for the cleavage fracture of a polycrystal was based on the attainment of a critical value of the tensile stress, $\sigma_{\theta\theta}$, in the unyielded grain. By an argument similar to that developed for the Petch model for yielding, a relationship was derived between the cleavage fracture stress and the material's grain size:

$$\sigma_F = \sigma_i + k_F d^{-\frac{1}{2}} \tag{7.9.1}$$

where k_F refers to the critical value of local tensile stress required to fracture the second grain. This theory makes the correct prediction, that the cleavage fracture stress is higher in fine-grained material, but gives no clear meaning to k_F and does not therefore adequately explain the experimental variation of fracture stress with grain size, as shown in *Figure 7.9.* For such results, k_F appears to be identical to k_Y for coarse grain sizes, but substantially greater than k_Y in finer grain sizes. Initially, the experimental values were 'corrected' and then appeared to fit the theory, but the procedure used cannot be regarded as satisfactory.

The meaning of the k_F value was investigated by Stroh[18,19]. He supposed that the dislocations at the head of the slip-band were squeezed together to produce a crack nucleus and calculated the magnitude of the local tensile stress, $\sigma_{\theta\theta}$, needed to spread this nucleus as a Griffith crack. This is a maximum at 70.5° to the line of the slip band. The model is drawn in *Figure 7.10c.* Certain modifications to the original Stroh analysis are needed to take account of the role of shear stresses in assisting crack growth and that the crack nucleus spreads under a non-uniform stress[20]. (Stroh also failed to allow for an incorrect factor of two in the first version of Griffith's equation.) When the corrections are made, Stroh's expression for the conditions under which the nucleus will spread may be written as[20]:

$$\tau_{\text{eff}} \equiv \tau_{\text{Y}} - \tau_{\text{i}} \geqslant \left[\frac{\pi\mu\gamma}{2(1-\nu)d}\right]^{\frac{1}{2}} \tag{7.9.2}$$

From this result, it may be clearly deduced that, if a crack nucleus can form, any increase in the length of the nucleus will lead to a decrease in the total energy of the system, provided that the surface energy encountered by the growing crack remains constant. This implies that cleavage fracture is *nucleation*-controlled, demanding simply that a number of dislocations sufficient to nucleate a crack may be squeezed together at the end of the slip-band. The critical number is given by:

$$n = \frac{\pi^2\gamma}{2\tau_{\text{eff}}b} \tag{7.9.3}$$

where b is the Burger's vector of a dislocation.

The important stress in causing cleavage fracture is therefore predicted to be the effective shear stress,

$$\tau_{\text{eff}} = \tau_{\text{Y}} - \tau_{\text{i}} = k_{\text{Y}}^{S} d^{-\frac{1}{2}} \tag{7.9.4}$$

from equation 7.8.3. This prediction does not then explain why cleavage fractures predominate at low temperatures [where the tensile yield stress is high, but $(\tau_{\text{Y}} - \tau_i)$ is little different from its value at room temperature] or why the tensile stress is so important in notched-bar fracture.

If tensile stress is to be the vitally important parameter indicated by the experimental results, fracture must be *growth*-controlled, rather than nucleation-controlled. Stroh's formula (7.9.2) shows clearly that this is impossible, assuming his nucleation mechanism, if the surface energy term, γ, remains constant.

7.10 Cleavage Fracture – Cottrell's Theory[7]

Cottrell proposed a dislocation mechanism for cleavage fracture, which allowed growth to be the controlling factor, by providing an easy nucleation process. If, as indicated in *Figure 7.11*, we have intersecting $\{101\}$ slip-planes in iron on which dislocations with Burger's vectors of type $\frac{a}{2}\langle\bar{1}\,\bar{1}\,1\rangle$ are gliding, we may obtain the reaction:

$$\frac{a}{2}[\bar{1}\bar{1}1]_{(101)} + \frac{a}{2}[111]_{(\bar{1}01)} \rightarrow a[001] \tag{7.10.1}$$

The resultant dislocation is formed with a reduction in energy and is a sessile edge dislocation (the line of intersection of the planes is [010]) with a Burger's vector normal to the cleavage plane (001). It therefore provides the first stage of crack nucleation. The relative motions of material above and below the slip-planes produce the effect of driving a wedge into the cleavage plane.

The importance of tensile stress in causing fracture may be deduced by considering the total energy of the system as more dislocations are fed into the nucleus to increase the wedging displacement in the [001] direction. Suppose that, for a displacement nb, produced by n pairs of slip dislocations, the length of nucleus is c. We recognise four contributions to the total energy per unit thickness:

(1) the Griffith energy of a crack of total length c under a tensile stress p:

$$U_1 = -\frac{p^2(1-\nu^2)}{E}\left(\frac{c}{2}\right)^2 \qquad (7.10.2)$$

(cf. equation 4.3.14 for a crack of length $2a$)

(2) the work done by the stress in forming the nucleus:

$$U_2 = -\tfrac{1}{2}p.n\boldsymbol{b}.c \qquad (7.10.3)$$

(the average displacement is $\frac{1}{2}n.\boldsymbol{b}$)

(3) the surface energy

$$U_3 = +2\gamma.c \qquad (7.10.4)$$

(4) the strain energy of the 'cracked' edge dislocation of Burgers vector $n\boldsymbol{b}$

$$U_4 = +\frac{\mu(n\boldsymbol{b})^2}{4\pi(1-\nu)}\ln\left(\frac{2R}{c}\right) \qquad (7.10.5)$$

where R is taken as the distance over which this strain field is effective and $(c/2)$ is taken as the core radius of the cracked dislocation. The equilibrium lengths for the crack are found by equating $\frac{\partial}{\partial c}(U_1 + U_2 + U_3 + U_4)$ to zero. A quadratic in c is obtained, which implies that either there are two stable crack lengths or no

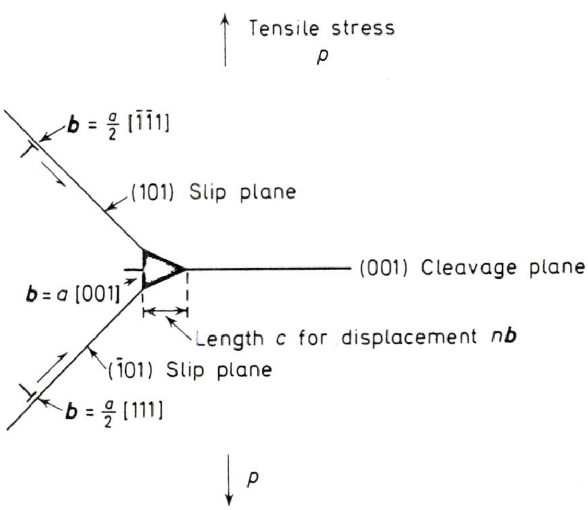

Figure 7.11 Cottrell's model for cleavage fracture[7]

real roots, in which case the energy decreases spontaneously. At the transition point, we obtain:

$$pnb = 2\gamma \tag{7.10.6}$$

or, substituting for $\boldsymbol{b}$ (= a[001]),

$$pna = 2\gamma \tag{7.10.7}$$

This may be taken as the critical condition where the curve of energy *v.* crack length decreases, but shows a horizontal point of inflection corresponding to a characteristic nucleus size. We may substitute for the displacement *na* the sum of the sliding displacements at the end of each slip band resolved in the [001] direction. If each slip band is one grain diameter *d* in length and the effective shear stress at the yield point is $(\tau_Y - \tau_i)$, the shear strain is $(\tau_Y - \tau_i)/\mu$ and the shear displacement is $d(\tau_Y - \tau_i)/\mu$. We may therefore write:

$$na = (\tau_Y - \tau_i)\frac{d}{\mu} = \frac{k_Y^S d^{\frac{1}{2}}}{\mu} \tag{7.10.8}$$

making the substitution

$$(\tau_Y - \tau_i) = k_Y^S d^{-\frac{1}{2}} \tag{7.9.4}$$

The value of the tensile stress needed to propagate a nucleus then becomes:

$$p \geqslant \frac{2\mu\gamma}{k_Y^S} d^{-\frac{1}{2}} \tag{7.10.9}$$

Cottrell applied this expression mainly to explain the experimental results on tensile specimens shown in *Figure 7.9*. Writing $p = \sigma_Y = 2\tau_Y$ for the value of tensile stress at the yield point, he obtained the expression:

$$\tau_Y k_Y^S d^{\frac{1}{2}} \geqslant \mu\gamma \tag{7.10.10}$$

or

$$\sigma_Y \geqslant \frac{2\mu\gamma}{k_Y^S} d^{-\frac{1}{2}} \tag{7.10.11}$$

as the stress required (on energetic grounds), for propagation, once a nucleus has formed. From *Figure 7.9* it is apparent that, for grain sizes coarser than d^*, the fracture is not growth controlled. It is *mechanism* controlled, because it is necessary to produce dislocations before they can form a crack nucleus. The fracture stress is then simply the yield stress. For grain sizes finer than d^*, yielding precedes fracture, but the crack nucleus will not spread until a stress

$$\sigma_F \geqslant \frac{2\mu\gamma}{k_Y^S} d^{-\frac{1}{2}} \tag{7.10.12}$$

is attained. (Strictly, the Cottrell argument as such applies only at d^*, because γ may increase and k_Y^S may decrease once the specimen has yielded.) The value of γ calculated from the results in *Figure 7.9* was about 20 J m^{-2}, i.e. about an order of magnitude greater than the surface energy of the lattice. Cottrell attributed this larger value to the extra work done in producing river lines or traversing grain boundaries.

The model was applied to explain notch fracture by writing

$$\sigma_Y k_Y^S d^{\frac{1}{2}} \geqslant \beta.2\mu\gamma \qquad (7.10.13)$$

where β is a factor [taken by Cottrell to be $\frac{1}{3}$ to conform with the Orowan general yield solution for deep external cracks (equation 7.4.1)] which relates the tensile yield stress to the maximum tensile stress below a notch. In fact, for a given notch geometry, β varies with temperature, because the plastic zone size at fracture increases with temperature[3]. The β factor is the reciprocal of the stress intensification, Q (see Section 7.4).

Cottrell's model therefore emphasises the role of tensile stress and explains effects of grain size ($d^{-\frac{1}{2}}$) and yielding parameters (τ_Y, k_Y^S) on fracture. Hardening, other than by decreasing grain size, is predicted to promote brittle cleavage fractures, by raising the value of tensile stress at the yield point. Initially, the Cottrell equation appeared to serve as a complete basis for the microstructural

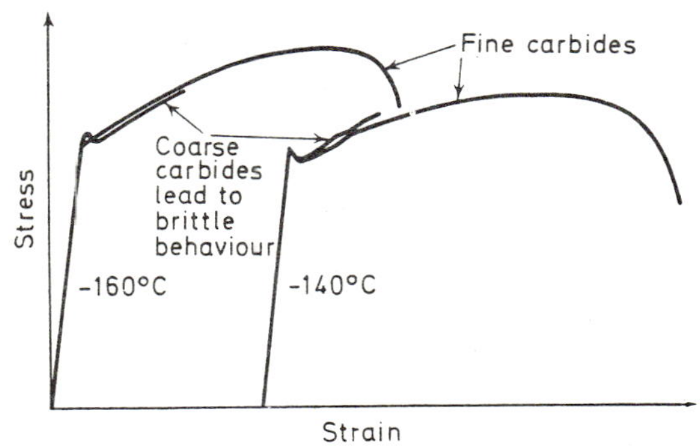

Figure 7.12 Schematic stress–strain curves for steels containing fine and coarse carbides (after McMahon and Cohen[21])

design of steels with improved resistance to cleavage, but one important variable was omitted from the model.

This was demonstrated by McMahon and Cohen[21] who tested, at various low temperatures, tensile specimens of identical yield and flow properties, but containing carbides of different sizes. Their results (*Figure 7.12*) showed that coarse carbides promoted cleavage, whilst fine carbides allowed the material to behave in a ductile manner.

7.11 Cleavage Fracture: Smith's Theory[22]

These experiments provided the starting point for an alternative model for growth-controlled cleavage fracture, to incorporate the effect of carbide particles.

The situation is drawn in *Figure 7.13*. Here, a brittle carbide of width C_0, at the grain boundary dividing adjacent grains is subjected to the concentrated stress ahead of a pile-up of length d. The system is subjected to a tensile stress σ. From

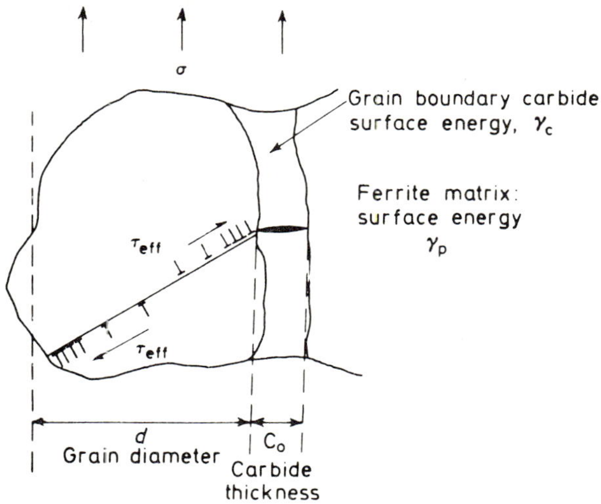

Figure 7.13 Smith's model for cleavage fracture[22]

a Stroh-type analysis (yielding a result of form similar to equation 7.9.2) it may be deduced that the carbide will be cracked by the pile-up if

$$\tau_{\text{eff}} = (\tau_Y - \tau_i) \geqslant \left[\frac{4E\gamma_c}{\pi(1-\nu^2)d}\right]^{\frac{1}{2}} \tag{7.11.1}$$

where γ_c is the surface energy of the carbide particle. If it is assumed that the work required to spread the crack from the carbide into the ferrite matrix of the unyielded grain (an 'effective' surface energy, γ_p) is greater than γ_c, we may deduce that fracture at the yield point is nucleation-controlled if

$$\tau_{\text{eff}} = (\tau_Y - \tau_i) \geqslant \left[\frac{4E\gamma_p}{\pi(1-\nu^2)d}\right]^{\frac{1}{2}} \tag{7.11.2}$$

If, however, τ_{eff} at yield lies between the limits set by equations 7.11.1 and 7.11.2 the change in energy as the crack length is increased may be examined in a manner similar to that for the Cottrell model to obtain as a failure criterion for growth-controlled fracture[22]:

$$\left(\frac{c_0}{d}\right)\sigma_F^2 + \tau_{\text{eff}}^2\left[1 + \frac{4}{\pi}\left(\frac{c_0}{d}\right)^{\frac{1}{2}}\frac{\tau_i}{\tau_{\text{eff}}}\right]^2 \geqslant \frac{4E\gamma_p}{\pi(1-\nu^2)d} \tag{7.11.3}$$

We notice that, if the dislocation contribution (the second term) is absent, equation 7.11.3 reduces to:

$$\sigma_F > \left[\frac{4E\gamma_p}{\pi(1-\nu^2)c_0}\right]^{\frac{1}{2}} \tag{7.11.4}$$

as for a Griffith crack of total length c_0 ($c_0 \equiv 2a$ in equation 4.3.18). This model therefore emphasises the importance, not only of yield parameters and grain size, but also of carbide thickness and indicates clearly that coarse carbides give rise to low fracture stresses. The predictions of equation 7.11.3 with respect to grain size are interesting. If τ_{eff} is written as $k_Y^S d^{-\frac{1}{2}}$ (equation 7.9.4), the equation predicts that σ_F is independent of grain size, other factors being equal. In practice, however, fine grains are associated with thin carbides, and values of σ_F are usually expected to be high.

7.12 Twin-initiated Cleavage Fracture

It is appropriate at this stage briefly to discuss the importance of mechanical twinning in the initiation and growth of cleavage cracks. In studies made to try to substantiate the Cottrell mechanism for nucleation (Section 7.10) in iron–silicon single crystals, it was found that a cleavage crack was initiated on the (001) cleavage plane, not by the intersection of slip bands, but by the intersection of twins[23]. Similar events have been observed in polycrystalline mild steel[2] (see *Figure 7.14*) and indeed, the value of k_Y obtained from the yield stress curve in *Figure 7.9* tends to suggest twin-initiated, rather than slip-initiated, fracture, even in this case.

In the body-centred cubic lattice, twinning occurs by the movement of partial dislocations of the type $a/6\langle\bar{1}11\rangle$ on $\{211\}$ planes. The twinning shear is

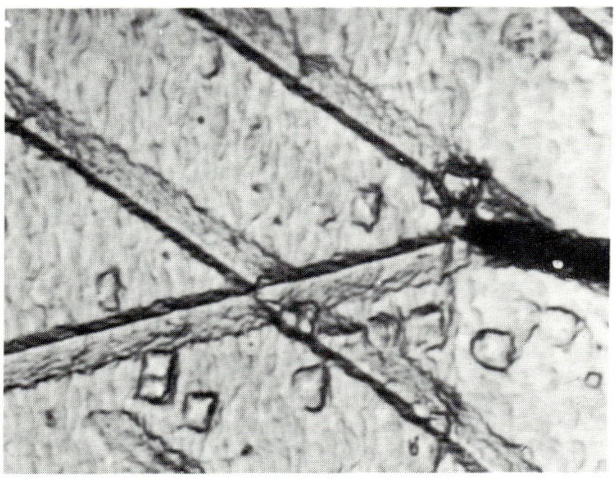

Figure 7.14 Initiation of cleavage microcracks by twins, in mild steel at 77 K. (*The section is almost a* (100) *plane and the edges of the etch pits show* ⟨010⟩ *directions. The twins show the traces of* $\{112\}$ *planes, and the crack, lying parallel to one set of etch pit edges, may be shown to represent the trace of the appropriate* $\{010\}$ *plane*) (x 1500).

0.707. By analogy with the slip dislocations, it is reasonable to suppose that the partials moving ahead of the growing twins might interact to produce a resultant displacement normal to the cleavage plane. Estimates of this displacement have been made by measuring the thicknesses of the intersecting twins and applying the known value of the twinning shear. Values of γ, calculated from the displacements are of the order $20\,J\,m^{-2}$, i.e. approximately an order of magnitude greater than the surface energy of the lattice.

In annealed and normalised steels deformed at low temperatures or high strain rates, where twinning is the operative mode of plastic deformation, twins may act as potent stress concentrators in the nucleation of cracks in brittle carbides. The friction stress resisting the motion of twinning dislocations is very small and is conventionally taken as zero. Smith's formula for the propagation of a crack from a grain boundary carbide (equation 7.11.3) then becomes:

$$\left(\frac{c_0}{d}\right)^2 \sigma_F + \tau_Y^2 \geqslant \frac{4E\gamma_p}{\pi(1-\nu^2)d} \qquad (7.12.1)$$

The implication of this result is that, even when cracks are nucleated by very large shear stress components, the total fracture event may still sometimes be controlled by the magnitude of the applied tensile stress. Experimental support for this prediction is given by results obtained on notched bars of various thicknesses, fractured at 77 K (see *Figure 7.15*)[24]. It is necessary to develop a critical

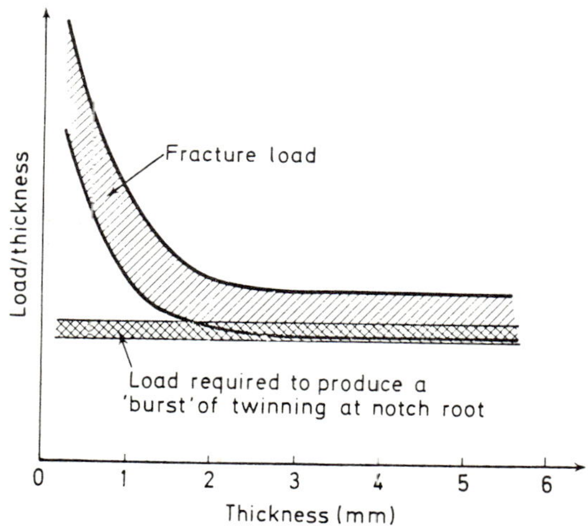

Figure 7.15 Variation of fracture load with thickness at 77 K

size of plastic zone beneath a notch before a 'burst' of twins is produced. In thick specimens, the tensile stress below the notch when the twins are formed is more than sufficient immediately to propagate any carbide cracks nucleated by the 'burst' and final fracture is coincident with the onset of twinning. In thin specimens, the triaxiality is low, and the tensile stress at the onset of twinning is insufficient to propagate the nuclei. It may be raised by increasing the size of the plastic zone and hence the applied load on the specimen. Consequently, the fracture loads for thin specimens are substantially greater than those for thick specimens. The temperature dependence of σ_F for twin-initiated fracture (equation 7.12.1) is predicted to be much greater than that for slip-initiated fracture because τ_Y is strongly dependent on temperature.

Even in coarse-grained mild steel, twin-initiated fracture would not be anticipated at temperatures above about 150 K, if the applied loading rates were equivalent to those normally obtained in standard fracture toughness testing or in the loading of a structure. Only if the external loading is carried out under impact conditions will the *initiation* of cleavage fracture at temperatures approaching ambient be associated with twins. Twinning is commonly associated with *propagating* cracks, however, because very high strain rates are produced ahead of the accelerating crack tip.

Since we are primarily concerned with the onset of crack extension in fairly slow tests, we shall not pursue these twinning mechanisms. In particular, the behaviour of mild steel is to be regarded as a model for more complex materials, which often do not deform by twinning. Slip-initiated fracture mechanisms are therefore more generally relevant.

7.13 The Fibre-loading Model

A further model, for crack nucleation[25] in a ferrite/carbide microstructure by slip dislocations, supposes that the carbide acts as a long thin fibre which is loaded elastically when the matrix around it deforms plastically, and which cracks when the matrix strain reaches a critical value. The energy release on cracking may be so great that the crack nucleus bursts into the matrix for a short distance. The final fracture is again growth-controlled and should therefore occur at a critical value of the applied tensile stress.

The implications of the fibre-loading model are that the thinnest carbides, aligned most nearly parallel to the tensile axis, are those which are the most likely to be cracked and that a significant amount of plastic strain must occur in the matrix before a fibre is loaded to its breaking point. The total fracture criterion becomes complicated because nuclei initiate more easily in thin carbides, yet the larger nuclei, formed in the thicker carbides, are able to be propagated by lower stresses. It is predicted that fractures produced in the local plastic zone ahead of a stress concentrator do not form at the plastic/elastic interface, but at some distance behind the interface, where the plastic strains are sufficient to form crack nuclei. One of the main virtues of the fibre-loading model, as opposed to simple tensile-stress controlled growth, appeared to be that fractures in 45° V-notched bars at low temperatures seemed to nucleate well behind the interface. However, recent finite-element analyses of the stress distributions in such plastic zones[9] show that the position of maximum tensile stress itself lies some considerable distance behind the interface (see Section 3.18), so that the observations do not, in fact, discriminate between the alternative carbide-cracking mechanisms.

7.14 Implications of the Cleavage Micro-mechanisms

The various micro-mechanisms proposed for the formation of cleavage cracks involve the nucleation of cracks by the high stresses developed locally at the ends of slip-bands and the propagation of the nuclei under the applied tensile stress to produce the final fracture. For fracture to be growth-controlled, rather than nucleation-controlled, it is necessary for the 'effective surface energy' to increase as the crack grows. Cottrell's model provides for this condition by means of a nucleation stage which is energetically favourable: Smith's model and the fibre-loading model postulate that the work required to crack a brittle carbide is substantially less than that needed to spread the nucleus into the surrounding ferrite matrix.

Another growth-controlled situation might occur in rather fine-grained material. Here, the growth of a crack through a single grain could be a relatively easy process compared with the difficulty of propagating a microcrack of length equal to a grain diameter into a second grain, where the cleavage plane has a different orientation. If the crack were to arrest at the grain boundary, the total fracture stress could be estimated by supposing that the fracture would continue to propagate when a new crack was nucleated in the second grain. The displacement in the second grain associated with nucleation is then given by equation 7.9.3 as:

$$nb = \pi^2\gamma/2\tau_{\text{eff}}$$

If this displacement is equated to that at the end of the microcrack in the first grain, we may calculate the fracture stress directly from equation 3.13.11, writing δ_{crit} as nb and taking for the crack length, $2a = d$, the grain diameter[26]. In a sense, we assume that the microcrack acts as a sensitive probe which samples the displacement in the adjacent grain. Values of γ calculated in this manner are, however, some two orders of magnitude greater than the surface energy. Although it does seem clear that, in some situations, the grain-boundary arrests provide the basic reason for growth-controlled fracture, it is difficult to see how they would cause the effects shown in *Figure 7.12* where the grain size was held constant, although the carbide thickness was varied.

At present, no single growth-controlled mechanism has been shown to cover all possible cases (particularly since grain size and carbide width are usually varied simultaneously in most practical heat-treatments) but Smith's model seems to fit most results for rather coarse-grained polycrystals. Results obtained by Oates[27] (see *Figure 7.16*) on a plain carbon steel and a manganese steel, having identical grain size, but containing different widths of grain-boundary carbides, support the quantitative predictions of the model closely, assuming a value for γ of $14\,\text{J m}^{-2}$. Fracture in the mild steel, which contained coarse carbides, could occur at low tensile stress levels and fracture at general yield was therefore obtained at relatively high temperatures, because the uniaxial yield stress did not need to be very high. The cracks were therefore nucleated by slip. In the manganese steel, on the other hand, fracture at general yield could occur only at very low temperatures where the uniaxial yield stress was high. Cracks were nucleated by twinning and the temperature dependence of the local fracture stress was consequently much increased (see Section 7.12).

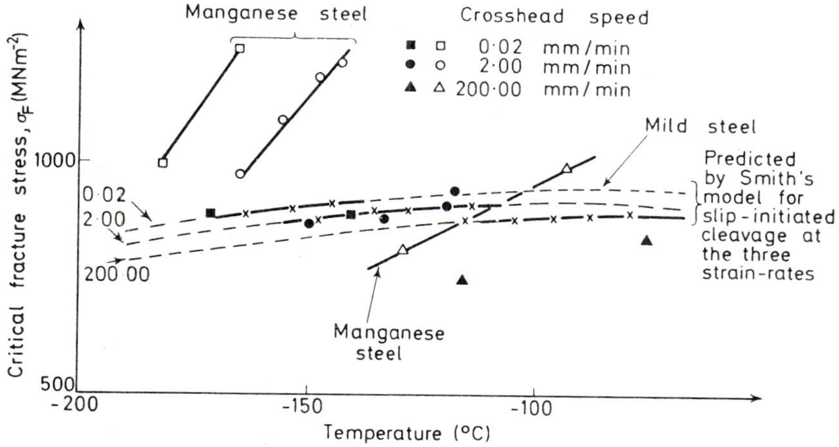

Figure 7.16 Variation of fracture stress with temperature for mild steel (filled points), compared with Smith's model (see also Figure 7.3). Results are included for twin-initiated fracture in a manganese steel (solid lines and open points) (after Oates[27])

The effect of grain size on local fracture stress in notched bars has been studied by Almond, Timbres and Embury[28] and by Knott[3]. Their results are shown in *Figure 7.17* and may be interpreted directly in terms of Smith's model, or of a modified version of it.

The general explanation of a critical tensile stress criterion for the fracture of notched bars of mild steel is that this represents the stress needed to propagate a cleavage crack nucleus. We expect that the magnitude of the stress depends on

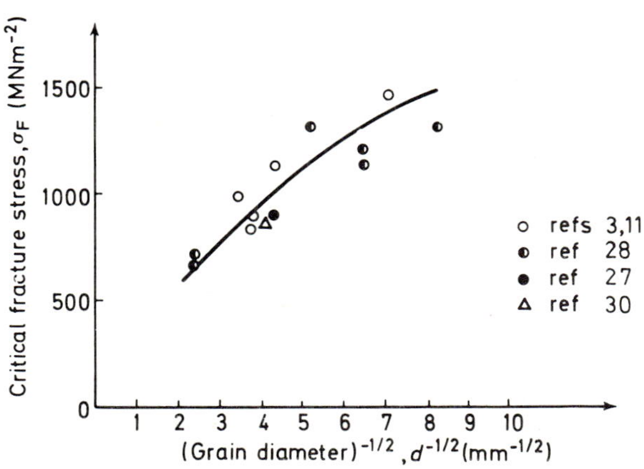

Figure 7.17 Variation of critical fracture stress, σ_F, with grain size, in mild steel (results from several authors)

the size of the nucleus and on the shear stress contribution from slip-bands or twins. Twins provide more effective stress concentrations and so for growth-controlled situations (notched bars, as opposed to uniaxial specimens), we expect the critical fracture stress to be lower for a constant nucleus size. The temperature dependence of the fracture stress depends on that of the 'effective shear stress' (or k_Y, see Section 7.9). If fracture is slip initiated, this is virtually independent of temperature: if it is twin-initiated, a strong dependence may be produced (e.g. *Figure 7.16*). In steels hardened by large precipitates or by forest dislocations, the friction stresses are long-range and thermal fluctuations are unlikely significantly to affect the k_Y term, so that the local cleavage fracture stress could again be relatively independent of temperature.

We therefore recognise a mechanical situation in which the macroscopic ductility of a notched specimen depends on the ability of the specimen to undergo general yielding before the tensile stress in the plastic zone exceeds the critical value needed to propagate a cleavage crack. A transition temperature, T_{GY}, is defined for the condition where fracture coincides with general yield. A combination of high uniaxial yield stress, large grain size, and coarse carbides leads to high transition temperatures. Usually, the metallurgical heat-treatments which produce hardening refine the grain size or carbide distribution, but simple hardening of the ferrite matrix in an annealed microstructure (e.g. by neutron irradiation hardening or by precipitation, as in the iron–copper–carbon system) ought to enable the increase in T_{GY} to be studied in an unambiguous manner. In general, an increase in yield stress may have two components: a temperature dependent component, $\Delta\sigma_i''$, arising from short-range forces (dislocation–interstitial interactions, solute clusters, jogs, etc.), and a temperature independent component, $\Delta\sigma_i'$, associated with long-range interactions (forest hardening, precipitates, etc.). If k_Y becomes temperature-dependent, re-analysis is necessary, because the model at present has been justified only for the two limiting conditions: τ_{eff} = constant (slip-initiated fracture) and $\tau_{eff} = \tau_Y$ (twin-initiated fracture). Complications might arise in, e.g. quench-aged material, where k_Y is temperature dependent and is a function of the state of ageing.

Almond and Embury[29] have shown clearly how the cleavage fracture resistance of a simple mild steel may be improved by altering the distribution of carbides. In the annealed condition, with coarse grain-boundary carbides, the cleavage fracture strength was 660 MN m^{-2}. In a quench-aged and fully spheroidised condition, the fracture stress was 900 MN m^{-2}. Even though the yield strength was increased by some 33% by the precipitation-hardening, the transition temperature *decreased* by some 40°C because the crack nuclei were so much more difficult to nucleate and propagate in the spheroidised condition.

7.15 Conclusions

From this extensive discussion of the cleavage fracture behaviour of mild steel, it is clear that the macroscopic fracture behaviour can be related to the micro-mechanisms of fracture and that these are in turn dependent on microstructure and flow characteristics. In certain cases, it has proved possible to exert metallurgical control on the microscopic fracture process, such that the macroscopic behaviour is improved. It is obvious that further detailed study of micromechanisms is necessary to establish technological limits on fracture resistance in more complicated materials. Some work of this sort will be discussed briefly in Section 8.10.

Recently, it has proved possible to use the cleavage fracture stress model directly to predict values of K_{IC} in mild steel at low temperatures. This will be described in Section 8.9. In the following chapter we examine the cleavage-fibrous transition in slow notch-bend tests and show how it relates to impact testing.

REFERENCES

1. Knott, J. F., *Effect of Second-Phase Particles on the Mechanical Properties of Steel*, 44, The Iron and Steel Institute (1971)
2. Knott, J. F. and Cottrell, A. H., *J. Iron Steel Inst.*, **201**, 249 (1963)
3. Knott, J. F., *J. Iron Steel Inst.*, **204**, 104 (1966)
4. Wilshaw, T. R., *J. Iron Steel Inst.*, **204**, 936 (1966)
5. Orowan, E., *Trans. Inst. Engrs Shipbuilders Scot.*, **89**, 165 (1945)
6. Hendrickson, J. A., Wood, D. S. and Clark, D. S., *Trans. Am. Soc. Metals*, **50**, 656 (1958)
7. Cottrell, A. H., *Trans. Am. Inst. Min. Metall. Petrol. Engrs*, **212**, 192 (1958)
8. Knott, J. F., *J. Iron Steel Inst.*, **205**, 288 (1967)
9. Griffiths, J. R. and Owen, D. R. J., *J. Mech. Phys. Solids*, **19**, 419 (1971)
10. Knott, J. F., *Proc. R. Soc.*, **A285**, 150 (1965)
11. Knott, J. F., 'Fracture 1969' *Proc. Second International Conference on Fracture*, 205, Chapman and Hall, London (1969)
12. Knott, J. F., *J. Iron Steel Inst.*, **205**, 970 (1967)
13. Griffith, A. A., *Phil. Trans. R. Soc.*, **A221**, 163 (1921)
14. Low, J. R., *Symposium on Relation of Properties to Microstructure*, 163, American Society for Metals (1954)
15. Hall, E. O., *Proc. phys. Soc.*, **64B**, 747 (1951)
16. Petch, N. J., *J. Iron Steel Inst.*, **173**, 25 (1953)
17. Cottrell, A. H., *Symposium on the Relation Between the Structure and Mechanical Properties of Metals*, 456, National Physical Laboratory, HMSO, London (1963)
18. Stroh, A. N., *Proc. R. Soc.*, **A223**, 404 (1954)
19. Stroh, A. N., *Adv. Phys.*, **6**, 418 (1957)
20. Smith, E. and Barnby, J. T., *Metal Sci. J.*, **1**, 56 (1967)
21. McMahon, C. J. and Cohen, M., *Acta Met.*, **13**, 591 (1965)
22. Smith, E., *Proc. Conference Physical Basis of Yield and Fracture*, 36, Inst. Phys. Phys. Soc., Oxford (1966)
23. Hull, D., *Acta Met.*, **8**, 11 (1960)
24. Knott, J. F., *J. Iron Steel Inst.*, **205**, 285 (1967)
25. Lindley, T. C., Oates, G. and Richards, C. E., *Acta Met.*, **18**, 1127 (1970)
26. Groom, J. D. G., *Ph.D. thesis*, University of Cambridge (1972)
27. Oates, G., *J. Iron Steel Inst.*, **207**, 353 (1969)
28. Almond, E. A., Timbres, D. H. and Embury, J. D., 'Fracture 1969', *Proc. Second International Conference on Fracture*, 253, Chapman and Hall, London (1969)
29. Almond, E. A. and Embury, J. D., *Metal Sci. J.*, **2**, 194 (1968)
30. Ritchie, R. O., Knott, J. F. and Rice, J. R., *J. Mech. Phys. Solids* (in press)

Eight

THE CLEAVAGE–FIBROUS TRANSITION, FIBROUS FRACTURE AND IMPACT TESTING

8.1 Introduction

In contrast to the previous chapter, which concentrated on the single theme of cleavage fracture, the present chapter deals with several topics. Initially, the cleavage/fibrous transition is treated in general terms. The precise conditions required for the initiation of fibrous fracture are discussed in detail, so that transitions in slow-bend and in impact can be related to a material's microstructure and flow properties. Transitions in notched impact testing and in dynamic tear testing are then described, to demonstrate the interaction of mechanical and metallurgical factors.

A model is developed which shows how the temperature dependence of K_{IC} in a mild steel microstructure may be explained in terms of the specific micromechanisms of fracture. The limitations of the cleavage/fibrous micromechanisms in providing a general basis for fracture toughness are discussed, with reference to some specific failure mechanisms in steels and aluminium alloys.

8.2 The Cleavage–Fibrous Transition

So far, we have considered two 'mechanical' transitions in notched-bar testing (Section 7.3). The ductility of a specimen increases as the temperature is raised because it is necessary to develop a larger plastic zone to increase the local tensile stress below the notch by plastic constraint. At the transition temperature, T_{GY}, fracture is coincident with general yield. Above T_{GY}, the stress intensification cannot be increased by further constraint, and the reason for the occurrence of cleavage fracture after general yield must be attributed to the contribution of strain-hardening in the region below the notch[1]. This increases the flow stress and we may write for the fracture criterion:

$$Q_{GY} \cdot (\sigma_Y + \Delta\sigma) = \sigma_F \tag{8.2.1}$$

where $\Delta\sigma$ is a function of strain, and where σ_F may also be increased by the general strain in the lattice (dislocation pile-ups are shorter, the effective surface energy is increased because the crack must propagate from sub-grain to sub-grain, and so on). The introduction of strain-hardening in a generally yielded specimen demands that the applied load is increased, and at a critical value, dependent on the geometry of the testpiece and on the strain-hardening properties of the material, gross-section yielding may occur. A typical value for deeply notched mild steel testpieces has been given as 1.25 times the general yield load. The relaxation of triaxial stresses produced by this yielding makes it virtually impossible to develop sufficient tensile stress to propagate a cleavage crack and the fracture load rises steeply with increase in temperature to the point where sufficient strain has been produced at, rather than far below, the notch root to initiate fibrous fracture. In such a situation, we expect the temperature at which gross relaxation occurs, T_W, to approximate closely to that at which evidence of fibrous fracture at the notch root is first obtained. We take this latter temperature to be the cleavage/fibrous transition temperature T_F. The fracture appearance transition temperature (50% fibrous, 50% cleavage) is, of course, substantially higher.

In general, the onset of fibrous fracture demands that a critical strain level is achieved at the notch root. Whether or not T_F is greater or less than T_W, or even T_{GY}, depends on the relative ease of producing fibrous fracture (strain at the notch root) compared with cleavage fracture (tensile stress below the notch root). As far as the macroscopic fracture behaviour (T_{GY}, T_W) is concerned, it is simply a question of whether the plastic zone size associated with the critical stress or strain can be accommodated before general yield, between general yield and gross yield, or only after gross yield. In the last case, T_F will usually be close to T_W, because the relaxation of triaxial stresses demands a rapid increase in general strain level. We have discussed, in previous sections, the factors controlling the micro-mechanisms of cleavage fracture. To decide how microstructure affects the position of T_F, it is now necessary to examine the micro-mechanisms of fibrous fracture at a stress concentrator.

8.3 The Initiation of Fibrous Fracture

In contrast to the large amount of theoretical and experimental information available on the production of cleavage fracture in notched bars, rather few studies have been made on the conditions leading to fracture initiation by fibrous mechanisms. The general behaviour usually involves the formation of voids around inclusions or second-phase particles and the subsequent growth of these voids in directions transverse, as well as parallel, to the applied tensile stress to give final coalescence. This growth stage may be affected by the hydrostatic component of the stress state, because the presence of transverse principal tensile stresses facilitates the lateral growth of the voids.

The situation in uniaxial tensile specimens is not particularly easy to analyse. The initiation of a void around a particle depends very much on the degree to which that particle is bonded to the matrix. For an inclusion, such as manganese sulphide in steel, the bonding is negligible and voids may 'initiate' and grow, at least in the direction of the applied tensile stress, at very low plastic strains. However, even though inclusions may not be bonded chemically, it is quite common for void initiation to be made difficult by the presence of 'tessellated' stresses (arising from the differences in thermal contraction between inclusion and matrix on cooling) which bind the matrix onto the particle[2]. If the stresses are tensile in the inclusions, it is possible that the inclusion will break before the interface separates. Other inclusions, such as copper oxide in copper, may wet (be bonded to) the matrix. Particles such as carbides or nitrides in steels are bonded quite strongly to the matrix and voids can be initiated only by the high local stresses produced when many prismatic vacancy loops are created at the 'north and south poles' of a particle. These loops derive from the 'glissile' types of loop left surrounding the particle as a result of the precipitate hardening mechanism[3]. Large plastic strains are then needed before voids initiate. The particles must be somewhat incoherent for this mechanism to be viable, because coherent particles are simply cut by the slip bands. A size effect on initiation can arise if particles are small and it becomes more easy for dislocations gliding through the matrix to cross-slip out of the particle's region of influence rather than tangle round it. Higher matrix strains are therefore needed to initiate a void. The effect is enhanced if the particles are more strongly bonded to the matrix when they are small.

There is consequently a large variation in the strain needed to initiate a void. In steels, voids are likely to initiate around inclusions at much lower strains than those at which they initiate around carbide or nitride particles and they initiate more easily around large, than around small, particles.

8.4 The Growth of Voids

The growth of voids to final coalescence has been the subject of several theoretical models. McClintock[4] considers the case of the expansion of cylindrical holes, with longitudinal axes parallel to the X_3 direction, subject to generalised plane strain. The holes have initial diameter $2r_2^0$ and spacing l_2^0 in the X_2 direction. Coalescence in the X_2 direction occurs when $2r_2 = l_2$ where r_2 and l_2 are the final values of radius and spacing. At any given instant, the 'damage', associated with partial void coalescence is given by:

$$d\eta_{32} = \frac{d\ln(2r/l)}{\ln(l_2^0/2r_2^0)} \tag{8.4.1}$$

A solution, assuming viscous flow properties, enables the damage rate to be calculated for a given equivalent strain, $\bar{\epsilon}^\infty$, when transverse stress components $\sigma_{11}{}^\infty$ and $\sigma_{22}{}^\infty$ are applied to the system. Approximating the effects of strain-hardening through the expression

$$\bar{\sigma} = \sigma_1 \bar{\epsilon}^n \tag{8.4.2}$$

the damage rate is given by:

$$\frac{d\eta_{32}}{d\bar{\epsilon}^{\infty}} = \frac{\sinh\left[(1-n)(\sigma_{11}^{\infty} + \sigma_{22}^{\infty})/(2\bar{\sigma}/\sqrt{3})\right]}{(1-n)\ln(l_2^0/2r_2^0)} \tag{8.4.3}$$

For the particular case where the ratios of stress components are held constant, this equation may be integrated to give for the fracture strain,

$$\bar{\epsilon}^{\infty} = \frac{(1-n)\ln(l_2^0/2r_2^0)}{\sinh\left[(1-n)(\sigma_{11}^{\infty} + \sigma_{22}^{\infty})/(2\bar{\sigma}/\sqrt{3})\right]} \tag{8.4.4}$$

If r_2^0 is large or l_2^0 is small, it is clear that the subsequent fracture strain will be small. If we substitute a typical value, $n = 0.5$, for the work-hardening rate and take values of $(\sigma_{11}^{\infty} + \sigma_{22}^{\infty})/(2\bar{\sigma}/\sqrt{3})$ equal to 1 and 2 respectively, it is clear that doubling the transverse stress component reduces the fracture strain by a factor of some two and a quarter. Reducing n from 0.5 to 0.2 in the latter situation further reduces the fracture strain, by approximately 17%. Thus, large volume fractions of voids, high transverse stresses and low work-hardening rates all promote low-strain fracture.

Similarly, for the radial expansion of a cylinder in the plane transverse to X_3, when subjected to a radial stress σ_{rr}, Rice[5] has obtained the expression

$$\frac{\dot{r}^0}{r^0} = \frac{1}{2}\dot{\epsilon}_{33}\left[\sqrt{3}\sinh\left(\frac{\sigma_{rr}^{\infty}}{\tau_Y}\right) - 1\right] \tag{8.4.5}$$

where $\dot{r}_0$ is the rate of increase of radius with time, $\dot{\epsilon}_{33}$ is the strain rate in the X_3 direction and τ_Y is the shear yield stress. Again, if the transverse stress, σ_{rr}^{∞}, is large, the rate of increase of void radius is high. The stresses ahead of a sharp crack tip are: $\sigma_{11} = 2\tau_Y(1 + \pi/2)$, $\sigma_{22} = \pi\tau_Y$. Taking an average value for σ_{rr}^{∞} as $\tau_Y(1 + \pi)$, we obtain, by substitution in equation 8.4.5

$$\frac{\dot{r}^0}{r^0} = 26.6\,\dot{\epsilon}_{33} \tag{8.4.6}$$

which suggests a very rapid rate of enlargement ahead of the crack.

Rice and Tracey[6] have considered the growth of an isolated spherical void in a remotely uniform stress and strain-rate field. The sphere has initial radius r^0 and the remote strain field comprises a tensile extension at the rate $\dot{\epsilon}$ in the X_3 direction, with contractions at the rate $-\frac{1}{2}\dot{\epsilon}$ in the X_1 and X_2 directions. This situation corresponds to a state of simple tension for an incompressible material. A Mises material is chosen for analysis. The relative rate of void expansion, $D = \dot{r}^0/\dot{\epsilon}r^0$, is shown as a function of σ^{∞}/τ_Y in *Figure 8.1* where σ^{∞} is the remote mean normal stress and τ_Y is the shear yield stress. For large values of σ^{∞}/τ_Y (high triaxiality) the shape change of the void is negligible compared with the dilatational growth, which can be expressed in terms of σ^{∞}/τ_Y in the analytic form:

$$D = \frac{\dot{r}^0}{\dot{\epsilon}r^0} = 0.283\exp\left(\sqrt{3}\,\frac{\sigma^{\infty}}{2\tau_Y}\right) \tag{8.4.7}$$

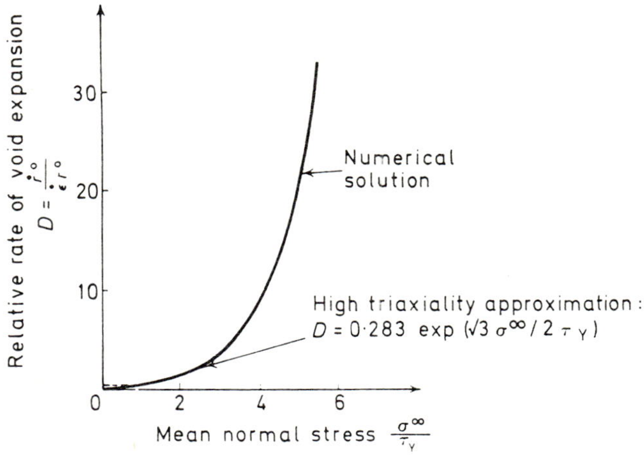

Figure 8.1 Dependency of relative rate of void expansion on mean normal stress (after Rice and Tracy[6])

This expression in fact fits the computed curve extremely well, even at low triaxiality, and is indistinguishable from it for values of σ^∞/τ_Y greater than 1.5 ($D > 1$). It is clear that the model indicates that even small increases in triaxiality produce very large increases in the rates of void expansion.

8.5 Thomason's Model

A rather different model is employed by Thomason[7] to describe void coalescence. As shown in *Figure 8.2* he takes an initially square array of square holes in a matrix whose flow behaviour is rigid/plastic. The deformation is assumed to be plane strain and is usually composed of two parts. If the voids are widely spaced, it is easier to deform the body as a whole, by gross-section yielding, than it is to produce internal necking between the voids. The effect of applying a tension is to draw the voids out in the X_1 direction and to bring their centres closer together in the X_2 direction. Eventually, the voids are spaced sufficiently closely for localised internal necking between them to become possible. Final void coalescence then occurs rapidly.

If the mean tensile stress in the X_1 direction, necessary to cause flow in the internal neck is σ_n, the general condition for the onset of coalescence is given by:

$$\sigma_n(1 - \sqrt{V_f}) + P < \sigma_{22} + 2\tau_Y \tag{8.5.1}$$

where σ_{22} is any tensile stress applied in the X_2 direction, P is the superimposed hydrostatic pressure and V_f is the volume fraction of voids. Coalescence cannot occur if $P > 2\tau_Y + \sigma_{22}$. For a uniaxial tension specimen, Thomason shows that the majority of the elongation comprises the uniform elongation, with only a small amount being attributable to the non-uniform internal necking.

The model is able to treat effects of volume fraction on ductility both in uniaxial tension and in the triaxial stress field ahead of a sharp crack. Up to the onset of coalescence, as given by equation 8.5.1, the change in geometry is given by:

$$\frac{a}{b} = \exp(2\epsilon_{11}) \ \sqrt{V_f}/(1-\sqrt{V_f}) \tag{8.5.2}$$

where ϵ_{11} is the uniform strain in the X_1 direction. This strain ϵ_{11} increases until the neck geometry a/b, and the corresponding constraint factor $\sigma_n/2\tau_Y$, satisfy the condition for the coalescence of cavities. Once internal necking has

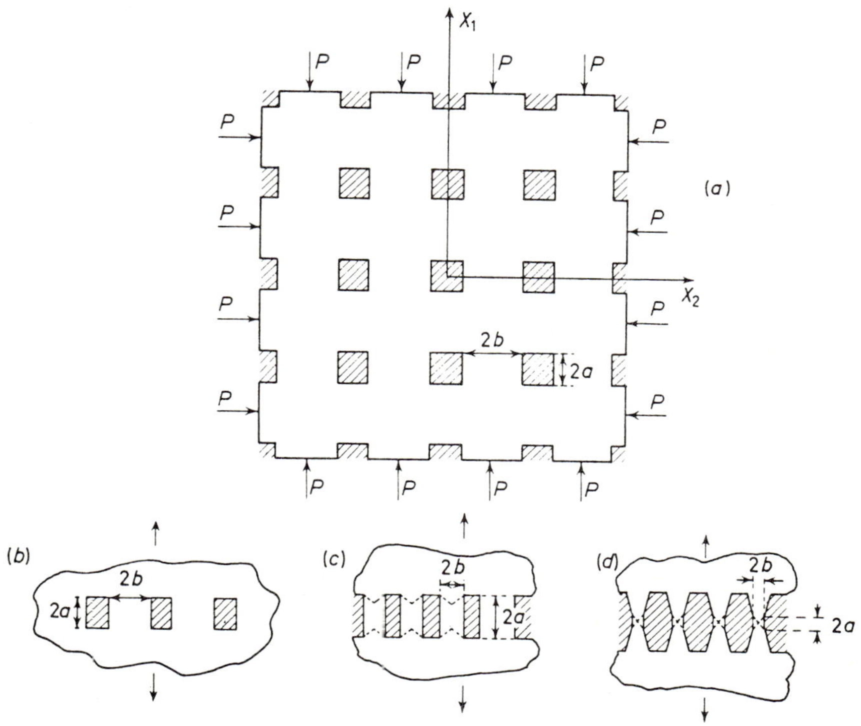

Figure 8.2 (a) Thomason's model for void coalescence. (b), (c), (d) Progressive stages of deformation

commenced, it is assumed that the plastic flow is such as to give the maximum rate of unloading. The total relative displacement between the two faces required to produce final fracture is of the same order as the inter-void spacing at the onset of internal necking which, for low volume fractions, implies that the macroscopic strain to the onset of coalescence may be taken as the total strain to fracture. Thomason refers to this strain as a post-instability strain, because he considers that a specimen must have been strained to greater than its UTS before any tendency to form internal necks exists. An application of his model to tensile tests is described in the following section.

The crack situation[8] is treated by taking a pseudo 'plane-strain' value of crack opening displacement, 2δ (equivalent to δ in earlier chapters) as:

$$2\delta = \frac{(1-\nu^2)K_I^2}{2E\sigma_m} \tag{8.5.3}$$

[cf. equation 3.13.12 with a substitution of $(1-\nu^2)$] where σ_m is the average value of the tensile flow stress in the plastic zone ahead of the crack, say $4k$, where k is a work-hardened shear flow stress (cf. preceding equation 8.4.6). Thus equation 8.5.3 may be re-written:

$$\delta = (1-\nu^2)(K_I/2k)^2/4(E/2k) \tag{8.5.4}$$

The plastic strain must decrease to zero at the plastic/elastic boundary. The shape of the plastic zone is taken as a wedge of width $2r_0$ at the crack tip and zero at the plastic/elastic boundary, $x_2 = 1$ (*Figure 8.3*). If the width at any point x_2 is $2r_x$, it is assumed that the displacement is given by:

$$\delta_{1(x)} = \delta\frac{r_x}{r_0}\left(1-\frac{x_2}{l}\right) \tag{8.5.5}$$

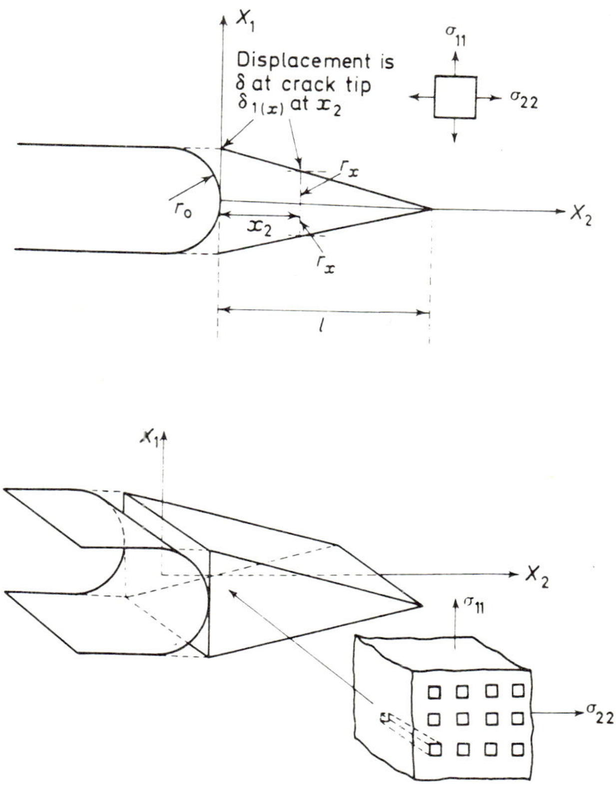

Figure 8.3 Relationship between void coalescence and fracture toughness

The logarithmic strain at x_2 is given by:

$$\epsilon_{\text{true}} = \ln(1 + \epsilon_{\text{nom}}) = \ln\left(1 + \frac{\delta_{1(x)}}{r_x}\right) = \ln\left[1 + \frac{\delta}{r_0}\left(1 - \frac{x}{l}\right)\right] \quad (8.5.6)$$

Using the theoretical relationship between post-instability fracture strain, volume fraction and transverse stress component, σ_{22}, derived for tension, it is possible to substitute in equations 8.5.2 and 8.5.4 to obtain:

$$\frac{K_{\text{IC}}}{\sqrt{3k}} = \left\{r_0\left[4.62\left(\frac{E}{\sqrt{3k}}\right)\frac{(e^{\epsilon_{\text{I}}} - 1)}{(1 - \nu^2)} + 8.81\tan\left(\frac{\pi}{2}\left(1 - \frac{V_{\text{f}}}{0.09}\right)\right)\right]\right\}^{\frac{1}{2}}$$

for $0.02 \leqslant V_{\text{f}} \leqslant 0.09$.

The main problems with this expression are that, first it is necessary to know the root radius of the crack tip, which must be related to the fracture displacement and that, secondly, the instability in the crack tip region is identified with that in uniaxial tension. The importance of high volume fractions in reducing crack-tip ductility is again apparent.

8.6 Experimental Observations on Fibrous Fracture

As indicated previously, the amount of definitive experimental work carried out on the fibrous fracture process is not large. In tensile tests, confusions may arise, if it is not clear whether the events are related to void initiation or to void growth. We may distinguish between two limiting situations.

For the case of a matrix containing only non-wetting inclusions, which does not produce interfacial thermal stresses on cooling, voids are present from the beginning of the test. Before internal necking can occur, however, they must be brought into close proximity. Some lateral movement is obtained during the uniform elongation of the specimen, but if the initial spacing is large, it will be insufficient to produce the critical spacing for internal necking between the voids. When a critical elongation (rate of work-hardening) has been reached, macroscopic necking of the specimen begins. This introduces radial tensile stresses in the transverse plane (see also Section 2.12), but, until the external neck deepens and these stresses become large, Thomason's model predicts that the internal necking process is primarily dependent on the void spacing. Experimental work by Baker[9] on the formation of voids around copper oxide particles in a copper matrix associated the onset of internal necking with the development of an external neck, but the technique used was such as to exaggerate the influence of even small radial stresses. Voids grow laterally under decreasing load in the triaxial stress field which exists in the centre of the neck. Whether the coalescence of voids at a particular point is a catastrophic process or not depends on the rate of release of elastic energy of the testing system. In a really hard (displacement-controlled) system, achieved perhaps by pulling a strong spring in parallel with the testpiece, it is possible to control the growth of the central 'cup' fairly well. As the cup becomes larger, the geometry changes to that of an internally cracked specimen, and the piece breaks by a combination of tension and shear on the 'cone' surface.

The other limiting type of particle is the carbide or nitride well-bonded to the matrix. Here, particularly if the particle is small, a large amount of matrix strain is necessary before voids initiate by the action of the concentrated stresses developed at the particle/matrix interface by the dislocation tangles left surrounding the particle.

The strain may indeed be so large that substantial macroscopic necking has occurred before initiation. Voids are then produced, more or less simultaneously, around particles which are now quite closely spaced, in a triaxial stress field. Under these conditions, void linkage is likely to be rapid and it may be difficult to arrest the central 'cup'. The fracture profile is similar to that in the previous situation, but the 'dimples' in the 'cup' are much shallower and more closely spaced. In such a catastrophic situation, it is difficult to distinguish between initiation and growth of the voids.

Most commercial steel microstructures contain a mixture of inclusions, fairly coarse carbides or nitrides (such as grain-boundary carbides or particles obtained by interactions of the interstitial elements with grain-refining elements such as Nb or Al) and perhaps fine carbides, produced by tempering a martensitic matrix. The sequence of events will then be as follows. Voids are formed round the inclusions at low strains and during the uniform elongation and early stages of necking draw closer together. Internal necking between the inclusions may then occur, with the carbides initially moving simply as part of the continuous matrix in this new 'macroscopic' (in terms of carbide size and spacing) neck. Eventually, the strains in the internal neck become so large that voids are initiated around the carbide particles. The final fracture surface is then composed of a number of rather large dimples, formed around the inclusions, with very much finer dimples, deriving from the carbides, where the inter-inclusion necks have been drawn out most extensively. The fairly large particles needed to restrict austenite grain growth (and hence situated on prior austenite boundaries) can form voids more easily than tempered carbides and it is sometimes possible to obtain a surface which shows dimples around inclusions, together with what appears to be a ductile intergranular fracture, caused by void linkage between the particles on the prior austenite boundaries.

The most commonly quoted experiments on the effect of volume fraction of particles on reduction of area in a tensile test are those carried out by Edelson and Baldwin[10]. Their results are given in *Figure 8.4* and show clearly that for a whole range of types and sizes (approx. 1–200 μm) of particles or voids in a copper matrix, volume fraction was the main single factor determining ductility. The work-hardening exponent, n, determined from an expression of the type: $\sigma = K\epsilon^n$, also decreased with increase in volume fraction of the dispersed phase. A similar effect has been found by Turkalo and Low[11] on the effect of volume fraction of carbide particles on the ductility of steels. Edelson and Baldwin discounted the argument that their results were similar for all types of inclusion because all were minimally bonded to the matrix, by reference to these effects in steel, but a closer comparison of the data reveals that the steels were, in fact, almost an order of magnitude more ductile for a given volume fraction (see *Figure 8.4*). The situation is very similar to that obtained for steels by Gladman[12] (*Figure 8.5*). Here, effects of particle shape are shown in addition to those of

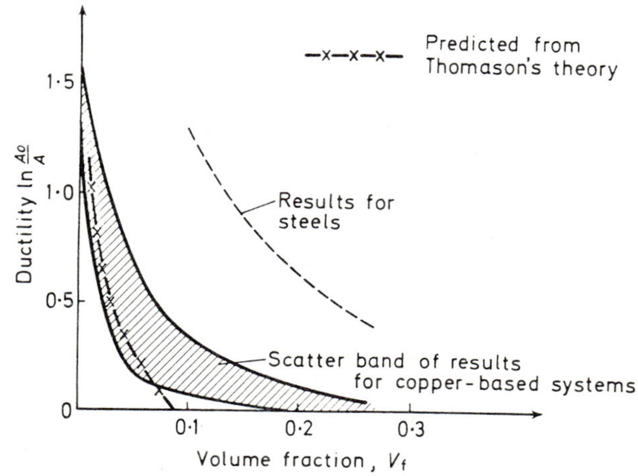

Figure 8.4 Dependence of ductility on volume fraction of second-phase particles. (Copper-based results from Edelson and Baldwin[10]; steel results from Turkalo and Low[11])

interface bonding. Disc-shaped sulphides are more detrimental to ductility than sulphides elongated parallel to the tensile axis. Pearlitic carbides, although plate-like, are less detrimental because void initiation occurs by particle cracking and this is stress dependent, so that a significant strain is required before voids

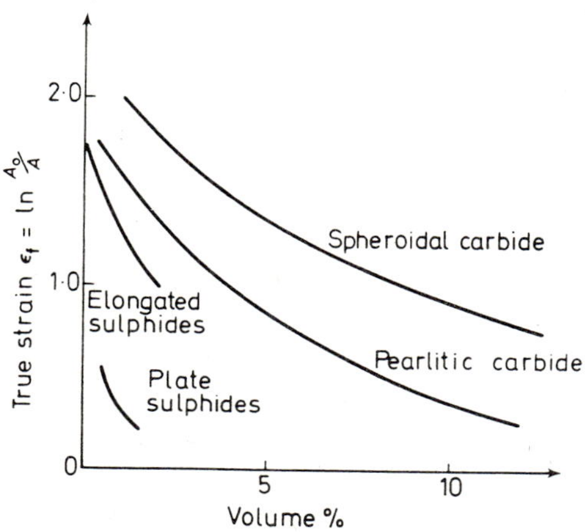

Figure 8.5 Effects of type and shape of a second-phase particles on the tensile ductility of steel (after Gladman, Holmes and McIvor[12])

initiate. Spheroidal carbides increase the ductility still further, because very large true strains (> 0.7) are required to initiate the voids.

The lack of bonding in Edelson and Baldwin's experiments may be further demonstrated by the good agreement obtained between the predictions of Thomason's model, derived for the growth of voids, with their results on the effect of volume fraction on ductility, see *Figure 8.4.* Further, results obtained by Palmer, Smith and Warda[13] indicate that, in internally-oxidised copper–silica alloys, where the interface bond is strong, the strains to fracture are very much increased.

The sequence of events leading to the initiation of fibrous fracture ahead of a crack-like stress concentrator is best described on the basis of the observations made on the mild steel/manganese sulphide system by Smith and Knott[14], which have been partly described in a previous chapter (Section 6.3). Here in normalised material, the effective gauge length of a fatigue crack was found to equal the mean (centre to centre) inclusion spacing (0.046 mm) and the fracture strain was some 80%. The growth of a fibrous crack by internal necking between inclusions (and not carbide particles) is shown in *Figure 8.6a.*

Observations of the dimple spacing on scanning electron micrographs of the fracture surface, which are essentially planar projections of rather rough surfaces, e.g. *Figure 8.6b,* gave a value of about 0.02 mm. By assuming that this was the final spacing at which internal necking between inclusions could start, it proved possible to estimate the 'uniform' and 'non-uniform' components of the total fracture strain. The general picture that emerges from this, and similar studies on other steels, is that the uniform component represents the strain necessary to draw inclusions sufficiently close together (a separation of some two to three particle diameters) to promote internal necking. The non-uniform component is that given by the necking process. The uniform strain may be similar to that at which macroscopic necking occurs in a tensile test, but this seems to be fortuitous. Two extremes in pre-cracked specimens may be recognised. In one case, represented by very closely spaced inclusions, such as type II manganese sulphide, the uniform strain is negligible and internal necking between inclusions, although small, is the sole contributor to the total strain. On the other hand, in very clean material, such as a vacuum-melted steel, the uniform strain needs to be very large to draw the inclusions together[15]. It may be so great that the general inter-inclusion strain field is sufficient to produce voids around matrix carbides or nitrides, which give fine-scale linkage between the inclusions. The total ductility will eventually be limited by this fine-scale microstructure.

The situation regarding the microstructural basis of fibrous fracture is therefore complicated. Recent experimental observations indicate that the very high rates of void expansion predicted by the modified visco-elastic theories are unrealistic, because particles tend to deform with the matrix (except at their poles) until the lateral separation has decreased to a critical value. Thomason's model expresses this in terms of plastic constraint and generally appears to represent the physical situation quite closely. However, the onset of macroscopic necking in a tensile specimen does not seem to be critically related to the ductile fracture events on the local scale, even in a tensile specimen (although the radial tensile stress in the neck promotes lateral void growth) and bears little relevance to the events near a stress concentrator.

The critical factors in determining ductility are therefore the critical distance of approach and the strength of the particle/matrix bonding.

To summarise, the critical crack tip ductility at the initiation of fibrous fracture depends on several factors. It decreases with increase in inclusion content (decrease of inter-particle spacing for a given size of particle), with decrease in

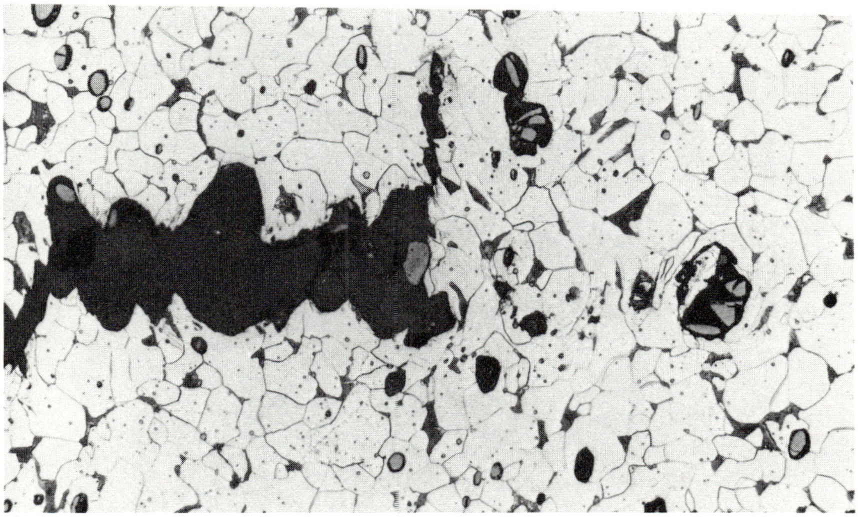

(*a*)

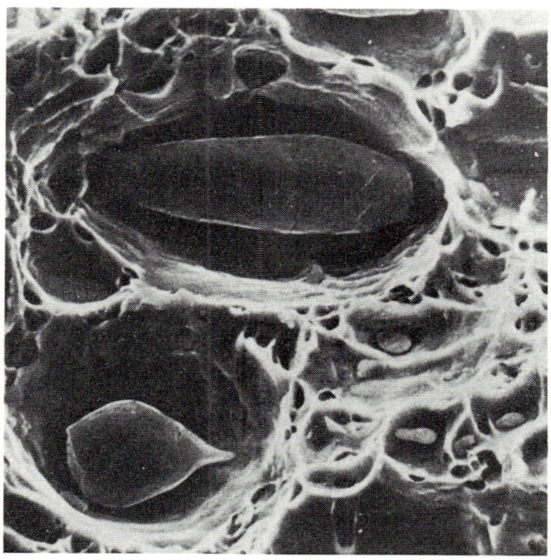

(*b*)

Figure 8.6 (*a*) *Section through growing fibrous crack showing formation of holes around inclusions, but not around pearlite* (× 240). (*b*) *Fracture surface, showing void formation around inclusions* (× 800) (*both courtesy of R. F. Smith*) work-hardening rate and with increase in transverse stress. The role of yield stress is not clear, although an increase in yield stress will increase the magnitude of the hydrostatic stress component around the stress concentrator.

8.7 Transitions in Slow-bend and in Impact

The general features of the cleavage/fibrous transition in slow notch-bend have already been described. The situation in small specimens is probably best illustrated by showing how the notch-tip strain or COD varies with temperature, in comparison with the general yield and fracture load curves (see *Figure 7.1*). Below T_{GY}, the critical COD value is low, because fracture initiates by cleavage mechanisms and the plastic zone is obviously smaller than the width of the specimen. Between T_{GY} and T_W, the COD increases more rapidly (and is directly proportional to the angle of bend of the specimen) but not until gross yielding has occurred, above T_W, does it rise really steeply. The fracture is cleavage until the COD is sufficient to produce fibrous fracture. The further increase in COD with temperature depends on the growth of the fibrous thumbnail prior to the onset of instability by cleavage fracture, until a limit is reached where the fracture is 100% fibrous.

A similar sequence is found in impact tests, where the hammer and tup have been instrumented to provide readings of dynamic loads[16]. In mild steel, the main effect of high strain-rate is to raise the yield stress at any given temperature, because less time is available for thermal fluctuations to lower the stress needed to glide dislocations through the matrix (the temperature dependent component σ_i'' of the friction stress, σ_i). A limit is reached at very high strain rates where all strain rate sensitivity of the yield stress is lost[17]. As might be expected the limiting strain-rate necessary to produce this 'cut-off' effect decreases with temperature. The limit may correspond in all cases to the onset of twinning as the mode of general plastic deformation, but this point does not appear to have been exhaustively investigated. In high strength steels, both the temperature dependence and strain-rate sensitivity of the yield stress are proportionately reduced, because far more of the friction stress derives from the temperature independent component σ_i' (long-range stress fields). The microscopic deformation modes of such steels under very high rates of loading are not well documented.

The effect of increasing the strain-rate on the various transitions at T_{GY}, T_W and T_F is to move them all to higher temperatures. Because the uniaxial yield stress at a given temperature is greater, a smaller plastic zone is required to develop a tensile stress below the notch sufficient to propagate a cleavage crack nucleus. The effect on T_{GY} is indicated schematically in *Figure 8.7*, assuming that the critical value of cleavage fracture stress does not vary with strain rate. The absolute shift of transition temperature, ΔT_{GY}, depends on its initial value, because the dependence of yield stress on temperature is not linear. If T_{GY} were initially at a low temperature where the yield stress curve was steep, a given increase in yield stress would produce a much smaller absolute shift in temperature than if T_{GY} were initially high. This point pertains particularly to the use that has been made of shifts in transition temperature to assess absolute amounts of grain-boundary embrittlement (see Section 8.10).

Although twinning can occur at low temperatures and high strain-rates, experiments on normalised and annealed mild steel indicate that slip-initiated cleavage fracture occurs at T_{GY}, even under impact conditions. Knott's analysis[18] of

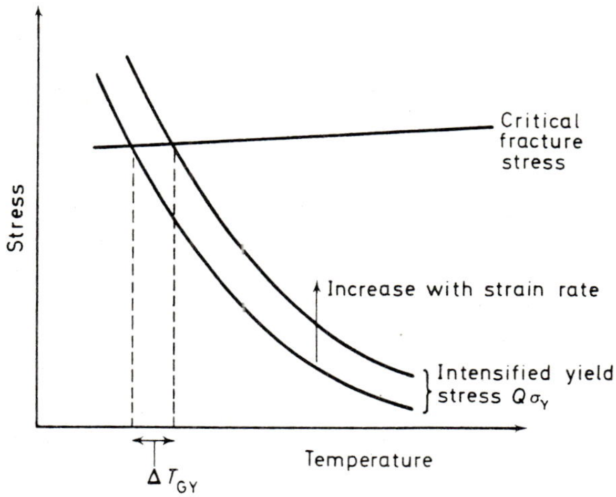

Figure 8.7 Schematic effect of increasing strain-rate on the ductile/brittle transition temperature, T_{GY}

Wilshaw's experimental results[19] showed that all could be explained in terms of a single critical cleavage stress, over a range of crosshead speeds spanning more than 10^6. Oates[20] has determined direct values of cleavage fracture stress for mild steel over a range of 10^4 (see *Figure 7.16*). Only for the highest strain-rate and lowest temperature did the slip-initiation model fail to apply. For a manganese steel of identical grain size, but containing finer grain boundary carbides, the temperatures were generally much lower and twins were involved in cleavage crack formation at T_{GY}.

The usual information gained in an impact test (e.g. *Figure 8.8*) is a composite value of the energy needed to cause fracture. At low temperatures, up to the initiation of fibrous fracture at T_F, this energy is fairly representative of the work required to initiate fracture, although it includes work expended in indentation of the specimen at the loading points. If this could be removed, e.g. by preloading a bar, before notching or fatigue cracking, so that the loading points were work-hardened and did not deform plastically when the bar was broken, it ought to be possible to relate the work done in a cracked, rather than notched, impact specimen directly to a 'J_{IC}' value (see Section 6.9) which could be related to K_{IC}. Because the yield stress decreases and the critical COD increases, the energy absorbed increases only weakly with temperature, until T_W is reached, when the relaxation in constraint allows such large displacements to occur that the energy rises sharply. Thus, the nil-ductility temperature (NDT) is close to T_W and, for specimens of constant geometry and similar work-hardening rates, comparative values of NDT will usually rate materials in the same order as one would obtain from accurate values of T_{GY}. The correlation would have to be applied very carefully in cases where the NDT values covered a wide range, because the difference between T_{GY} and NDT varies with the steepness of the yield stress/temperature curve.

It is clear that T_W, and hence NDT, varies substantially with specimen geometry. Shallow notches allow gross relaxation to occur rapidly after, or even before, constrained general yielding (Section 7.6). In bend specimens which are relatively deep compared with their thickness (B) it is possible to obtain stress relaxation by through-thickness yielding, rather than by gross-section yielding.

The other transition temperature commonly used in conventional impact testing is the fracture appearance transition temperature (FATT), representing 50% fibrous, 50% cleavage fracture (*Figure 8.8*). This is greater than the value,

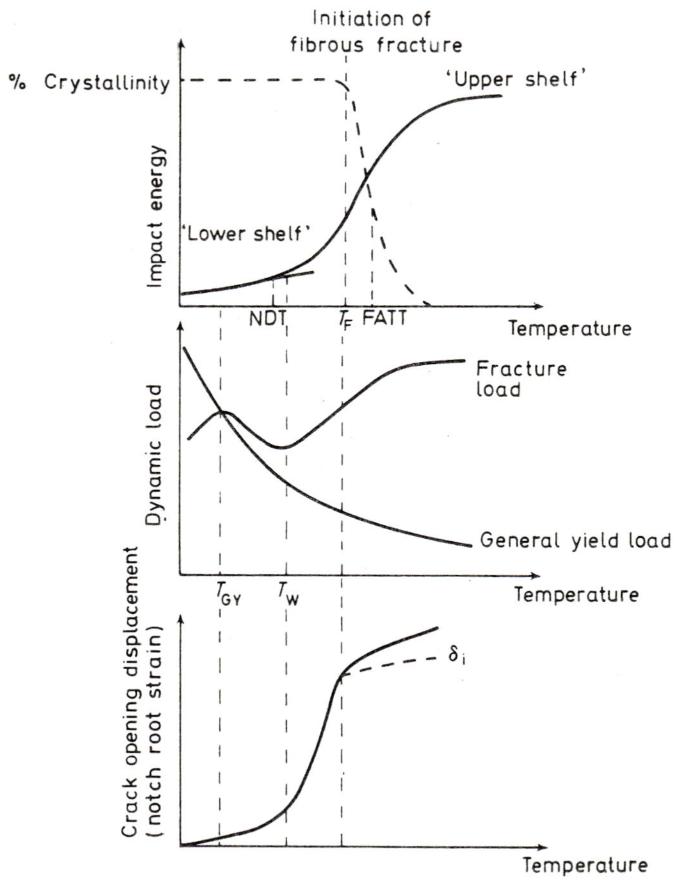

Figure 8.8 Schematic relationship between quantities in impact testing

T_F, that we have taken to indicate the temperature at which the first trace of a fibrous thumbnail is seen around the notch root on the fracture surface. The amount by which it is greater depends critically on the growth characteristics of the fibrous fracture and on the criterion which governs the transition from fibrous to cleavage fracture.

In detail, the growth characteristics are complicated, because the impact loading of a specimen containing a growing crack is variable. The time element is, however, virtually removed, because the loading is fast. We can approximate the increase in energy to the extra displacement involved in growing the crack to a given length (see *Figure 6.7*) and conclude that the reason that the energy increases is simply because the length of fibrous crack needed to promote cleavage increases with temperature. The magnitude of the energy increase obviously depends on the form of the COD/crack growth curve and we may generally expect that , if the notch root has not been work-hardened before testing, a fibrous fracture that initiates at low COD will show little increase of COD with growth.

The reason for the change from fibrous to cleavage fracture has not been fully established. Initially, it was thought that the fibrous fracture accelerated as it grew, so that the local strain-rate in the region ahead of the advancing crack tip increased and eventually reached a level which elevated the yield stress, and hence the tensile stress, in that region sufficiently to allow cleavage cracks to propagate. An example of the change from fibrous to cleavage fracture is shown in *Figure 8.9*. However, *displacement-controlled* tests by Smith[14] have shown

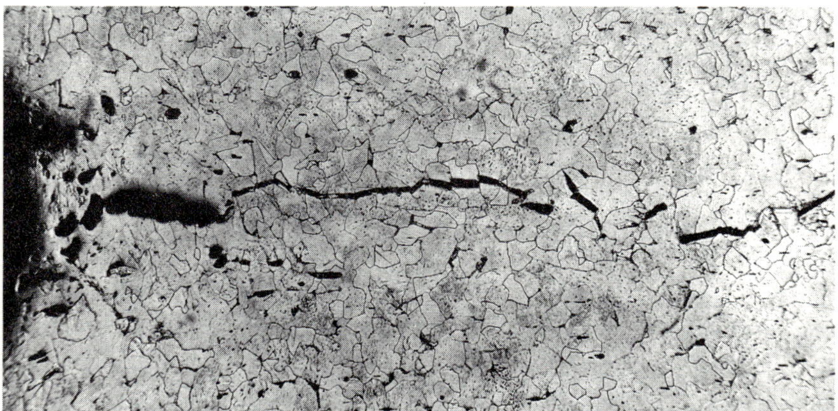

Figure 8.9 Change in fracture mechanism from fibrous to cleavage as a crack grows (mild steel, room temperature) (x 42)

that the tendency of a growing fibrous fracture is initially to slow down, because the tunnelling forward of the plane-strain fracture in the centre of the piece is restrained by the need to develop high crack tip displacements before the side ligaments (shear lips) can fracture (see Section 6.4). In his tests, it seems that the effect of a growing fibrous fracture is to increase the constraint, and hence the stress intensification Q (see Section 7.6), in the unfractured part of the specimen, because this section deforms under deep-notch, rather than shallow-notch, conditions. It is not clear that the same results would be obtained from load-controlled tests, where accelerations might well be obtained. The actual form of the loading programme during an impact test is therefore very important.

Once a growing crack situation has been obtained, it is not possible to relate the total energy values to J_{IC}, because the displacement is produced both by plastic work and by increased compliance. The COD value at the original notch tip for this situation is a combination of a general base level displacement, δ_i, obtained when the first element of fibrous fracture was initiated at the notch, together with incremental displacements representing the integrated effect of increasing the base level displacement to δ_i for all other elements which fracture. At present, experimental curves of the type shown in *Figure 6.7* are the best way of visualising the situation. The increasing COD cannot be easily related to the increasing energy value at this stage, except by experimental, or empirical, correlation, because it is no longer proportional to the angle of bend of a specimen.

To summarise, we may expect to recognise general features of impact transition curves as follows. The 'lower shelf' energy level is associated with cleavage fracture and is therefore expected to be relatively high for material of fine grain size and fine carbide distribution. The transition, T_{GY}, is not determinable as such, but the NDT will, in general, relate to T_W or T_{GY} in a comparative manner, as a measure of the particular balance of yield stress and cleavage fracture stress for a given material which gives rise to failure at a specified value of specimen deformation (e.g. gross-section yield). The NDT is therefore critically dependent on specimen geometry. The temperature, T_F, indicates when fibrous fracture is first obtained: a COD measured at T_F would give the value of δ_i at T_F directly. This may not be the same value of δ_i as that obtained at room temperature: it is, in general, lower. The FATT is representative of 50% fibrous fracture. Its closeness to T_F relates to the conditions for changing from fibrous to cleavage fracture: we may expect that these are also geometry dependent; markedly so, if the effect arises from notch-deepening and constraint. The positions of T_F and FATT depend on how easy it is to initiate and grow a fibrous fracture as opposed to a cleavage fracture. The 'upper shelf' level depends entirely on the characteristics of fibrous initiation and growth. In terms of microstructure, low 'upper shelf' levels are produced by high contents of closely-spaced, non-wetting inclusions, by low rates of work-hardening and by high yield strength levels.

The impact transition curve for a clean, annealed mild steel will therefore exhibit a low 'lower shelf' and a high NDT (T_W, T_{GY}), because cleavage fracture is easy to produce; a steep transition, governed by the gross relaxations at NDT (T_W); and a high 'upper shelf' value. Additions of sulphide inclusions to the steel will reduce the 'upper shelf', but will not affect the transition temperature. A similar effect can be obtained by testing a constant inclusion content in different orientations[14] (see *Figure 8.10*), because the interparticle spacing is reduced in the transverse orientation. The transition curve for an equally clean medium strength steel is far less discontinuous in nature. The lower shelf is higher (because the refinement of the microstructure more than compensates for the increase in yield strength), the NDT is lower, and the upper shelf is lowered by the increased yield stress and decreased work-hardening rate.

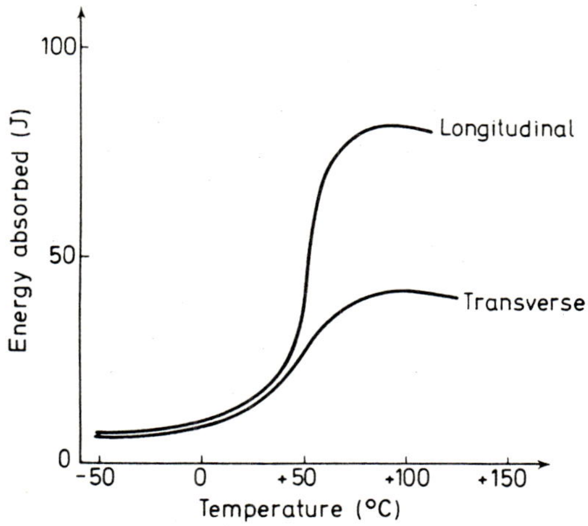

Figure 8.10 Effect of testpiece orientation on the 'upper shelf' level. Data for a free-machining mild steel. Note that the transition temperature is virtually unaltered (after Smith and Knott[14])

All these comments are clearly semi-quantitative at best. Some energy values may be related to more useful parameters (COD, J_{IC}) in certain cases, but regions of applicability have to be clearly defined. The two transition temperatures, NDT and FATT, are likely to vary with geometry in ways which just cannot be estimated from conventional impact tests. This major fault of the method has led to the establishment of more realistic impact testing techniques.

8.8 'Drop-weight' and 'Dynamic Tear' Testing

The main work which has been done to extend impact testing to geometries which are more representative of service application has been carried out by Pellini and co-workers at the Naval Research Laboratory[21–23]. The standard design of dynamic tear (DT) specimen is that of a beam containing a sharp crack-like stress concentrator which has been deliberately embrittled, for example, by using a Ti-embrittled electron beam weld as a crack starter in order to remove the 'initiation stage' of fracture, by subjecting the specimen to a fast-running crack when the weld breaks. Testpieces are of full-plate thickness (from 16 to 305 mm) and the standard dimensions for different thicknesses are indicated in *Table 8.1*. It is clear that the variable ratios of notch depth to specimen depth and depth to thickness dimensions cannot be regarded as entirely satisfactory, in the light of our previous discussion on the importance of these parameters (Sections 7.5 and 7.6). The older drop-weight-test (DWT) specimens similarly provided an embrittled crack starter, but are different in detailed design. The impact loading system ranges from a balanced impact tester for small specimens to vast drop-weight facilities (simply, large masses descending on guide rails from a fixed height) of capacities up to 1 MJ (75×10^4 ft. lbs), for the really large pieces.

Early tests on drop-weight (DWT) specimens served to estimate the position of the 'nil-ductility temperature' (NDT), which defines the temperature below which the fracture resistance is so low that brittle plane-strain (cleavage) fractures can be initiated from small flaws under the high strain-rate conditions in the test (cf. Section 8.7). The standard definition of the position of the NDT temperature in DWT tearing has been shown to correspond to a ratio of dynamic fracture toughness, K_{ID}, to yield stress, σ_Y, of 0.5[21]. In terms of specimen thickness requirements (see Section 5.12) this implies that specimens less than 16 mm thick cannot be used to establish fracture toughness values corresponding to the NDT temperature.

Table 8.1 DIMENSIONS OF DYNAMIC TEAR (DT) TESTPIECES

Designation	*Thickness, B* (mm)	*Depth, W* (mm)	*Brittle weld or Notch depth, a* (mm)
$\frac{5}{8}$ in (16 mm)	16	41	13
1 in (25 mm)	25	120	44
2 in (50 mm)	50	203	76
3 in (76 mm)	76	203	76
6 in (152 mm)	152	305	76
12 in (305 mm)	305	381	76

In low-strength steels, the steepness of the transition from low to high toughness values over a narrow temperature range enabled a simple philosophy to be developed with respect to design against fracture. This stated that, if the operating temperature were more than 33.3 °C (60 °F) higher than the NDT, the resistance of the material to fracture would be such that fracture would occur only at stress levels greater than the general yield stress. To design against fracture at low temperatures, steels with correspondingly lower NDT temperatures would have to be used. This procedure might be expensive in terms of alloying elements and would be unduly conservative if the operating stresses were substantially less than the yield stress.

The use of higher-strength materials with less sharply defined energy transitions and the general desire to design as efficiently and economically as possible have led to a broadening of the original philosophy based on the NDT temperature. In dynamic tear (DT) testing, the whole transition curve is obtained for what is regarded as the worst case of service loading: the running-crack situation. The transition curves vary in position with specimen thickness up to a limiting thickness (75 mm in a reactor pressure vessel steel) and two curves are drawn in *Figure 8.11* to represent the behaviour of thin (16 mm) and thick (> 75 mm) specimens, with energy normalised by dividing by the fracture area. It is important to note that the NDT does not vary with thickness for a given size of starter crack (weld), because even a 16 mm specimen can provide sufficient constraint to give brittle fracture: if the crack size is scaled with thickness, however, the NDT increases with specimen size. The transition curve for the thick specimen is taken as the limiting transition temperature range (LTTR) for the brittle–

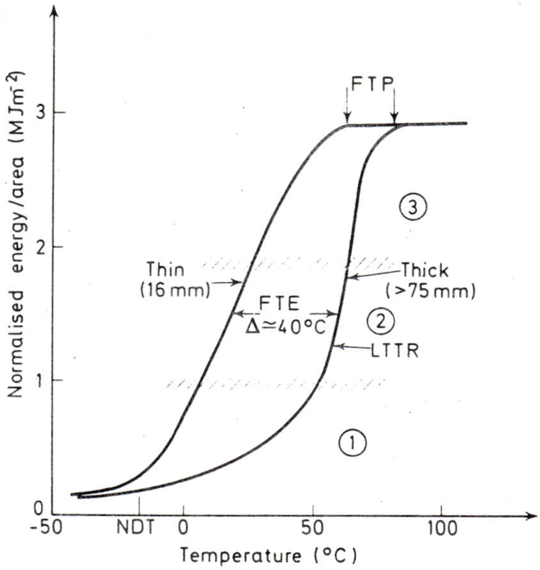

Figure 8.11 Dynamic tear test curves (after Lange[22])

ductile transition, because the combination of high strain rate (dynamic testing) and large thickness (> 75 mm) is regarded as the most severe condition likely to be met in service.

It is conventional to describe (index) the transition curves in terms of temperatures and energies. The NDT temperature has already been defined and provides the main indexing point, since it does not vary with thickness. The temperature at which the 'upper-shelf' (100% fibrous fracture) is first obtained is known as the fracture transition plastic, or FTP. The temperature corresponding to an energy which is the mean of the upper and lower shelf values is known as the fracture transition elastic, or FTE. Certain other modes (regimes) of behaviour are indicated on the diagram (*Figure 8.11*). In regime 1, fracture occurs under linear elastic plane strain (dynamic) conditions; in regime 2, loss of constraint leads to a rapid increase of toughness with temperature: the initial running crack may arrest, re-initiate by a fibrous mechanism and re-propagate by cleavage; in regime 3, material ductility is so high that the initial crack arrests and the rest of the total fracture is fibrous: at the FTP, the length of arrested crack running from the weld is essentially zero. The transition from regime 1 to regime 2 represents the plane-strain limit for dynamic conditions and, for convenience, the FTE is taken as a slightly conservative estimate of this transition.

The DT transition curve between the NDT and the FTP is equated to the Robertson crack arrest curve. In the Robertson test[24], a large plate is subjected to a uniform stress and a running crack is propagated through the plate, either in a temperature gradient or at constant temperature. For a given stress level, it is found that a critical temperature can be determined, above which the crack arrests. The complete series of critical temperatures *v*. applied stress levels defines a crack arrest curve. In relation to the DT curve, the FTE represents the Robertson crack arrest temperature (CAT) for an applied stress equal to the yield stress. The early design philosophy in low strength steels of maintaining an operating temperature of NDT + 33.3 °C (60 °F) implied, from the general shapes of the DT transition curves found experimentally, that the operating temperature coincided with the FTE and that crack propagation could not occur for service stresses less than the yield stress.

The comparison of the transition curves for thick and thin specimens (*Figure 8.11*) reveals that the main shift in position is due to an expansion of regime 1, i.e. the thicker specimens can maintain plane strain conditions to higher temperatures. Once the basic curves have been obtained, it should be possible to estimate the LTTR curve in regime 1, for a fairly similar type of steel, by making measurements on 16 mm specimens and adding some 40 °C (70 °F) to the FTE. The NDT does not vary with thickness and the curve can be indexed to these two points.

It is possible also to relate the general forms of the transition curves to the stress levels required to propagate flaws in service. These relationships are based mainly on service experience and some of the force of the DT approach is lost because it is found that, in practice, the sizes of critical flaws which initiate brittle fractures are usually such that fracture must have started under static, rather than dynamic conditions. The summary information is presented in terms of a fracture analysis diagram, as indicated in *Figure 8.12*, which plots stress *v*. temperature. Curves equivalent to the DT energy transition curves are indexed by means of the NDT and FTE. The NDT corresponds to the temperature at which fracture may be initiated by a small flaw (less than 25 mm long) under local dynamic loading of yield stress level, or to that at which a long crack propagates under minimal applied stress. The FTE represents the temperature at which long cracks can continue to propagate under yield stress loading, as in the Robertson test. Loadings greater than yield stress are not considered in detail, because these correspond to failure by gross plastic collapse rather than by fracture. Below the NDT, failure stress/flaw size relationships can be calculated using linear elastic fracture toughness techniques. These agree fairly well with service experience to give some indication of the sizes of flaws likely to propagate under applied stresses of, say, $\frac{3}{4}$, $\frac{1}{2}$ and $\frac{1}{4}$ of the dynamic yield stress at the NDT. Curves starting from these three points at the NDT may then be drawn parallel to the transition curve appropriate to a particular specimen thickness to indicate the stress level required to propagate a flaw of given size at a given temperature. The aim is to give results in a temperature region in which linear elastic calculations cannot be used and yet in which the economical factors have demanded operating conditions, of stress above, and of temperature below, the LTTR.

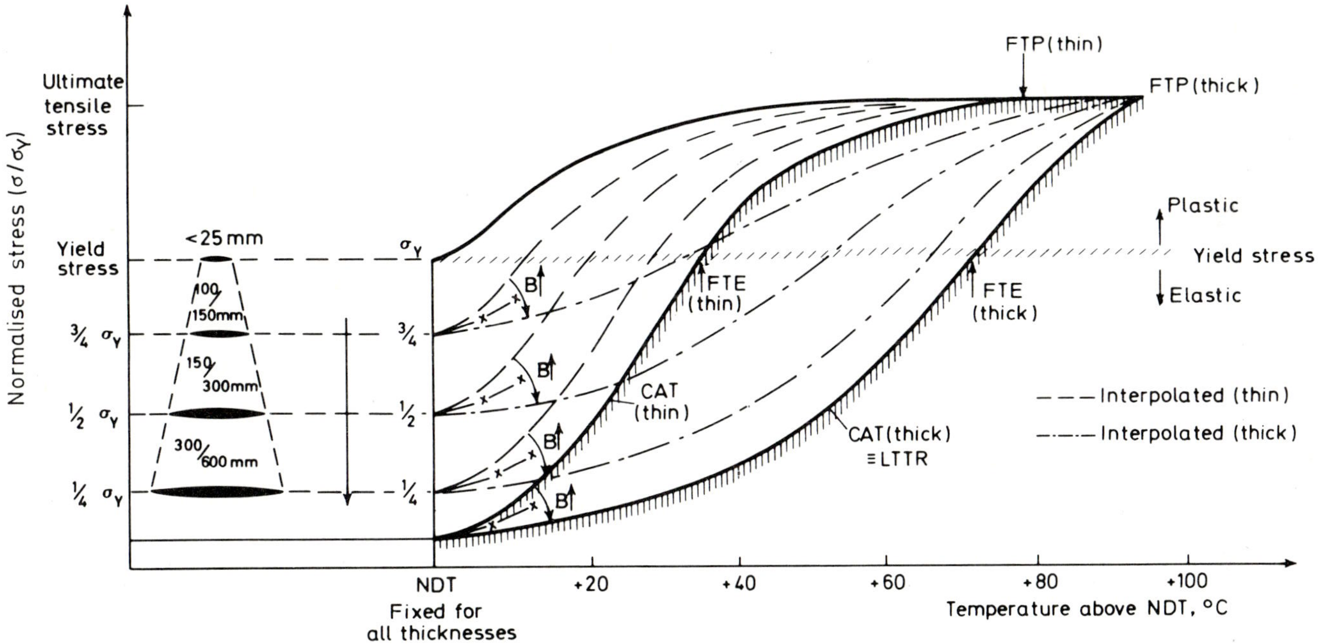

Figure 8.12 *The fracture analysis diagram (FAD) (after Loss*[23]*)*

Despite the empirical nature of many of the correlations which are made, dynamic tear testing can often adequately define the likely behaviour of steels in service and has been used, for example, in the assessment of materials for nuclear pressure vessels. It concentrates, however, on the dynamic 'metallurgical' transition from cleavage to fibrous fracture and is not easily able to treat unstable propagation by fast fibrous fracture. The importance of this is demonstrated by the behaviour of pneumatically loaded, as opposed to hydraulically loaded, vessels. In the hydraulically loaded vessel, as in the Robertson test or the drop-weight test, the applied stress on the system drops rapidly as the crack propagates and a clearly defined transition is obtained. In a pneumatically pressurised vessel, even ductile fracture (accompanied by strains of the order unity and local bulging of the walls of the vessel) propagates faster than the pressure can be reduced by leakage and a catastrophic situation can occur. With very long flaws, catastrophe can be produced by low applied stresses at temperatures higher than the LTTR[25]. The lengths of these flaws are such that they would not normally be found in plate material. Emphasis is therefore placed on the need to examine sub-critical crack growth mechanisms (by fatigue, stress corrosion, ductile tearing, etc.) in such situations.

Transitions in impact testing, whether conventional notched bar testing or DT testing, are obviously dependent on specimen geometry, but it is clearly possible to examine effects of geometrical variables, as for slow-bend testing, so that the position of a particular transition is explained simply in terms of the conditions for initiating cleavage or fibrous fracture. These are then related to the micro-mechanisms of fracture which we have discussed previously. The main factors are yield stress, stress intensification and microstructure with regard to cleavage fracture: strain concentration, strain gradient and microstructure with regard to fibrous fracture. The impact transitions must take account of the effects of high strain-rate on yield stress and work-hardening rate.

We shall now attempt to relate the variation of linear elastic fracture toughness, K_{IC}, with temperature to the micro-mechanisms of fracture in a typical mild steel microstructure.

8.9 A Model for K_{IC} in Low-strength Steel

The previous sections have been concerned with the micro-mechanisms of cleavage and fibrous fracture in notched bars and how these are related to the ductile/brittle transition. It is now of interest to examine the extent to which these micro-mechanisms can explain the temperature dependence of K_{IC} in low-strength steel, where the cleavage/fibrous transition occurs under fully plane strain conditions, i.e. where the position of the transition is controlled directly by the metallurgical factors, rather than by any stress relaxation caused by gross-section or through-the-thickness yielding. Typical results for a pressure-vessel steel have been given in *Figure 5.22b*, where it is seen that K_{IC} increases steeply, from some 60 MN m$^{-3/2}$ at −20°C to over 150 NM m$^{-3/2}$ at +20°C.

Unfortunately, no information on local cleavage fracture stress, grain size, carbide or inclusion distribution is available for steels on which large-scale toughness tests have been carried out, and it will be necessary to try to deduce behaviour from results obtained on simpler materials.

Recent experimental determinations of K_{IC} in an annealed high nitrogen steel (of 60 μm grain size) are presented in *Figure 8.13*. An 'effective elastic crack' correction (see Section 4.6) has been applied to the experimental K_Q

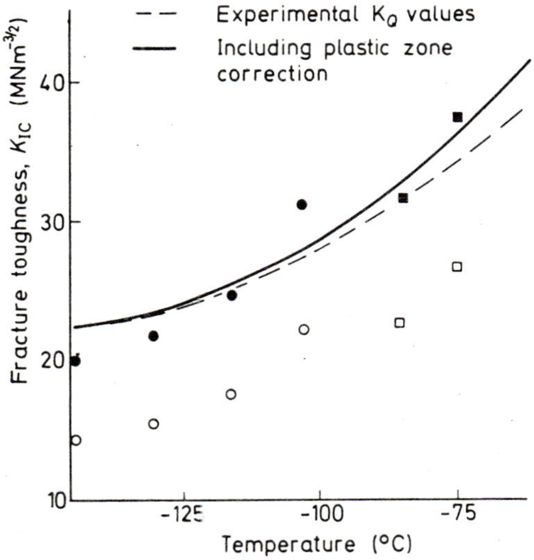

Figure 8.13 Experimental variation of fracture toughness with temperature, compared with theoretical predictions (see text). Filled points assume a critical distance of two grain diameters: open points a distance of one diameter. ○● *calculated using Figure 3.18a;* □■ *using Figure 3.18b (after Ritchie, Knott and Rice*[26]*)*

values, but this makes very little difference to the temperature dependence of the fracture toughness. The results are appropriate only to low temperatures, because the specimens were of small dimensions. The cleavage/fibrous transition did not set in until very much higher temperatures, well after general yield.

We postulate that the temperature dependence of K_{IC} is of the form that it is because the plastic zone at each temperature must be of a size such that the local tensile stress is equal to the cleavage fracture stress, as in notched bars. Only recently, with the development of finite element stress analyses which give the elastic/plastic stress distribution ahead of a sharp crack, has it become possible to test this hypothesis in a quantitative manner. Typical stress distributions were drawn in *Figure 3.18* for non-hardening material and for various work-hardening exponents with $\sigma_Y/E = 0.0025$ (the average value over the temperature range of interest).

Consider the curve for $n = 0.1$ (corresponding to the average low temperature work-hardening rate). At low temperatures, where the yield stress is high, little stress intensification is needed and the critical fracture stress can be met at a point, not too far behind the plastic/elastic interface (*Figure 8.14a*). The absolute size of plastic zone (value of K) is small because the yield stress is high. As the temperature is raised, more and more stress intensification is required and the failure point moves closer to the maximum in the curve. It does not necessarily move with respect to the crack tip, because the maximum moves away from the tip as the tip blunts (*Figure 8.14b*).

We suppose that the critical stress at any temperature must be attained at some *fixed* distance ahead of the tip, governed by the microstructure of the material[26]. The association of cleavage fracture with grain boundary carbide cracking mechanisms, as described in Chapter 7, leads us to choose one or two grain diameters as the characteristic distance. Smith's model (Section 7.11), which has been used successfully to relate the critical value of local fracture stress to microstructure in notched bars, assumes that a microcrack grows into the ferrite matrix from a carbide under the action of a uniform tensile stress. In the plastic zone ahead of the macroscopic crack tip the stress is non-uniform and it is plausible that insufficient stress is generated across the second grain to propagate the carbide crack nucleus if the critical stress predicted by Smith's theory is achieved only at the first boundary.

The high-nitrogen steel, for which the fracture toughness results are shown in *Figure 8.13*, possessed a grain size of 60 μm. Notched-bar tests, carried out over a range of low temperatures, enabled the critical local fracture stress to be determined as 850 MN m^{-2} for this grain size. Taking the figure 850 MN m^{-2} to be independent of temperature, and using values of uniaxial yield stress measured over the appropriate temperature range in subsidiary tensile tests, it is then possible to calculate the critical size of plastic zone, and hence the fracture toughness, as a function of temperature. The results of these calculations are shown in *Figure 8.13*, assuming that the stress 850 MN m^{-2} is achieved at distances ahead of the crack tip of 60 μm and 120 μm respectively. Agreement with the experimental results is obviously very close if the latter figure is assumed.

If the value of approximately two grain diameters is taken to be of general significance, it is possible to examine the validity of the cleavage crack propagation model with regard to the prediction of fracture toughness in other steels. In particular, we examine the pressure-vessel steel data shown in *Figure 5.22b*. Unfortunately, no information on the microstructure of this steel was included with the toughness results. We assume that, at the NDT, a condition has been reached where the maximum stress intensification is operating. The temperature is higher than for the high nitrogen steel, and it is appropriate to take n as 0.2. From the experimental values of K_{IC} (75 MN m$^{-3/2}$) and σ_Y (530 MN m^{-2}) at the NDT, it is possible to deduce that the critical distance is some 30 μm. This implies a grain size of 15 μm, which is reasonable for a steel of the appropriate composition and heat-treatment. The local value of fracture stress is found to be about 2600 MN m^{-2}, which is considerably greater than the value of some 1600 MN m^{-2} which would be typical of an unalloyed, normalised steel of the same grain size (*Figure 7.17*). The carbide distribution in the pressure-vessel steel

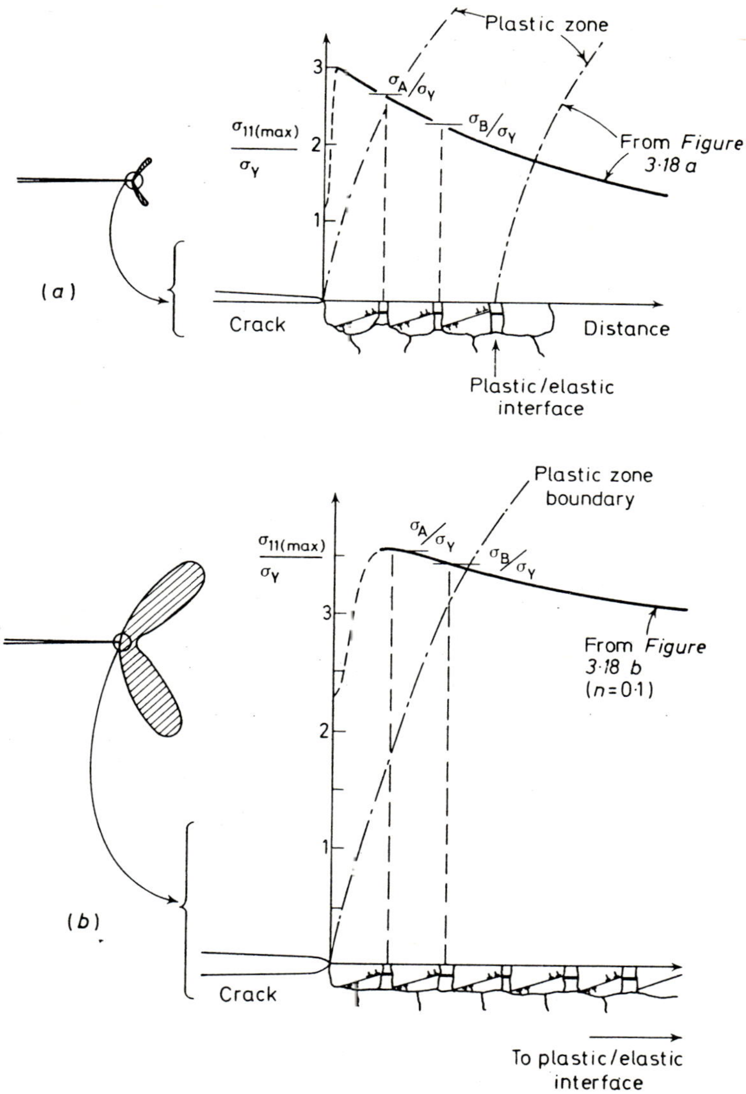

Figure 8.14 (*a*) *Situation at low temperature. The critical cleavage fracture stress is attained at the first* (*or second*) *grain boundary ahead of the crack. Because the yield stress is high, the stress intensification* (σ_A/σ_Y *or* σ_B/σ_Y), *need not be high and the critical event can occur close to the plastic/elastic interface. The plastic zone size* (*and hence* K_{IC}) *can therefore be small.* (*b*) *Situation at higher temperature. A high stress intensification is now needed, because the yield stress is lower. The plastic zone size at fracture* (*and hence* K_{IC}) *must be larger* (*see Figure 3.18*)

is, however, likely to be very much finer than in the plain carbon steel and this will give rise to a much higher critical stress. The effect is observable in Oates's results (*Figure 7.16*), where, for coarser grain sizes, a manganese-containing steel had a much higher fracture resistance because the carbides were more finely distributed.

Taking the value of 2600 MN m^{-2} as the critical value, we examine the situation at −150°C, where σ_Y = 800 MN m^{-2} and at −100°C, where σ_Y = 600 MN m^{-2}. The required stress intensification factors (of 3.25 and 4.3) imply critical values of K_{IC} of 32 and 60 MN m$^{-3/2}$ compared with the experimental values of just under 40 and just over 50 MN m$^{-3/2}$. These results cannot be regarded as completely convincing, because σ_F and n may vary with temperature, but give sufficiently close agreement to warrant further investigation of the application of the cleavage model to the calculation of K_{IC}.

At the high temperature end, the mode of crack extension should change from cleavage to fibrous. At the NDT, where K_{IC} = 75 MN m$^{-3/2}$ and σ_Y = 530 MN m^{-2} as before, the value of COD is calculated as approx. 0.05 mm. This is close to the critical value for the initiation of fibrous fracture found in free-machining mild steel (Section 6.4) and suggests that the transition in fracture initiation occurs just above the NDT when sufficient strain to produce internal necking between inclusions has been developed at the crack tip.

The sequence of events in precracked specimens therefore appears to be very similar to that in notched specimens, except that the cleavage fracture is controlled by a critical microstructural distance, taken for the present as about two grain diameters. Effects of root radius on fracture toughness in a mild steel of grain size approx. 50 μm show that the critical distance there is again about two grain diameters[27] and, using this argument to apply to high-strength steels, where the cut-off occurs at about 6 μm (e.g. *Figure 5.15*), we may deduce that the limiting fracture toughness for a steel of yield strength 1400 MN m^{-2} is about 40 MN m$^{-3/2}$, in agreement with the experimental results (assuming an average stress intensification of 4). The general development of fine grain size, fine carbide distribution and widely spaced inclusions will therefore produce high toughness, although the refinement of microstructure is limited with regard to cleavage resistance, because the critical distance is much reduced. Before a fully quantitative model for K_{IC} in steels can be developed, it is obviously necessary to investigate further the micro-mechanisms of fracture in fine carbide dispersions, to devise suitable fracture criteria for these microstructures.

8.10 Other Fracture Modes

The modes of fracture used in the explanation of the general temperature dependence of K_{IC} in steel were cleavage, at low temperature, and fibrous, at high temperatures. An alternative form of low temperature brittle fracture can occur in alloy steels if minor impurity elements (Sn, Sb, P, As, etc.) have segregated to prior austenite grain boundaries in amounts sufficient to induce intergranular failure. Such segregation may be present in an as-quenched or lightly tempered steel, or may occur during slow cooling after tempering at higher temperatures. The 'degree of embrittlement' is conventionally assessed by measuring the difference in impact transition temperature between an embrittled and unembrittled steel, but this shift cannot be related directly to the segregation phenomena, because the fractures at different transition temperatures take place in materials of different yield stress[28]. Similarly, the shift, as a measure of embrittlement, can be misleading if, for example, results on a high yield strength martensite are compared with those on a lower strength bainite.

A recent detailed investigation into temper-brittle fractures has shown that they behave in a way similar to that for cleavage. In notched bars, fracture appears to occur at a critical tensile stress, which varies slightly with temperature, and which is lower than the critical cleavage stress, because the segregating species form brittle boundary films. Coarse carbides (e.g. in bainites, rather than martensites) reduce the fracture stress, but the effect of grain size is more complicated, because it also affects the degree of grain-boundary coverage by impurities. In principle, however, it would seem feasible to be able to relate the critical fracture stresses obtained in notched-bar tests to K_{IC}, as for cleavage fractures.

The low-energy fracture of aluminium alloys in thick section involves the formation of dimples around intermetallic particles or inclusions[29]. Large iron- and silicon-bearing particles fracture at low strains and form the sites for primary voids (rather like non-wetting inclusions in a steel), but the linkage between these large voids is enhanced by the formation of secondary dimples around smaller particles at higher strains: these smaller particles being associated with the age-hardening or the grain-refinement of the alloys. Higher-toughness alloys therefore demand pure components. The fracture appears to be displacement-controlled. Aluminium alloys do not exhibit any transitions in impact or toughness results over a range of low temperatures. The yield stress increases slightly with decrease in temperature and, for displacement-controlled behaviour, one would expect that the toughness at low temperature would be somewhat greater than at room temperature, because it is necessary for the applied load to produce the same displacement in harder material. There is, indeed, some evidence that the toughness is higher at low temperature.

8.11 Conclusions

It is apparent that different types of material will fracture in different ways and that many more investigations need to be carried out before the relationships between micro-mechanisms and toughness are understood. In general, coarse, brittle particles will crack at low stresses and non-wetting inclusions allow voids to form at low strains. Refinement of the microstructure is apparently the best way to achieve high toughness, but the metallurgical properties need to be developed in conjunction with a logical appreciation of the role of engineering design. As shown in *Table 8.2* the same working stress can be achieved with greater safety from fast fracture, not by improving the material's toughness at a

Table 8.2 PERMISSIBLE DEFECT SIZES IN 3% Cr–Mo–V STEEL AT OPERATING TEMPERATURE (350 K)

	Critical defect size for applied stress/proof stress of		
0.2% *Proof stress*	0.5	0.57	0.61
800 MN m^{-2}	35 mm	25 mm	21 mm
860 MN m^{-2}	17 mm	12.5 mm	11 mm
980 MN m^{-2}	9 mm	7 mm	6 mm
Working stress 490 MN m^{-2}			

given high strength level, but by using a numerically smaller 'safety factor' in material of lower yield strength. The aim of the Fracture Mechanics approach is to replace the degree of ignorance in conventional design factors by a quantitative parameter which is a direct measure of the material's fracture resistance. Numerically smaller factors used properly can mean that the true factor of safety is much increased.

REFERENCES

1. Knott, J. F., *J. Iron Steel Inst.*, **204,** 104 (1966)
2. Brooksbank, D. and Andrews, K. W., *J. Iron Steel Inst.*, **206**, 595 (1968)
3. Ashby, M. F., *Phil. Mag.*, **14,** 1157 (1966)
4. McClintock, F. M., *Int. Jnl Fracture Mechanics*, **4,** 101 (1968)
5. Rice, J. R., *Fracture – An Advanced Treatise* (edited by H. Liebowitz) Vol. 2, 191, Academic Press, New York (1968)
6. Rice, J. R. and Tracey, D. M., *J. Mech. Phys. Solids*, **17,** 201 (1969)
7. Thomason, P. F., *J. Inst. Metals*, **96,** 360 (1968)
8. Thomason, P. F., *Int. Jnl Fract. Mech.*, **7,** 409 (1971)
9. Baker, C., *Ph.D. Thesis*, Cambridge University (1963)
10. Edelson, B. I. and Baldwin, W. M., *Trans. Am. Soc. Met.*, **55,** 230 (1962)
11. Turkalo, A. M. and Low, J. R., *Trans. Am. Inst. Min. Metall. Petrol. Engrs,* **212,** 750 (1958)
12. Gladman, T., Holmes, B. and McIvor, I. D., *Effect of Second-Phase Particles on Mechanical Properties of Steel*, 68, Iron and Steel Institute, London (1971)
13. Palmer, I. G., Smith, G. C. and Warda, R. D., *Conf. Proc. Physical Basis of Yield and Fracture*, 53, Institute of Physics and the Physical Society (1966)

14. Smith, R. F. and Knott, J. F., *Conf. Proc. Practical Application of Fracture Mechanics to Pressure-vessel Technology*, 65, Institute of Mechanical Engineers (1971); Smith, R. F., *Ph.D. Thesis*, Cambridge University (1972)
15. Chipperfield, C. G., Smith, R. F. and Knott, J. F., *Third International Congress on Fracture*, Munich (April 1973)
16. Almond, E. A. and Embury, J. D., *Mat. Sci. Jnl*, **2,** 194 (1968)
17. Wood, D. S. and Clark, D. S., *Trans. Am. Soc. Metals*, **43,** 571 (1951)
18. Knott, J. F., *J. Mech. Phys. Solids*, **15,** 97 (1967)
19. Wilshaw, T. R. and Pratt, P. L., *Proc. First International Conference on Fracture*, 973, Sendai, Japan Society for Strength and Fracture of Metals (1966)
20. Oates, G., *J. Iron Steel Inst.*, **207**, 353 (1969)
21. Loss, F. J. and Pellini, W. S., *Practical Fracture Mechanics for Structural Steel*, Paper J, UKAEA/Chapman and Hall, London (1969)
22. Lange, E. A., *Naval Res. Lab. Report* 7046 (1970)
23. Loss, F. J., *Naval Res. Lab. Report* 7056 (1970)
24. 'The Robertson Crack Arrest Test,' *Br. Weld. J.*, 387 (Aug. 1968)
25. Nichols, R. W., *Proc. R. Soc.*, **A285,** 104 (1965)
26. Ritchie, R. O., Knott, J. F. and Rice J. R., *J. Mech. Phys. Solids* (in press)
27. Tetelman, A. S., Wilshaw, T. R. and Rau, C. A., *Int. Jnl Fract. Mech.*, **4**, 147 (1968)
28. Geniets, L. C. E. and Knott, J. F., *Metal Sci. Jnl*, **6**, 69 (1972)
29. Tanaka, J. P., Pampillo, C. A. and Low, J. R., *Am. Soc. Test. Mat.*, Spec. Tech. Pub. No. 463, 191 (1970)

Nine

APPLICATION OF FRACTURE MECHANICS TO CRACK GROWTH BY FATIGUE OR STRESS-CORROSION MECHANISMS

9.1 Introduction

In many practical applications, it is clear that the sizes of defects needed to produce catastrophic, brittle fractures under service stresses are so large that the defects could always be detected and eliminated, by repair work, before the piece entered service. Nevertheless, catastrophic failures can occur, if initially small defects grow to critical lengths during operation, by 'subcritical crack growth' mechanisms, such as fatigue or stress-corrosion. The success of conventional linear elastic fracture mechanics in characterising the final fracture event in terms of parameters which can be related to the applied stresses on a structure has led to attempts to characterise subcritical crack growth by similar methods. The present chapter contains descriptions of the experimental observations which have been used to provide a basis for a fracture mechanics approach to fatigue-crack propagation and to stress-corrosion crack growth and of the models which have been developed to explain the experimental results.

9.2 Fatigue Failure

A fatigue failure is one which occurs after a number of cycles under alternating stresses whose peak amplitudes are less than the strength of the material in a uniaxial tensile test. The complexity of interpretation of the measurements used to determine resistance to fatigue failure is reduced if we concentrate on limiting situations, which may be regarded as initiation-controlled and propagation-controlled.

The initiation-controlled situation is found typically in circumstances where smooth shafts, rotating in bearings, are subjected to eccentric loadings, which produce a bending moment. A region of the surface of the shaft is then subjected alternately to tensile and compressive maximum fibre stresses (see *Figure 1.2*) as it rotates. If these stresses locally exceed the elastic limit, alternating plastic deformation is produced in the surface grains. In practice, this deformation is not fully reversible and gives rise to two effects. First, the redundant dislocation density within a grain forms a debris structure or cell structure which produces 'fatigue hardening' and which is characteristic of the plastic strain amplitude, the temperature and the ability of the dislocations to cross-slip. Secondly, non-reversible flow at the surface produces 'extrusions', which pile up material on the surface, and associated 'intrusions', which act as embryonic cracks along the active slip-bands (see e.g. Plumbridge and Ryder[1] and Ham[2]). The general hardening of the grain facilitates localisation of deformation along a slip-band and the lack of reversibility of plastic flow at the surface is enhanced by an environment, such as air, from which gaseous species may adsorb onto freshly exposed slip steps, making the reversal of flow at such positions more difficult. The intrusion propagates along the active slip-band, inclined roughly at 45° to the direction of the maximum principal stress, as a 'stage I' crack, until it reaches a length significant with respect to the geometry of the piece, such that the crack tip stress field becomes dominant. It then propagates as a 'stage II' crack, normal to the maximum principal stress until it becomes so long that the piece separates by a fast tensile fracture. The fracture surface during stage II propagation shows a sequence of 'ripples'.

This situation is initiation-controlled at low plastic strain amplitudes because the growth of the crack from intrusion to stage II is a rapidly accelerating process, as the net section decreases in size and stresses become intensified around the growing crack tip. For low amplitudes, some 90% of the total fatigue life is expended in developing the initial intrusion. The important point is that, since fairly substantial amounts of alternating plastic flow are needed to form a crack-like profile in an initially smooth surface, the alternating stress needed to initiate fatigue fracture is high. This implies that the stress applied to the specimen is high, so that a crack, once initiated, propagates rapidly.

Conventional fatigue testing has concentrated primarily on the behaviour of smooth specimens. Testpieces are machined to provide parallel-sided gauge lengths or 'hour-glass' (waisted) profiles and are tested in plane bending, rotating bending or uniaxial compression–tension ('push–pull') or tension–tension. Information is presented in terms of an S–N curve, which shows the dependence of the number of cycles required to cause failure on the applied alternating stress (see *Figure 9.1*). Materials generally exhibit one of two types of behaviour. In mild steel, and in other strain-ageing materials, a sharp 'fatigue-limit' is obtained: below a critical value of applied stress, specimens appear to last indefinitely. The importance of strain-ageing is demonstrated by the effect known as 'coaxing': a specimen fatigued for a sufficiently long time at stresses below the fatigue limit is subsequently found to possess an improved resistance to fatigue failure, because its flow stress has been increased by strain-age-hardening. Non-ageing materials do not show a sharp fatigue limit and it is conventional to define an 'endurance limit', which is the stress required to cause failure in 10^8 cycles. It is important to note, although the point is seldom stated explicitly, that conventional fatigue or endurance limits are nearly always greater than the macroscopic uniaxial yield stress, although less than the UTS. An important exception occurs in the case of fine-grained mild steel: here, the combination of short dislocation pile-ups in the most easily activated grains and strongly locked dislocations in neighbouring grains can produce intrusions at stresses as low as some 80% of the macroscopic lower yield stress[3]. The yield stress of individual grains has, of course, been exceeded. In such fine-grained steel, the fatigue limit represents the stress at which heavy fatigue damage and intrusions are just able to spread from grain to grain. The difference between this and the simple strain-ageing effect is

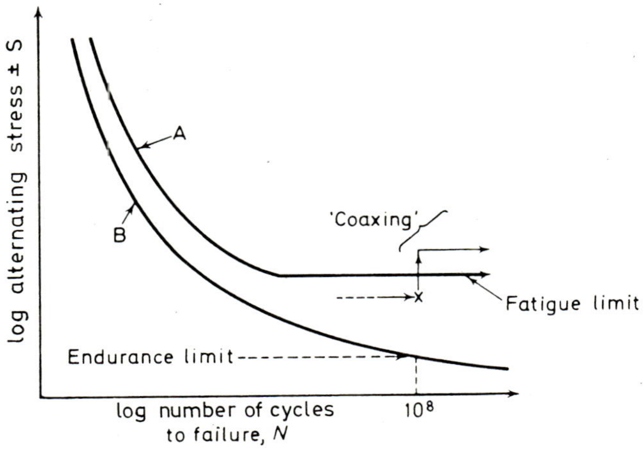

Figure 9.1 'S–N' Fatigue curves. Curves of type A are typical of mild steel and alloys which strain-age. Curves of type B are typical of non-ageing alloys

shown by the fact that a sharp fatigue limit persists at low temperature in fine-grained mild steel, but not in coarse-grained mild steel.

The importance of plastic strain range, $\Delta\epsilon_p$, on fatigue fracture in smooth specimens is clearly demonstrated by the so-called Coffin–Manson[4,5,6] relationship between $\Delta\epsilon_p$ and the number of cycles to failure, N_f. In its most general form, the relationship is given by:

$$\Delta\epsilon_p \, . \, N_f^{C_1} = C_2 \tag{9.2.1}$$

where C_1 is a constant, often having a value close to 0.5, and C_2 is a constant, approximately equal to the true strain to fracture in a tensile test (about unity). The extent to which this equation is obeyed by a wide range of metals is illustrated by *Figure 9.2*. There is also a marked effect of mean stress on the endurance limit (*Figure 9.3*), which indicates that, once the tensile yield stress has been exceeded and alternating plastic strain is possible, a mean tensile stress accelerates the fatigue fracture mechanisms.

The behaviour in such tests has been divided into three parts: intrusion formation, stage I propagation along 45° slip-bands and stage II propagation normal to the applied stress. It is clear that large plastic strain amplitudes will produce large slip steps on the surface and perhaps substantial surface buckling and folding. The general scale of upheaval is much increased and it is reasonable to suppose that fewer cycles are needed to form a definite crack embryo. Similarly, the larger alternating shears are likely to produce more rapid crack advance during stage I propagation along the active slip-band. If the plastic strain amplitude is very large, such that failure occurs in less than 100 cycles, for example, stage II propagation may start after the first few cycles. The magnitude of the strain needed to produce this situation is large. From equation 9.2.1 a life of 100 cycles requires that $\Delta\epsilon_p$ is about 10%, i.e. a plastic movement of ± 0.5 mm of the ends of a gauge length of 10 mm.

A model has been developed to explain the mechanism of stage II propagation

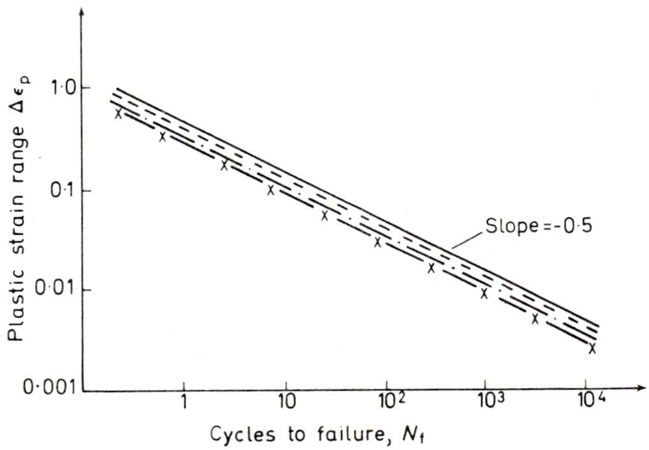

Figure 9.2 Coffin–Manson relationship. ——— C/Mn *steel;* – – – Ni/Cr/Mo *alloy steel;* – · – Al–Cu *alloy*; – x – Al–Mg *alloy*

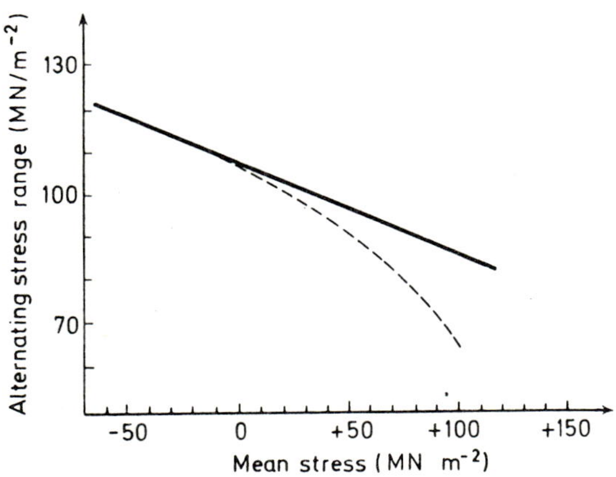

Figure 9.3 Effect of mean stress on fatigue strength at 10^8 *cycles in two aluminium alloys*

in terms of plastic flow processes[7,8]. The sequence of events is depicted in *Figure 9.4*. As an initially sharp crack is loaded in tension, shear deformation is concentrated in bands inclined at some 70° to the crack plane (see *Figure 9.4a* and Section 3.15). With further straining, the initial crack is subjected to a simple opening, until a point is reached where physical separation occurs in the

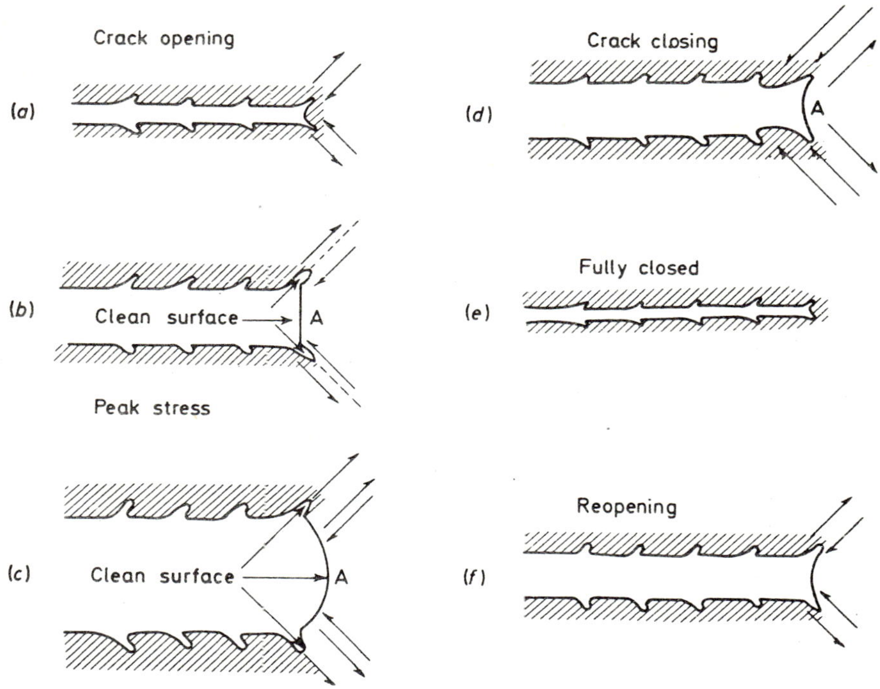

Figure 9.4 Schematic representation of fatigue crack advance by plastic blunting process

slip bands. At peak tensile strain, the profile contains two 'ears' at the end of the crack (*Figure 9.4b,c*). In the earlier version of the model the portion of the crack front between the ears was apparently squeezed forward on closure, so that the tip of the new crack protruded beyond the tips of the ears. Advance of the point A on closure, however, seems physically unreasonable, unless it is possible, in a specimen undergoing very large reversed plastic strains, to concede that lateral movements of exterior surfaces are possible and to regard A as a fixed hinge, about which the upper and lower faces rotate (counter clockwise and clockwise respectively) on closure. More recent versions tend to show advance of point A during the tensile stroke, i.e. between *Figure 9.4b* and *9.4c*. In either case, it is difficult to understand why the tensile deformation is not identically reversed on compression, to give the original profile, as in *Figure 9.4a*. Alternative possibilities are that cracking or voiding occurs ahead of the original crack tip (rather than in the 'ears') when the tensile stress or strain reaches a critical value (see Chapters 7 and 8) or that adsorption of gaseous matter on the freshly exposed surfaces at the tip prevents re-welding on closure, so that an increment of crack advance depends on the amount of freshly exposed surface (crack opening), on the environment (availability of species) and on the time of exposure (frequency). This latter alternative may prove to be the more realistic model at low plastic strain amplitudes. At high amplitudes, the opening and closing, with 'ear' formation, provides a good explanation for the formation of the 'ripples' observed on the fracture surface and confirms the one-to-one relationship between number of ripples and number of fatigue cycles.

These events have been concerned with smooth pieces in which, unless the applied plastic strains are very large, the total failure is governed by the initiation of a small slip-band crack in the smooth surface. It is necessary to develop alternating plastic flow in the surface layers before any cracks can be initiated and commercial treatments to prevent fatigue in components such as crankshafts therefore rely primarily on the hardening of the surface, so that plastic flow is difficult to produce. Treatments such as flame-hardening, induction-hardening or case-carburising simply increase the yield stress: shot-peening and case nitriding (or cyaniding) also introduce residual compressive stresses in the surface layer, which then needs a yet larger stress amplitude to be applied before cracks are initiated (see *Figure 9.3*). One problem, particularly with shot peening, is that the surface treatment may so increase the surface roughness that much of the benefit of the treatment in terms of preventing crack initiation is lost. It may be necessary to grind away some of the hard surface layers to remove flaws which would otherwise serve as preferential sites for initiation. In all hardened smooth pieces, initiation control is even more important, because the initiation stress is so high that subsequent events proceed very rapidly.

9.3 Fatigue Crack Propagation

The opposite extreme to the case of initiation-controlled fatigue failure in smooth components is found in situations where fatigue stresses operate on structures which contain large stress concentrators. These may be defects in welds or features of the general structural design, as in the Mark I Comet aircraft, where the sharp corners of windows in the fuselage were sufficient to allow cracks to initiate and grow under normal service stresses (pressurisation/depressurisation cycles, wind gusts, vibrations, bumpy landings, etc.). If a piece is fully heat-treated after the introduction of a stress concentrator, it is reasonable to suppose that the same sequence of events (i.e. development of dislocation sub-structure, localisation of slip, formation of cracks in slip bands) will be followed at the root of the concentrator as would occur at the surface of a smooth specimen. However, if the stress-concentration factor is large, the macroscopic applied stress needed to activate these events can be small and the inception of a stage II crack does not necessarily lead rapidly to final fracture. It will propagate at a steadily accelerating rate, but the rate of advance may be so small for such long periods of time that it would be possible to rely on periodic inspections of a structure to guarantee integrity. This approach would necessitate both a knowledge of the material's terminal toughness and a knowledge of how the crack growth rate was related to the stresses likely to be encountered in service.

In fairly large pieces, containing substantial stress concentrators, it is possible for stage II propagation to occur at stresses sufficiently lower than the general yield stress for the stress analysis to be treated as linear elastic. In conventional fatigue testing, it is clear that the alternating plastic strain amplitude, $\Delta\epsilon_p$, is the main factor controlling the number of cycles required to produce failure. For the quasi-linear-elastic situation, where the plastic zone associated with the propagating crack is very much smaller than the length of the crack or the width of the uncracked ligament, we expect that any crack tip plastic strain will be proportional to the size of the plastic zone or to the COD. In turn, these are given by:

$$r_Y = \frac{K^2}{2\pi\sigma_Y^2} \qquad \text{(equation 3.12.2)}$$

$$\delta = \frac{K^2}{\sigma_Y E} \qquad \text{(equation 3.13.12)}$$

in plane stress. The situation becomes complicated when reversed plastic strains at the crack tip are involved and will be discussed later in Section 9.6: at present, it is sufficient to indicate that the variable parameter taken to be equivalent in function to $\Delta\epsilon_p$ is ΔK, which is defined simply as the difference between the maximum and minimum stress intensities encountered during each cycle:

$$\Delta K = K_{max} - K_{min} \qquad (9.3.1)$$

It is anticipated that the increment of crack growth per cycle, or 'crack growth-rate' da/dN, should be a monotonic function of the instantaneous value of the 'alternating stress intensity', ΔK.

A simple functional relationship of the form:

$$\frac{da}{dN} \propto \Delta K^m \qquad (9.3.2)$$

was introduced by Paris to cover early experimental observations of fatigue crack growth rates, made on a variety of materials. Generally, he found that the *average value* of the exponent, m, taken over a wide range of growth rates, was close to 4 for nearly all materials: some high strength steels, however, occasionally exhibited values as high as 10. The variance between these slopes will be discussed in Sections 9.5 and 9.6.

The existence of a fully established and understood relationship between crack growth rate and ΔK would greatly facilitate the assessment of growth rates in a structure, because relatively simple laboratory tests could be used to provide information which could be used directly, knowing the service stresses. The existence of inexplicably high exponents, coupled with marked effects of mean tensile stress on growth rate in some alloys, but not in others, leads to a lack of confidence in the general application of fracture mechanics to fatigue crack propagation and retards the adoption of what are, potentially, very useful diagnostic techniques. Further problems involve crack propagation in thin sheets and under non-uniform alternating stresses (overloading, random fatigue, etc.). Before the attempts that have been made to meet these problems are described, it is worthwhile first to discuss some of the experimental difficulties involved in obtaining the basic crack growth-rate data.

9.4 Experimental Methods

Equation 9.3.2 relates the crack growth rate, $\mathrm{d}a/\mathrm{d}N$, to the instantaneous value of alternating stress intensity, ΔK. Experimental techniques are therefore concerned mainly with the control and measurement of ΔK and with the measurement of growth rate under controlled conditions of temperature and humidity. The experiments study the way in which crack length increases with number of cycles. For testpieces of CTS or SEN bend or tension geometry (see Section 5.9), this implies that the compliance (Section 4.3) of the testpiece increases as the crack grows. This change in compliance not only means that the local crack tip value of ΔK changes as the crack grows, even though the applied alternating stress remains constant, but can also produce severe testing difficulties, if the fatigue machine is of the resonance type, because resonance may be lost as the testpiece progressively becomes a weaker spring. Attempts have been made to propagate cracks under conditions of constant ΔK, either by manually reducing the applied stress amplitude as the crack grows or by designing a testpiece geometry whose compliance remains constant as the crack grows. A design for a CTS type of 'constant K' specimen is shown in *Figure 9.5*. Generally, however, the change in ΔK with crack growth has been assumed to occur so gradually that steady-state conditions, pertaining to the instantaneous value of ΔK, are always maintained at the crack tip.

Several forms of measurement of crack length have been employed. In thin sheet specimens, visual observation of the crack length on the side faces may be perfectly satisfactory, but surface measurements are unreliable when made on thicker pieces, because the fatigue crack front tends to bow forward in the centre, in a manner somewhat analogous to the tunnelling forward of a plane-strain 'pop-in' (Section 5.6) or a fibrous thumbnail (Section 6.4). The effect is not usually so large, however. In these situations, techniques are employed, which reflect the average crack size: in particular, use has been made of ultrasonics[9] or the change in d.c. resistance of the testpiece[10,11] as the crack grows.

An application of ultrasonics to fatigue crack propagation in specimens of a type similar to the CTS testpiece (a WOL specimen: see Section 5.9) has been made by attaching a probe to the top edge of a specimen[9] (see *Figure 9.6*). The probe is moved along this top surface so that a constant relationship between the crack tip and the position of the probe (as judged from the magnitude of the crack tip reflection signal) is maintained. Then the movement of the probe gives the amount of crack advance. Systems can be devised so that the movement of the probe is linked automatically to that of the crack and also gives continuous read-out.

Probably the most popular method of monitoring crack growth is the Electrical Potential method[10,11]. Here, a constant d.c. current is passed through the specimen and the change in resistance of the specimen as the crack grows is detected by measuring the change in potential across the open mouth of the

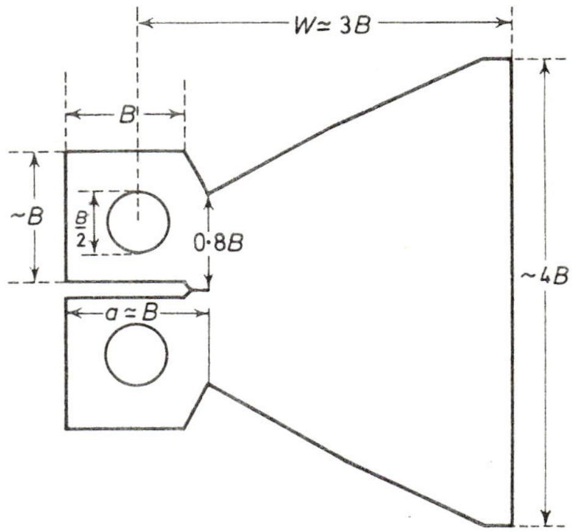

Figure 9.5 Schematic design for 'Constant K' testpiece of thickness B (dimensions approximate)

starter notch. Optimum positions for attachment of current leads and potential measuring probes are indicated in *Figure 9.7* for 'uniform current' configurations in SEN bend and CTS testpieces, together with the distributions of equipotential lines corresponding to these configurations as the crack grows (*Figure 9.8*). It is possible to provide theoretical calibrations for the potential distributions in cracked pieces, by solving, for the appropriate boundary conditions, the Laplace equation

$$\nabla^2 \phi = 0 \qquad (9.4.1)$$

where ϕ is the steady electric potential. For 'uniform current' applied to a sheet

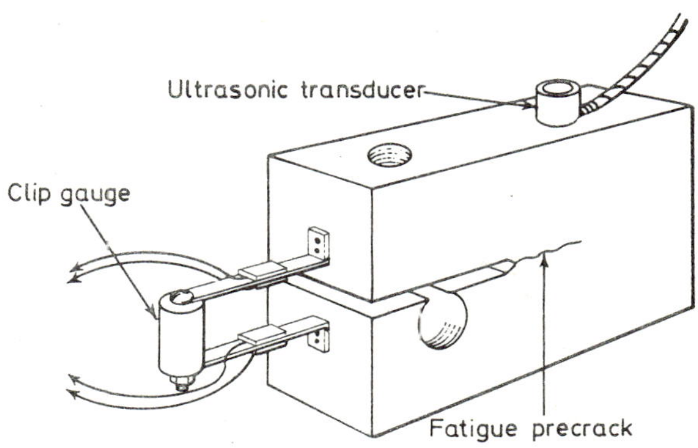

Figure 9.6 WOL-type specimen for fatigue crack growth-rate measurements

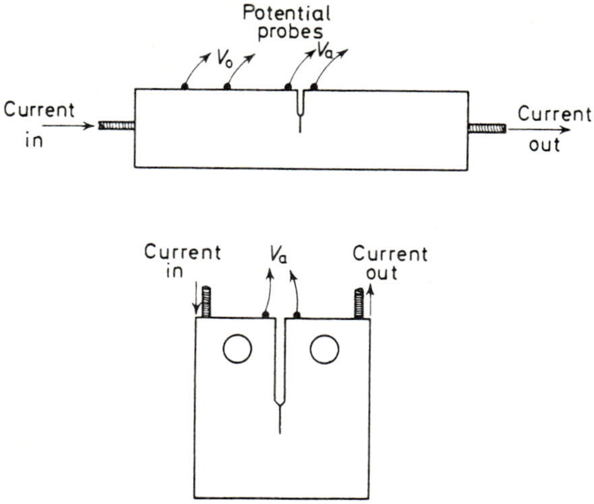

Figure 9.7 Optimum positions for current leads and measurement probes on SEN bend and CTS testpieces (see also Ritchie, Garrett and Knott[11])

Figure 9.8 Distribution of equipotential lines in SEN bend and CTS testpieces for uniform current configuration. (a) Uncracked SEN bend testpiece. (b) Cracked SEN bend testpiece. (c) Uncracked CTS testpiece. (d) Cracked CTS testpiece (see also Ritchie, Garrett and Knott[11])

of width W containing an edge crack of length a (or a sheet of width $2W$ containing a central crack of length $2a$), a mapping procedure leads to the choice of a potential function[12] of the form:

$$\phi = \mathrm{Im}\left[C \cos^{-1} \frac{\cos(\pi z/2W)}{\cos(\pi a/2W)}\right] \qquad (9.4.2)$$

where z is the complex variable: $z = x_1 + ix_2$; x_1 and x_2 are defined, as for the Westergaard analysis, along the crack and normal to the crack (see Section 3.7) and the uniform lengthways potential gradient is $C\pi/2W$. The boundary conditions are that the exterior surfaces and the crack are perfect insulators ($\partial\phi/\partial x_1 = 0$ when $x_1 = 0$ or W; $\partial\phi/\partial x_2 = 0$ when $x_2 = 0$ and $x_1 \leqslant a$) and that the uncracked ligament ($x_2 = 0, a \leqslant x_1 \leqslant W$) is an equipotential line. It is then possible to predict the potential difference between any two points across the mouth of the crack, for any length of crack. The theoretical calibrations have been applied primarily to relatively simple configurations, such as the centre-cracked and edge-cracked plate. For more complicated testpiece geometries it is generally necessary to produce direct experimental calibrations of the change in potential with crack length. Although attempts have been made to apply the theoretical analyses directly to real testpieces some of the assumptions made have not been justified rigorously. By optimising the system and using very low noise ($< \pm 0.1\ \mu$V) and low drift (less than 0.05 μV for the duration of a test) amplification, it has proved possible experimentally to measure crack lengths to within 0.1 mm in different specimens and changes in crack length of less than 10 μm (about half a grain diameter in normalised steel). Such sensitivity demands very close control of the constant current and of testing temperature (thermal e.m.f.'s are highly significant). The rise in specimen temperature due to the passage of current is of the order of 1–2°C, but this is allowed for by passing the current through a specimen for a sufficiently long period before the fatigue test begins.

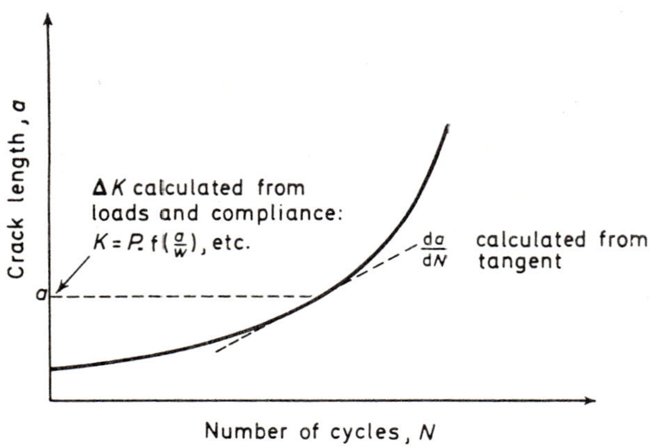

Figure 9.9 Basic data: crack length v. number of cycles

Further methods of measuring crack growth involve the change of compliance as the crack grows or a.c. resistance changes. The compliance change may be detected by a clip-gauge mounted across the crack mouth, but the measurement is not particularly sensitive and is ambiguous if substantial plastic flow accompanies crack extension. The a.c. methods would appear to hold promise, particularly if the 'skin effect' is utilised, but have not been used extensively. Stress-wave emission techniques are not really suitable for monitoring continuous crack growth, but may prove to be very useful in conjunction with, e.g. the electrical potential technique if discrete bursts of cracking occur (see Section 9.6).

The electrical potential method additionally offers a means of propagating cracks under constant alternating stress intensity, because the feedback signal, which is proportional to crack length, can be made to reduce the stress applied by a servo-controlled machine so as to maintain constant ΔK conditions automatically.

9.5 Experimental Results and Crack Growth Models

The basic information obtained from conventional crack propagation studies is a graph of crack length versus number of cycles, for a constant amplitude of alternating applied stress, as shown in *Figure 9.9*. Equation 9.3.2 indicates that the incremental crack growth per cycle is expected to depend on the alternating stress intensity through a simple power-law relationship. The information given in *Figure 9.9* is therefore processed as follows. At any given crack length, a, it is possible to calculate ΔK directly from the maximum and minimum loads applied to the specimen, using the compliance function: $Y = f(a/W)$ (*Table 5.2*); appropriate to the instantaneous specimen geometry (a/W). The compliance functions are often extended to high (a/W) values to cater for fatigue crack growth in geometries which would not be accepted as 'valid' for fracture toughness tests: for example, for notch-bend specimens with (a/W) values greater than 0.6, K_I is given by[13]:

$$K_I = \frac{\bar{Y}M}{(W-a)^{\frac{3}{2}}B} \tag{9.5.1}$$

where $\bar{Y}$ is equal to 4 to within 1% error. At this same crack length, the value of (da/dN) is determined from the gradient of the curve in *Figure 9.9*, either manually or by fitting a polynomial function to the whole curve, or, better, to the section of immediate interest, and numerically calculating the slope of this function at crack length a.

It is then possible directly to test the power-law relationship:

$$\frac{da}{dN} \propto \Delta K^m \tag{9.3.2}$$

by plotting a graph of $\log(da/dN)$ *v.* $\log \Delta K$. A typical curve of this type is drawn in *Figure 9.10*.

Early analyses of such crack growth-rate data determined the average value of the slope as 4 for a wide range of materials. With improved techniques and closer examination of the data, however, it has become more clear recently that the growth-rate curves are usually not straight lines, but are composed of three regions. At low values of crack extension (region A, *Figure 9.10*), the graph rises rather rapidly; over most of the ΔK range experienced in the test, the slope, or exponent m, is usually about 2–3 (region B);[14,15,16] and, at high ΔK values, such that K_{max} approaches K_{IC}, the slope steepens again (region C). We consider first the behaviour in region B, which, in ductile metals, may be regarded as representing pure stage II propagation.

The discussion in Section 9.2 suggested that the most important parameter in determining the rate of crack advance by plastic processes was likely to be the

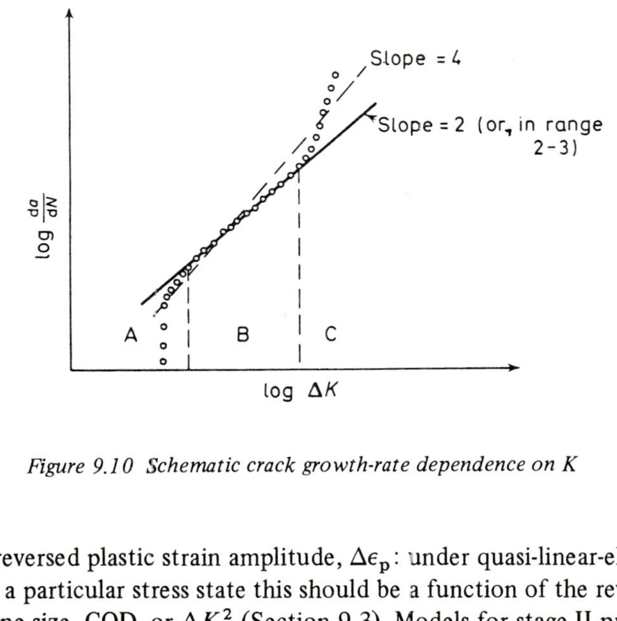

Figure 9.10 Schematic crack growth-rate dependence on K

range of reversed plastic strain amplitude, $\Delta\epsilon_p$: under quasi-linear-elastic conditions and a particular stress state this should be a function of the reversible plastic zone size, COD, or ΔK^2 (Section 9.3). Models for stage II propagation suppose that the crack increment per cycle is a proportion of the instantaneous COD. This proportion represents the amount of non-reversibility, caused perhaps by the adsorption of species from the environment onto freshly-exposed slip steps at the crack tip and preventing re-welding of the new surfaces on closing the crack. The radius of the reversible plastic zone can be deduced by assuming that the local flow stress is twice the monotonic yield stress (or, more logically, twice the fatigue-hardened flow stress) and using the standard relationship (equation 5.10.2) to obtain, for plane strain conditions:

$$r_f = \frac{\Delta K^2}{6\pi(2\sigma_Y)^2} \qquad (9.5.2)$$

The reversible COD is proportional to r_f and so to ΔK^2 also. In general terms, therefore, a dependency of crack growth rate, da/dN, on ΔK^2, appears to arise simply from any model dependent on reversed plastic opening at the crack tip. In terms of a physical model, the critical factor is, however, more likely to be the alternating plastic strain amplitude in the region immediately ahead of the tip. In plane strain, a given value of $\Delta\epsilon_p$ will be associated with a particular value of reversed COD or plastic zone size, but different factors of proportionality between these parameters are to be expected in plane stress (see Section 9.7).

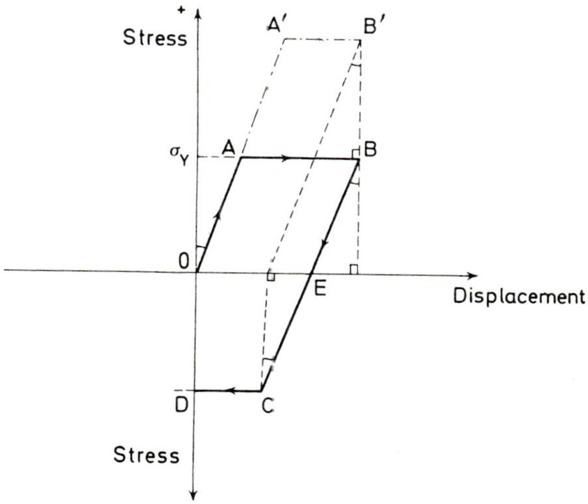

Figure 9.11 Model for reversed plastic flow. The material yields in tension at a stress σ_Y and extends plastically from A to B. On unloading, the stress is at first removed elastically to zero (BE) and then loads elastically in compression (EC) until reversed plastic flow is produced. The amount of this reversed flow at zero overall displacement (CD) is given by the amount of flow corresponding to a material with a flow stress $2\sigma_Y = OA' = BC$ (i.e. $CD = A'B'$). At zero displacement there are still positive dislocations present, giving a compressive stress. The Bauschinger effect is ignored

The reason for using a value of twice the monotonic flow stress to represent the stress required to produce reversed plastic flow may be understood by reference to *Figure 9.11*, which shows a schematic stress–displacement curve for the region immediately ahead of the crack tip. The curve is non-hardening, because, at the point of crack advance under steady-state conditions, saturation hardening has already been achieved. As a specimen is unloaded from the peak of the tensile stroke, the local stresses at the crack tip unload elastically and then become compressive. Some reverse plastic flow is obtained when the compressive stresses become equal to the flow stress in compression. We may estimate the amount of *reversed* flow by equating it to the amount of tensile flow that would have occurred had the specimen had a flow stress equal to the tensile unloading plus the compressive loading, i.e. to twice the tensile flow stress. When the displacement is zero, the plastic flow has not fully reversed and compressive stresses exist in the crack tip region. This ignores the Bauschinger effect.

Alternatively, one may consider the equilibrium of dislocations in the inverse pile-ups ahead of the crack tip (see Section 3.14 and *Figure 9.12*). At the maximum of the tensile stroke, the balance of forces is such that a dislocation is just prevented from moving any further forward against the friction stress, σ_i, (or yield stress, σ_Y, in a non-hardening material), and the back stresses of more

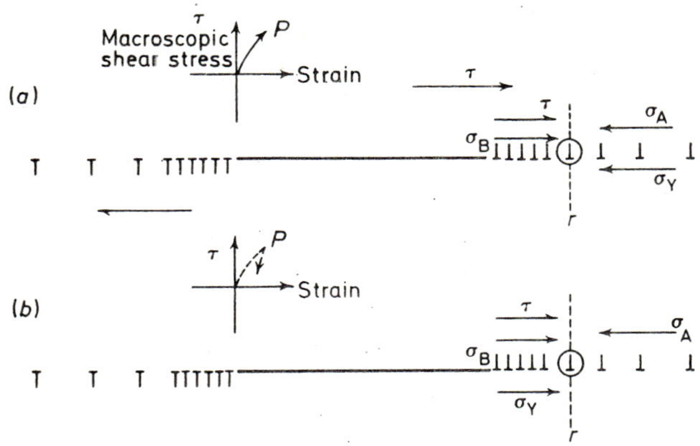

Figure 9.12 Dislocation model for reversed plastic flow. At the peak tensile stress, the rth dislocation is in equilibrium under stresses

$$\overset{\tau}{\longrightarrow}\;\underset{\sigma_B}{\longrightarrow}\qquad \overset{\sigma_A}{\longleftarrow}\;\underset{\sigma_Y}{\longleftarrow}\ \text{as in (a).}$$

As the load begins to decrease, so that the dislocation will want to move to the left, the equilibrium becomes

$$\tau\ (\text{decreasing}) \longrightarrow,\ \sigma_B \longrightarrow,\ \sigma_Y \longrightarrow \qquad \overset{\sigma_A}{\longleftarrow}\ \text{as in (b).}$$

The instantaneous difference between (a) and (b) is $2\sigma_Y$ and the amount of reversed flow is estimated, as in Figure 9.11 by calculating the amount of forward flow for a yield stress, $2\sigma_Y$

advanced dislocations, σ_A. The applied shear stress, τ, and stresses from less advanced dislocations, σ_B, are trying to move it forward. Thus

$$\tau + \sigma_B = \sigma_Y + \sigma_A \tag{9.5.3}$$

On unloading, the dislocation tries to move in the opposite direction. The force opposing reversed motion as unloading begins is $(\tau + \sigma_B)$ plus the friction stress, σ_Y. The force promoting reversed motion is the back-stress, σ_A, only. The net resistance to motion is then $(\tau + \sigma_B) + \sigma_Y - \sigma_A$, or using equation 9.5.3, $2\sigma_Y$. Again, the amount of reversed flow may be estimated for a non-hardening material by calculating the crack tip opening displacement for a material of flow stress $2\sigma_Y$.

In plane strain, the dependency of $(\mathrm{d}a/\mathrm{d}N)$ on ΔK^2 is not only predicted for a model involving the attainment of a critical value of $\Delta\epsilon_p$, but also for situations in which a fixed proportion of the alternating plastic displacement is prevented or even accentuated. So far, we have considered a relatively passive environment, such as air, in which gas might adsorb onto a clean surface and prevent complete reversibility of local slip mechanisms, much as it does on the surfaces of smooth specimens. There is, indeed, evidence that growth rates are much reduced when testing is carried out in a vacuum and that the usual ripples, characteristic of stage II propagation, are absent. Alternatively, it is possible that the environment is aggressive and actually dissolves some of the clean surface, physically advancing the crack tip by a stress-corrosion mechanism (see Section 9.10). Again we might expect that the effect would be proportional to the amount of fresh surface produced, i.e. to the crack opening. Of course, the actual amount of opening in a particular material under a particular stress cycle depends on the cyclic stress–strain properties (the 'fatigue hardening') of the region immediately ahead of the crack tip. The relationship between $\Delta\epsilon_p$ and ΔK and hence between da/dN and ΔK may then depend on the cyclic work-hardening exponent, as it will if the crack advance is controlled by alternating local stress rather than alternating plastic strain[17].

As a base-line for fatigue crack growth in plane strain, a 'log-log' graph with slope $m \cong 2-3$, seems to be the best representation of the more recent experimental results and plausible reversed plastic flow mechanisms. The exponent, 4, which has been commonly quoted, represents an average value, including some contributions from regions A or C (*Figure 9.10*), to which we now turn our attention.

Region A represents the initial stages of crack growth in a specimen. Consider the ideal situation of a sharp starter crack in a material which has been fully heat-treated after this pre-cracking. At extremely low values of ΔK, the local stress intensity is just not high enough to produce an amount of dislocation activity at the crack tip sufficient to develop any significant local slip-band cracking, i.e. the local stresses are below the local endurance limit. Estimating the scale of this activity is difficult but if a value of 1 μm were taken for the reversed plastic zone size (this corresponding to the size of the dislocation cell structures observed close to the surface of a propagating crack[18], to about one-twentieth of the grain size of normalised mild steel or to an associated reversed crack tip opening of about 150 Å) it could be deduced, from equation 9.5.2, that the absolute threshold value of ΔK would be less than 1.5 MN $m^{-3/2}$, assuming that the fatigue-hardened flow stress is about 200 MN m^{-2}. Very low threshold values, in the range 1–2 MN $m^{-3/2}$ have been observed and the absolute limit does seem, therefore, to depend simply on creating sufficient activity in the crack tip region. In normalised mild steel, values of ΔK lower than about 6 MN $m^{-3/2}$ confine the reversed plastic zone to less than a grain diameter: there is evidence here that the fracture surface shows a number of grain-boundary facets[19], perhaps because individual grains are hardened, leaving the boundaries as rather weak paths.

In the majority of published results, however, region A does not represent a genuine threshold. For a fully heat-treated pre-cracked specimen to which an arbitrary value of ΔK has been applied, it is not reasonable to suppose that crack propagation will ensue immediately. The local alternating plastic strain amplitude must produce local saturation hardening before cracking initiates. This hardening gives a situation where further crack tip strain can be accommodated only by intensely localised shear along a slip-band: in annealed material, the high work-hardening rate means that the strain is accommodated by general deformation in the crack tip region. There will therefore be an initiation period for any value of ΔK applied, although the higher that ΔK is, the shorter this period will be. The situation is directly analogous to that of a conventional fatigue test, where observations show that higher alternating stresses (plastic strain amplitudes) produce initiation after fewer cycles. The effect will be observable on curves of crack length versus number of cycles in terms of the number of cycles preceding definite crack growth. On the conventional graph of $\log\left(\frac{da}{dN}\right)$ *v.* $\log \Delta K$, however, this absolute initiation period is lost. Until a new crack initiates at the starter crack tip, (da/dN) is of course zero, so that $\log(da/dN)$ is $-\infty$. If the value of ΔK has been sufficient to create steady-state saturation hardening conditions, once crack extension has initiated, region A in an ideally homogenous material will consist simply of a vertical line, whose abscissa is $\log(\Delta K)$, and which rises until it intercepts the base-line growth curve, region B. Different initial values of ΔK will give different positions for this line.

The argument presupposes that the region ahead of the crack tip is initially in a fully stress-relieved condition and that microstructural inhomogeneities do not affect the behaviour. In many experiments, prestrain introduced by machining starter notches or by pre-cracking techniques has not been removed by subsequent heat-treatment and the initial stages of crack growth may be able to proceed more rapidly through this worked region. Region A may then rise vertically from $-\infty$ and show a region of steep slope followed by a decrease until steady-state conditions are obtained. Again, in steels tested at very low initial ΔK values, where the reversed plastic zone size is less than a grain diameter, the incidence of grain-boundary fracture can affect the form of region A, apparently increasing the exponent, m (equation 9.3.2) and presumably elevating the whole level of that part of the growth-rate curve. This situation is similar to the events to be described in the following section.

This first region can be said to be a 'threshold' only if it is clear that lower values of ΔK produce no growth whatsoever. The indications are that the threshold levels of ΔK are very small, although it is possible that they are increased somewhat in strain-ageing materials, by analogy with the sharp fatigue limits observed in conventional *S–N* curves. In sharply notched specimens of strain-ageing material, it is certainly possible to obtain non-propagating cracks. Additionally, the threshold level appears to be sensitive to the testing environment.

In region C of the crack growth-rate curve (*Figure 9.10*) higher slopes are again observed. Problems of interpretation which arise with respect to this region are two-fold. First, the high value of K_{max}, associated with high ΔK values, may mean that the total plastic zone size, if not the reversed plastic zone size, has become large with respect to the specimen dimensions, so that values of K

obtained from the elastic compliance function are suspect. Secondly, these same high K_{max} values may have produced so much tensile stress or strain at the crack tip that the total 'fatigue' crack growth contains contributions from 'static' or monotonic modes of fracture in addition to the normal ripple mechanism, so that the overall rate is accelerated.

Eventually, the fatigue crack reaches a length such that final fast fracture is produced. The terminal value of K_{max} at this point is sometimes found to equal K_{IC}, but is often found to exceed K_{IC}. This enhancement of toughness by fatigue has not been fully investigated, although it is clearly recognised in terms of permissible fatigue levels for precracking standard toughness specimens (see Section 5.13 and *Figure 5.16*). It does help to explain why substantial patches of monotonic fracture may be seen on regions of a fatigue crack surface, where K_{max} was close to, or greater than, K_{IC}. The occurrence of monotonic modes can also help to explain anomalies observed in region B of the crack growth-rate curve.

9.6 Monotonic Modes of Fracture during Fatigue Crack Propagation

Two main problems encountered in trying to develop a single model for fatigue crack propagation behaviour are, first, that some materials, particularly high-strength steels and aluminium alloys, seem to possess growth rates which are much more sensitive to ΔK than are those in the majority of other alloys, and, secondly, that in some cases, the growth rates are markedly sensitive to the *mean* stress level, where ΔK is constant. Results obtained for a variety of alloy steels, where growth rates were measured by ultrasonic techniques[20], are indicated in *Table 9.1*. It can be seen that some very high values of m can be found (up to 10) and that high values tend to be associated with low fracture toughness.

An important indication of the reasons for this behaviour is given by recent studies made by Ritchie on the effect of mean stress on fatigue crack propagation

Table 9.1

Steel specification	*0.2% Yield stress* MN m^{-2}	K_{IC}, MN $m^{-3/2}$	*Exponent,* m
A1S1 1045	260	55	4
A1S1 1144	545	66	5
A1S1 4140	455	62	10
ASTM A533 Grade B Class 1	475	210*	2.2
ASTM A216 WCC grade	335	170*	3
ASTM A469 Class 4	520	100*	2.7
ASTM A470 Class 8	650	60	6.7
ASTM A471 Class 4	790	220*	1.4

* Apparent toughness based on extrapolation of existing K_{IC} data.

in a medium strength (0.2% proof stress 750 MN m^{-2}) alloy steel, which was tested in a temper-embrittled condition and in the unembrittled state after tempering[14]. In these two conditions, uniaxial yield, flow and fracture properties were identical, but the embrittled steel was subject to grain-boundary fracture in notched tests. The curves of crack length *v.* number of cycles and associated crack growth-rate curves for different mean stress levels (as characterised by the ratio

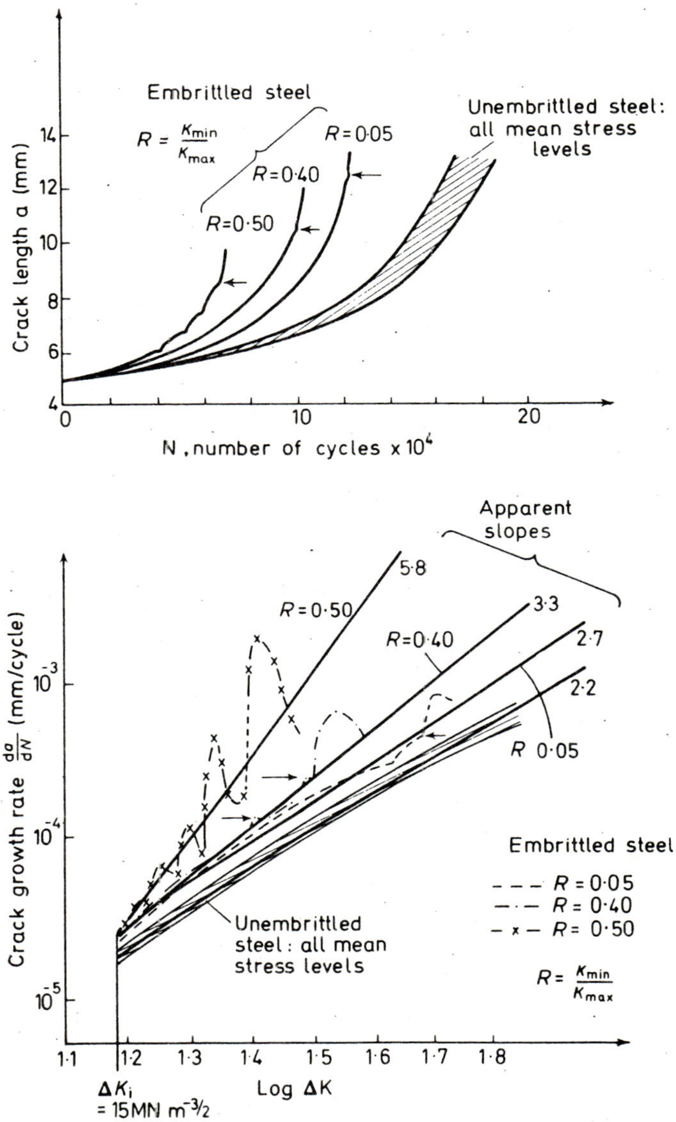

Figure 9.13 (*a*) *Effect of mean stress on fatigue crack length in low-alloy steel* (ΔK_i = 15 MN $m^{-3/2}$). (*b*) *Effect of mean stress on crack growth-rate and apparent slope of growth-rate curve. Apparent slopes obtained by least mean squares regression analysis* (*after Ritchie and Knott*[14])

$R = K_{min}/K_{max}$) at the same initial ΔK value (15 MN $m^{-3/2}$) are indicated in *Figures 9.13a* and *9.13b*. For the unembrittled condition, there is no apparent effect of mean stress on growth rate and the average gradient of region B of the growth-rate curve is close to 2, indicating that growth occurs mainly by a reversed plastic flow mechanism. The fracture surface shows typical fatigue striations or ripples (*Figure 9.14a*).

(*a*)

(*b*)

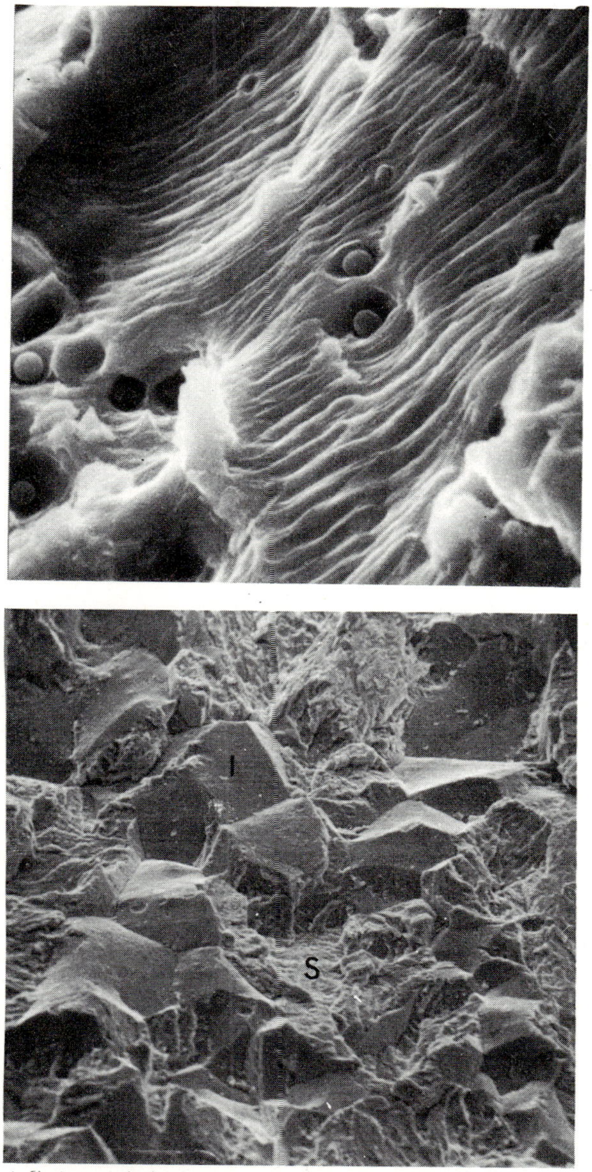

Figure 9.14 (a) Fatigue striations typical of an unembrittled alloy steel (× 4000). (*b*) *Intergranular facets (I) and striations (S) in temper-embrittled alloy steel* (× 350) (*courtesy of R. O. Ritchie*)

In the embrittled steel, very different behaviour is observed. The graph of crack length versus number of cycles is markedly affected by the mean stress level: higher mean stress both steepening the curve, and producing sharp discontinuities in it (*Figure 9.13*). The discontinuities could be associated with rather large 'bursts' of brittle intergranular fracture and the total 'fatigue' fracture surface additionally showed a large number of isolated intergranular facets (e.g. *Figure 9.14b*). The incidence of grain-boundary fracture had clearly increased the overall crack growth-rate. A further effect was observed on examining the crack growth rate [log $(\mathrm{d}a/\mathrm{d}N)$ *v.* log ΔK] curves, *Figure 9.13b.* In embrittled material at high mean stress (i.e. $R = 0.50$), the 'bursts' of intergranular fracture give rise to a curve, which, in the fine detail resolved by very sensitive crack-monitoring equipment, has a saw-tooth appearance, composed of accelerated rates during each burst superimposed on the normal straight line (which, in turn, has been raised slightly by the incidence of small amounts of intergranular fracture). A conventional treatment of such a saw-tooth curve, using fewer data points or less sensitive equipment, would tend to regard the jagged distribution of points as scatter and would fit, by least mean squares regression analysis, a single straight line to all points, as indicated in *Figure 9.13b.* The effect would be to make the data appear as if the gradient of the graph of log$(\mathrm{d}a/\mathrm{d}N)$ *v.* log ΔK, i.e. the exponent m in equation 9.3.2, had also been increased by mean stress, in addition to the fact that the absolute value of $(\mathrm{d}a/\mathrm{d}N)$ at a given ΔK value had been increased.

It is plausible that an effect of this type on a more refined scale (smaller, but more numerous, 'bursts') might be responsible for the high and variable values of m observed generally in steels. The brittle cracking need not be intergranular: in some ferritic steels, it is clear that cleavage fractures can contribute to the overall growth rate in an analogous manner. It has, in fact, been possible to show that the onset of a critical 'burst' of cleavage fracture during fatigue occurs at a critical stress intensity corresponding to the attainment of a critical local tensile stress, at a boundary carbide ahead of the crack tip[21], closely similar to that required to produce total cleavage fracture under monotonic loading at low temperatures (see Section 8.9). In other materials, it is possible that microvoid coalescence adds to the overall growth rate[22]. In aluminium alloys, contributions are to be expected from fracture processes involving the brittle, intermetallic second-phase particles: either fracture of the particles themselves, or voiding around them.

The occurrence of monotonic fracture modes also helps to explain otherwise anomalous values of crack growth rate, obtained by measuring striation spacings on a fracture surface and assuming a one-to-one relationship between the number of striations and the number of cycles. At high ΔK values, it is often found that the 'striation-spacing' growth-rate is substantially less than that obtained using monitoring techniques. This is clearly to be expected, if the total 'fatigue' crack growth contains components of monotonic fracture.

The form of the monotonic mode will obviously vary from material to material, but, in all cases, it is to be expected that the effects will be most clearly discernible if the material's fracture toughness is low, because it is in such a material that monotonic fracture can be produced at a low value of K_{max}. Little data on effects of mean stress on growth rate as a function of fracture toughness exist, but *Figure 9.15* shows the results of a large number of determinations of *m*, plotted versus K_{IC}. High values of *m* are obtained only in materials of low toughness, as expected, and, in high-toughness materials, *m* has a value close to 2.

It is generally concluded that the basic mechanism for fatigue crack propagation in region B is some form of reversed plastic flow, such that (d*a*/d*N*)

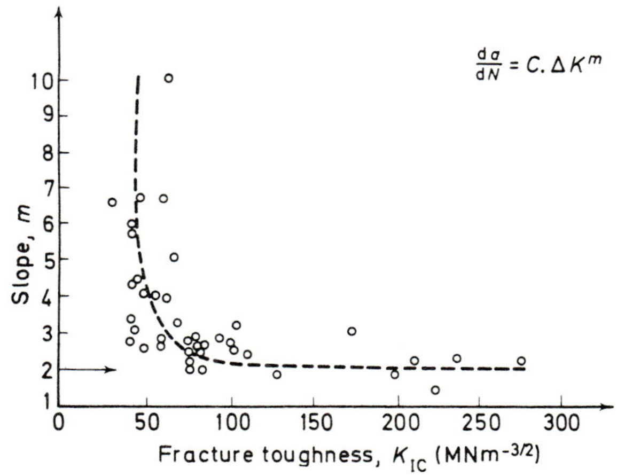

Figure 9.15 Variation of apparent slope, m, with monotonic *fracture toughness,* K_{IC}, *for many steels and aluminium alloys (for sources of data see Ritchie and Knott*[14]*)*
Ritchie and Knott[14]*)*

depends in many cases on ΔK^2. However, in detail, the behaviour must be related to cyclic stress–strain behaviour and the value of the cyclic work-hardening exponent will affect the power, *m*, to which ΔK is raised[17]. In many experimental observations, high values of *m* may be attributed to the occurrence of additional monotonic fracture modes. These will occur predominantly in low toughness alloys and are discrete processes, so that it may be possible to detect them as a series of pulses, if stress-wave emission equipment is used in conjunction with the normal crack-monitoring device. The occurrence of monotonic modes will greatly enhance the effect of mean tensile stress on crack growth rate, because they react to the increased value of K_{max}.

From a metallurgical point of view, it is encouraging if high growth rates and exponents can be related to the same microstructural parameters as those which lead to poor fracture toughness, rather than to some mysterious feature of the fatigue propagation itself. There is then hope, not perhaps of improving resistance to crack propagation beyond that which is currently available in the toughest materials, but of ensuring that structural materials in general do not permit cracks to propagate at anomalously high rates. For example, by correct heat-treatment after tempering, the properties of the otherwise temper-embrittled alloy steel are markedly improved (*Figure 9.14*). It is obviously beneficial if the same measures improve both toughness and resistance to fatigue crack growth: not only will the cracks propagate more slowly, but they will also have to grow to greater lengths before final catastrophic failure occurs.

It is necessary to pursue these concepts in more detail, but it does seem that an understanding of the factors governing propagation is being gained. The results so far have, however, been concerned primarily with the central regions of thick specimens and a further problem arises when the effect of testpiece thickness on crack propagation is considered.

9.7 Effect of Thickness on Propagation

Some confusion exists in the literature as to the predominant effect of thickness on propagation and to the reasons for the effects that have been observed[23,24]. The present section will describe some of the main observations and will attempt to give a coherent explanation for them.

First, it should be made clear that brittle monotonic fracture modes, as described in the previous section, are not expected to occur in thin pieces, because the hydrostatic stress state, needed to develop high crack tip tensile stresses or strains (Sections 7.4 and 8.6), will not be present: it will be relieved by through-the-thickness yielding (Section 2.11) when the size of the plastic zone is comparable with the sheet thickness. If monotonic modes contribute to a high rate of propagation in thick specimens, therefore, the rate in thin pieces will be slower.

The same conclusion holds if flat, or square, fatigue crack propagation occurs in thin sheet (true plane stress conditions), even if the propagation in thick specimens occurs by a reversed plastic flow mechanism. The effect has been shown clearly in thin bend specimens where the deformation is forced to remain in-plane[25]. A similar state of affairs is apparent if one examines the fracture surface profile in a specimen of moderate thickness. Here, the fatigue in the mid-thickness of a plate (under plane strain conditions) tunnels forward more rapidly than that at the edges. Such behaviour can be explained if crack advance depends on the attainment of a critical amount of plastic strain amplitude (giving saturation hardening, slip-band cracking, etc.) in the region immediately ahead of the crack tip, rather than on the reversible plastic zone size or COD as such. For a given value of stress intensity, the size of plastic zone in plane stress may be larger than that in plane strain but the strain is spread more diffusely throughout it, so that more cycles are needed to develop the critical strain level immediately in front of the crack tip. The effect of thickness is therefore analogous to that on the initiation of fibrous fracture (see Section 6.4). It is possible for a model for propagation under these conditions to be related to prevention of reversibility of the opening displacement, but the relationship between $\Delta\delta$ and $\Delta\epsilon_p$ would not be the same in plane stress as it is in plane strain.

The situation is, in any case, confused by the behaviour of precracked thin sheet tensile specimens. Here, the crack initially propagates on a plane normal to the applied alternating tensile stress. As it grows, the plastic zone at the crack tip becomes larger. At a critical size, related to the sheet thickness, the fracture plane changes to 45° slanting through the thickness and the crack growth rate accelerates markedly. This mode of fatigue should be regarded as Mode III antiplane strain (see Sections 2.11 and 5.4) rather than plane stress and will occur in situations where elastic buckling and bending of the sheet allow the parts immediately above and below the crack tip to be displaced laterally out-of-plane. The reversed plastic strain amplitude in the narrow pocket of slip on the 45° plane can remain concentrated, rather than diffuse. The relationship between crack tip displacement and $\Delta\epsilon_p$ will be numerically different from that in plane stress or in plane strain.

The events therefore depend on loading configuration in addition to the geometry of the piece. If out-of-plane displacements occur by buckling or by, for example, the bulging of a thin-walled pressure vessel, cracks may propagate more rapidly in thin section than in thick, unless brittle monotonic modes are highly significant in the thick piece: if out-of-plane displacements are restrained, the cracks should propagate more slowly. Similarly, the static fracture toughness of a thin sheet is much increased if buckling is restrained (see Section 5.4 and Irwin, Kies and Smith[26]).

Two types of model are used to describe fatigue crack propagation: one, based on concepts derived from experience with stage II propagation in generally yielded specimens, referring to re-welding of freshly exposed faces *at* the crack tip and thus depending on $\Delta\delta$; the other, relating to the accumulation of fatigue damage *ahead* of the crack tip, and depending on $\Delta\epsilon_p$. In a given state of stress, if we regard the region ahead of the tip as a miniature tensile specimen cyclically hardened by an amount dependent on $\Delta\epsilon_p$, it is clear that a constant relationship between $\Delta\delta$ and $\Delta\epsilon_p$ will be obtained under steady-state conditions. However, the slower growths obtained in plane stress tend to indicate that $\Delta\epsilon_p$ is the controlling factor and we are led to seek some form of localised slip-band crack in the fatigue-hardened zone ahead of the crack tip as the main means of crack extension. The role of an aggressive environment in promoting crack advance could be related to preferential chemical dissolution at the active slip band: the role of a passive environment might be, through adsorption on the generally opened clean crack tip surface, to enable deformation to be localised to a narrow slip-band, so that a single shear crack, rather than a number of slip steps is produced. Events obviously need to be studied in materials of different slip character and fatigue hardening behaviour to decide whether it is $\Delta\epsilon_p$ or $\Delta\delta$ that is the ultimate factor in controlling propagation. Since ΔK, in fact, is a measure of the crack tip alternating stress intensity rather than strain amplitude, it seems reasonable that the exponent, m, in equation 9.3.2 is dependent on the relationship between the cyclic stress and the cyclic strain, i.e. on the cyclic work-hardening exponent[17]. This might explain the variation in m over the range 2–4, but higher values are attributed to the additional modes of monotonic fracture described in Section 9.6.

9.8 Overload Effects and Random Fatigue

Once an understanding of the propagation of fatigue cracks under constant, or 'steady-state', ΔK conditions has been gained, the next problem to be faced is how to characterise the rate of growth of cracks in situations where ΔK changes rapidly. These may exist, in association with constant amplitude applied stress, when the cracks are propagating very rapidly or the stress gradient is very steep, but are more commonly found in practical applications, where the fatigue loads themselves are variable. Although service loadings can be recorded on magnetic or digital tape, (using strain gauges, load cells, or accelerometers attached to the component of interest) and reproduced in servo-controlled machines, so that actual components can be tested in the laboratory, general material assessment and quality-control testing on less sophisticated machines demands that any differences between variable-amplitude testing and constant amplitude testing should be fully investigated. The aim is to be able to predict from simple data the total life under a known, complex loading system.

Endeavours of this sort have met with some success, when dealing with situations in which a specimen is subjected to general alternating plastic strain. The total life appears to be predicted adequately by a 'cumulative damage' law due to Miner[27], which states that failure occurs when the sum of the number of cycles, n^i, at each stress (or plastic strain) range expressed as a fraction of the number of cycles required to cause failure at this stress range, n_F^i, becomes unity, i.e.

$$\Sigma \frac{n^i}{n_F^i} = 1 \tag{9.8.1}$$

In some versions, the figure 1 may be replaced by a constant.

Similar laws do not appear to hold for cracks propagating under quasi-linear-elastic conditions, although the base-line data have usually been obtained under steadily varying, rather than constant, alternating stress intensity. The application of a single overload to a constant amplitude cycle in fact, initially decreases the anticipated propagation rate over subsequent cycles. Contributions to this effect could occur from several factors, such as the blunting of the crack tip during overload, the production of high residual compressive stress on reducing the overload, the destruction of the characteristic 'steady-state' dislocation sub-structure, or from the 'crack-closure' mechanism[28] (the closing of the crack faces whilst the applied stress is still tensile, due to the residual stresses produced by the plastically deformed regions behind the crack tip). The point to make, however, is that at present, prediction of quasi-linear-elastic behaviour under variable or random amplitude fatigue loading from basic constant amplitude crack growth-rate data, differs widely from experience with specimens subjected to general plastic strains.

The quantitative treatment of subcritical crack growth by fatigue mechanisms therefore seems amenable to treatment by linear elastic methods for basic growth-rate information, although this may have to be modified for realistic application to service conditions. In subsequent sections, we shall see how fracture mechanics may be applied to problems involving subcritical crack growth by stress-corrosion mechanisms.

9.9 Stress–Corrosion Cracking

Up to now, we have discussed fracture events and mechanisms for crack propagation, which are essentially mechanical in nature, although it was clear that fatigue crack growth processes cannot be fully explained unless some interaction with the testing environment is considered. Situations involving stress corrosion include a very broad range of material–environment interactions, many of which are specific to the single system under study. It is not possible to give comprehensive coverage to all possible interactions in the present work and descriptions will be confined mainly to the general use of the fracture mechanics approach[29] and to experimental methods.

A definition of stress-corrosion cracking may be taken as 'a cracking process caused by the simultaneous action of corrodent and stress'. This is, wisely, a general statement, which ignores the specific reaction leading to the formation or extension of cracks. It therefore includes both anodic dissolution processes, which preferentially remove heavily stressed material by chemical means, and cathodic processes such as the evolution of hydrogen, which can diffuse into a material and cause fracture ahead of a crack tip by hydrogen embrittlement mechanisms.

Traditional stress-corrosion testing is concerned with the initiation of cracks. Testpieces are usually in the form of thin strips or wires and are stressed by simple loading with weights or by being clamped in various configurations which produce high bending stresses on parts of their surface. Some testpieces may be profiled to maintain the macroscropic stress constant as the crack grows. Testing variables are the chemical nature, pH and temperature of the environment and the maximum stress level applied. In some cases, the testpiece is held at an anodic potential to accelerate the corrosion processes. Information is presented as graphs of applied stress *v.* time to failure for each testing condition. The times to failure are usually dependent on the times required for initiation because the applied stresses are relatively high and a crack, once initiated, propagates rapidly.

It is important that yielding should be activated on some part of the surface to produce localised stress corrosion rather than general chemical dissolution[30]. The initial attack usually produces some form of pitting and the pits can act as stress concentrators, so that yield is produced locally under the action of the applied stress. The corroding environment must also be such that metal is dissolved rather than a passivating film formed. If the microstructure contains anodic phases, these are likely to be dissolved preferentially and, if these are associated with soft precipitate free zones at grain boundaries, the alloy tends to show pronounced cracking in the grain-boundary regions. In other alloys, the corrodent may produce pitting in slip bands. There is evidence that altering the character of the slip deformation in the grains by varying the state of ageing of precipitate-hardening particles can alter the resistance to stress-corrosion cracking: the sharp slip bands associated with an underaged condition promoting more rapid attack.

Resistance to stress-corrosion cracking in an initiation-controlled situation generally depends on the preservation of an impervious protective film on a metal surface. This may be an oxide, a paint or polymer coating, or a cladding treatment, such as that given to high strength aluminium alloys, which are clad with a dilute alloy, less prone to stress corrosion. The resistance of an alloy to stress corrosion is deemed to be sufficient if there are no obvious signs of cracking in laboratory tests unless the applied stress is greater than some 75% of the yield stress.

However, it is not always possible to rely on protective coatings in service. One case in point is that of an undercarriage forging on an aircraft. This is a redundant member of the structure for all airborne purposes and so it is usually made out of high-strength material to reduce its weight to a minimum; it can be painted, but is difficult to clad, because of the complex shape. On landing on a wet runway, it is quite possible for sharp stones to be flung up and penetrate the coating, under wet conditions, which do not allow a protective oxide film to form. The protection of the component then depends on whether or not this crack nucleus grows rapidly under the operative service conditions to a length such that fast fracture can occur. If it grows slowly, the presence of the crack can be detected by the normal periodic inspections of the aircraft. If it grows rapidly, and is not detected by cursory examination during flight stops, the aircraft is likely to collapse on a subsequent landing or take-off. A further example of a major application where stress-corrosion crack growth was critical was found in a recent failure of a large turbo-generator forging where cracks grew under wet steam conditions[33].

We therefore now consider how fracture mechanics is employed to characterise stress corrosion, in such a situation, where a crack is present from the very beginning of the stress corrosion test.

9.10 The Application of Fracture Mechanics to Crack Growth by Stress-Corrosion Mechanisms

The first investigations of stress-corrosion cracking in precracked specimens simply employed testpieces of the standard fracture toughness geometries (*Figure 5.12*) or the constant compliance geometry (*Figure 9.5*), subjected to dead loading, in the appropriate environment. Typical results (e.g. Brown[29]) obtained from such tests are presented in *Figure 9.16*, where a graph is drawn of the total time to failure for a particular initial stress intensity, K_i. The intercept at zero time is, of course, K_{IC}.

The first point studied was the terminal toughness, i.e. to discover whether the amount of crack growth under the initial load, corresponding to K_i, was just sufficient to make the instantaneous value of K at fracture equal to the conventional fracture toughness. In many broken alloys, the region of stress-corrosion crack growth was clearly distinguishable from that of the final fracture, so that the stress-corrosion crack length at final fracture could be measured directly on the fracture surface. In some cases, good agreement between K_{IC} and the calculated values (denoted by K_{ISC} to indicate that the 'pre-crack' is a stress-corrosion

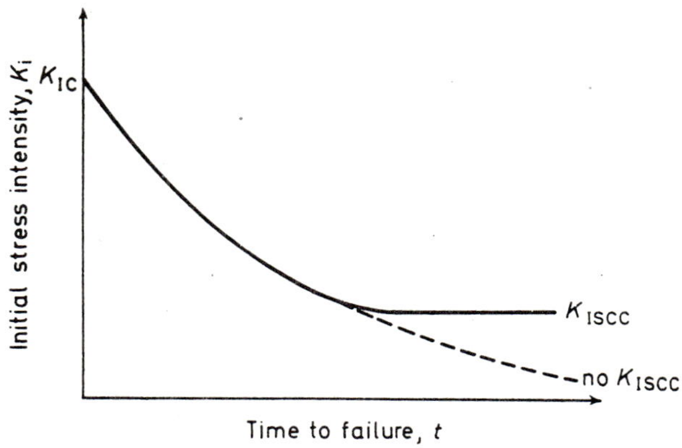

Figure 9.16 Typical behaviour in precracked stress-corrosion tests

crack) was obtained: in others, branching of the stress-corrosion crack led to the occurrence of higher values for K_{ISC}.

In a number of alloys, a definite limiting value of K_i was found, below which a specimen appeared to last indefinitely (see *Figure 9.16*). This limit, termed K_{ISCC}, is not, however, a feature of all alloys; at least, not at the comparatively high values of K observed (approx. 10–15 MN m$^{-3/2}$ for medium-strength steel).

It is clear that, if a definite K_{ISCC} limit exists, resistance to stress-corrosion failure can be guaranteed by ensuring that service stresses do not produce values of K greater than K_{ISCC} around flaws of the size likely to be encountered in service. If no K_{ISCC} limit exists, or if it is so small that service stresses would have to be held at an unrealistically low level, design against stress corrosion must rely on the assurance that crack growth rates in service will be so slow that the cracks can be detected and eliminated during the periodic inspections. Slip activity on the scale envisaged for the fatigue 'threshold' limit (Section 9.5) would, of course, imply extremely low values of K_{ISCC}.

A graph of the form shown in *Figure 9.16* does not by itself give direct information on the dependency of the rate of growth on K, although the final failure line represents the integrated effect of crack growth to constant terminal values (K_{ISC}). The general lack of precision in the data points means that various functional relationships between growth rate (da/dt) and K could be derived, each of which would appear to give an equally good fit. It is necessary to measure the growth rates directly.

Techniques[29] considered for the measurement of growth rate have been generally similar to those used in fatigue testing (see Section 9.4). The electrical potential method has again been used widely, although there are serious doubts as to the wisdom of measuring an absolute value of the rate of crack propagation by passing some 30 A through the specimen, because it is felt that this may well affect the anodic dissolution rate at the crack tip. Nevertheless, rates measured by the electrical potential method are generally considered to agree well with those determined in other ways. Measurements of the change in testpiece compliance as the crack grows have also been used under stress-corrosion conditions. The values of K are often small, so that the plastic zone size associated with the growing crack is not large and changes in compliance can, experimentally, be related to changes in crack length unambiguously.

Experimental determinations of the way in which growth rate depends on K give results which can vary markedly with the metal–environment system studied. Both exponential and linear relationships have been obtained and *Figure 9.17* shows results[31] for a high-strength aluminium alloy, subject to intergranular stress corrosion, in which a linear dependence and a complete lack of dependence on K can be obtained by modifying the structure of the grain boundary region by heat-treatment, whilst maintaining the yield stress and matrix microstructure constant.

In the light of this wide variation of experimental results, it is clearly not possible to produce a single, unified theory to explain stress-corrosion crack growth, particularly since the crack tip processes may be anodic dissolution in

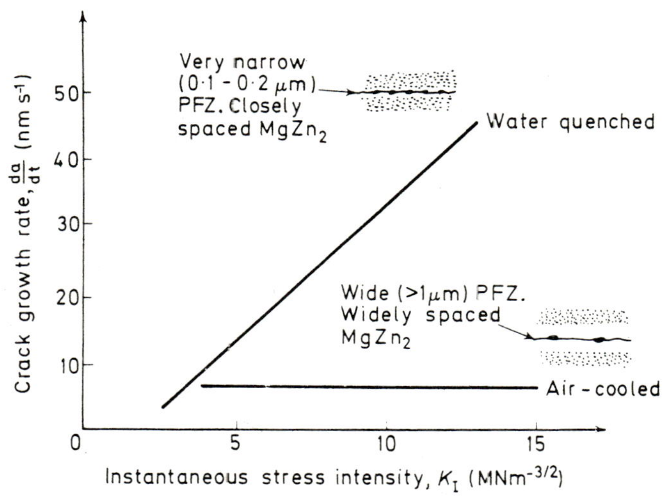

Figure 9.17 Variation of stress-corrosion crack growth-rate with stress intensity. (Pure Al–Zn–Mg *alloy, water-quenched or air-cooled before ageing) (after Shipp*[31]*)*

some cases and cathodic hydrogen evolution in others. The operative process can be detected by the sign of the potential which must be applied to a specimen to accelerate or retard the stress-corrosion process: if the process is anodic dissolution and the specimen is made more anodic, the rate will accelerate, and vice versa. The difficulty of establishing a viable model is increased by two further factors. First, the pH of the corrodent at the crack tip may not be equal to that of the bulk solution. This has been demonstrated in an elegant manner by removing a corroding specimen from solution and freezing it; the specimen may then be broken in half whilst cold and the pH of the liquid frozen at the crack tip is determined as it melts. Highly acidic crack tip environments are often found, even when the bulk solution is neutral. The second factor affecting a quantitative model is that the dissolving crack tip surface may be partially filmed or that, at low K values, access of the corrodent to the crack tip may be blocked by corrosion product. In some cases, this effect may be responsible for the K_{ISCC} limit.

One model which has been developed to explain stress-corrosion crack growth assumes that propagation normally proceeds by internal necking between inclusions, spaced at regular intervals ahead of the tip of the starter crack[32]. The rate of decrease of area between inclusions is then supposed to be the result of two factors: one due to the simple (plastic) Poisson's ratio effect as tensile strains are produced ahead of the crack tip; the other due to the dissolution processes. Internal necking (crack advance) is predicted to occur when an instability, as in the uniaxial tensile test, is produced in the contracting ligament. The theory, as put forward, makes a number of simplifying assumptions, which do not justify its reproduction in quantitative detail, but the general physical picture seems to be reasonable for some alloys. Even the intergranular stress-corrosion fractures, whose growth rates are indicated in *Figure 9.17*, showed fine-scale dimples centred around the grain boundary ($MgZn_2$) particles, at least if the particles were widely spaced. In this case, the lack of sensitivity of growth rate to stress intensity could be well explained[31] by the fact that, because the particles were so far apart, most of the life was spent in slow dissolution of large areas of precipitate-free grain boundary: the role of stress then became significant only when the growing crack tip came close to another particle. If the (readily soluble) anodic particles were closely spaced, however, stress played a significant part in the rupture of the regions between particles and a marked dependence on K was obtained.

In other systems, internal necking cannot be contemplated as a reasonable model for the crack growth and the cracks may follow 'cleavage' planes or active slip planes. In all cases, the calculation of a dissolution rate will be difficult, because the crack tip surface is likely to be partially filmed. The role of stress has not been fully clarified and confusions arise between behaviour in plane strain and in plane stress, where resistance to stress corrosion appears to be much improved. Naïvely, since the size of the plastic zone is larger in plane stress, it might be expected that more dissolution would occur under such conditions. However, as with fibrous fracture (Section 6.4) and fatigue fracture (Section 9.7) it may be that the important factor is the concentration of strain immediately ahead of the crack tip, so that it may be the confinement of strain to a small region, in the plane strain stress state, that produces the faster crack growth.

Since so many variables can contribute to stress-corrosion fracture mechanisms, it is clear that tests should be carried out under conditions which closely simulate those in service. The acceleration of stress corrosion, e.g. by raising the temperature of the corrodent or applying a potential to the specimen, is a matter which should be treated with great caution. If a material shows a definite K_{ISCC} limit, it may be that the limit is not affected by an alteration in conditions, but this fact can only be established experimentally. It defeats the whole object to measure growth rates under other than service conditions: corrections for effects of temperature, assuming a constant activation energy for the process, are again extremely dubious, unless experimental confirmation has been obtained.

9.11 Sub-critical Crack Growth – Conclusions

It is apparent that many attempts are being made to characterise subcritical crack growth in terms of critical values of parameters which can be related to design stresses. In many cases, the threshold, below which a sub-critical mechanism will not operate (ΔK threshold, K_{ISCC}), is so low that it may be unrealistic to consider that a component can be stressed so that crack growth will not occur. The minimum stress level is, of course, strongly dependent on the size of any defect that may exist in a structure and emphasis is therefore placed yet again on the need to design structures free from macroscopic stress raisers and to employ fabrication techniques and materials which minimise internal defects. It must be possible to measure the sizes of any defects that are present accurately by the non-destructive-testing methods used. If reliance is to be placed on slow crack growth rates, the periodic non-destructive-testing must also be thorough and capable of detecting any growth that has occurred in service.

In terms of understanding the mechanisms of sub-critical cracking, so that laboratory data can be applied to service situations with confidence, more progress has been made with models of fatigue crack growth, where purely mechanical considerations are involved, than with those for stress-corrosion, which is so critically dependent on the interaction between a particular alloy and a particular environment. Even so, there are still many unsolved problems in fatigue, concerning effects of mean stress or variable stress amplitude. In general, it would appear that development of high static fracture toughness gives good resistance to fatigue crack propagation. Such alloys are not necessarily passive in service environments and it seems likely that, unless environmental or potential control can be exerted, there are many circumstances where a material's resistance to stress-corrosion crack growth will prove to be the ultimate factor in deciding whether or not it will be selected for a particular application.

REFERENCES

1. Plumbridge, W. J. and Ryder. D. A., *Met. Rev.*, No. 136, 119 (1969)
2. Ham, R. K., *Canad. Metall. Q.*, **5**, 161 (1966)
3. Oates, G. and Wilson, D. V., *Acta Met.*, **12**, 21 (1964)

4. Coffin, L. F., *Appl. Mater. Res.*, **1**, 129 (1962)
5. Manson, S. S., *U.S. Nat. Advis. Cttee. Aeronaut.*, Tech. Note 2933 (1954)
6. Coffin, L. F., *Internal Stresses and Fatigue,* Elsevier, Amsterdam (1959)
7. Laird, C. and Smith, G. C., *Phil Mag.*, 7, 847 (1962)
8. Laird, C., *Am. Soc. Test. Mat.*, Spec. Tech. Pub., No. 415, 139 (1967)
9. Wessel, E. T., *Practical Fracture Mechanics for Structural Steel*, Paper H, UKAEA/ Chapman and Hall (1969)
10. Jack, A. R. and Price, A. T., *Int. Jnl Fract. Mech.*, **6**, 401 (1970)
11. Ritchie, R. O., Garrett, G. G., and Knott, J. F. *Int. Jnl Fract. Mech.*, **7**, 462 (1971)
12. Gilbey, D. M., and Pearson, S., *Royal Aircraft Establishment Technical Report* 66402 (1966)
13. Wilson, W. K., *Engng Fract. Mech.*, **2**, 169 (1970)
14. Ritchie, R. O. and Knott, J. F., *Acta Met.*, **21**, 639 (1973)
15. Richards, C. E. and Lindley, T. C., *Engng Fract. Mech.*, **4**, 951 (1972)
16. McEvily, A. J., *Proc, U.S. Air Force Conf. on Fatigue and Fracture of Aircraft Structures and Materials,* AFFDLTR–70–144, 451 (Sept., 1970)
17. Tomkins, B., *Phil. Mag.*, **18**, 1041 (1968)
18. Klesnil, M. and Lukás, P., *Phil Mag.*, **17**, 1295 (1968)
19. Birkbeck, G., Inckle, A. E. and Waldron, G. W. J., *J. Mater. Sci.*, **6**, 319 (1971)
20. Clark, W. G. and Wessel, E. T., *Am. Soc. Test. Mat.*, Spec. Tech. Pub., No. 463, 160 (1970)
21. Ritchie, R. O. and Knott, J. F., *Mater. Sci. Engng* (in press)
22. Griffiths, J. R., Mogford, I. L. and Richards, C. E., *Met. Sci. Jnl*, **5**, 150 (1971)
23. Jack, A. R. and Price, A. T., *Acta Met.*, **20**, 857 (1972)
24. Griffiths, J. R. and Richards, C. E., *Mater. Sci. Engng,* **11**, 305 (1973)
25. Ritchie, R. O., *Ph.D. Thesis,* Cambridge University (1973)
26. Irwin, G. R., Kies, J. A., and Smith, H. L., *Proc. Am. Soc. Test. Mater.*, **58**, 640 (1958)
27. Miner, M. A., *Trans. Am. Soc. Mech. Engng, J. appl. Mech.*, **67**, A-159 (1945)
28. Elber, W., *Am. Soc. Test. Mater.*, Spec. Tech. Pub., No. 486, 230 (1971)
29. Brown, B. F., *Met. Rev.*, **129**, 171 (1968)
30. Hoar, T. P., *Proc. Second Int. Conf. on Metallic Corrosion*, 14, Houston, Texas (National Association of Corrosion Engineers) (1966)
31. Shipp, R., *Ph.D. Thesis*, Cambridge University (1971)
32. Krafft, J. M., and Mulherin, J. H., *Trans. Am. Soc. Metals*, **62**, 64 (1969)
33. Kalderon, D., *Proc. Instn. mech. Engrs,* **186**, 341 (1972)

FURTHER READING

ASTM, Special Technical Publications No. 381 (1965), No. 410 (1966), No. 415 (1967), No. 463 (1970), No. 486 (1971), No. 514 (1972)

ASTM, Tentative Method E399-70T.

BSI, *Methods for Plane Strain Fracture Toughness* (K_{IC}) *Testing*, Draft for Development No. 3 (1971)

BSI, *Methods for Crack Opening Displacement* (COD) *Testing*, Draft for Development No. 19 (1972)

Cottrell, A. H., *Mechanical Properties of Matter*, Wiley, New York (1964)

Dieter, G. E., Jr., *Mechanical Metallurgy*, McGraw-Hill, New York (1961)

Dugdale, D. S. and Ruiz, C., *Elasticity for Engineers*, McGraw-Hill, New York (1971)

Hill, R., *The Mathematical Theory of Plasticity*, Clarendon Press, Oxford (1960)

Practical Application of Fracture Mechanics to Pressure-vessel Technology, Institute of Mechanical Engineers (1971)

Liebowitz, H. (Ed.), *Fracture – An Advanced Treatise*, Academic Press, New York (1968)

Love, A. E. H., *A Treatise on the Mathematical Theory of Elasticity*, 4th Edn, Cambridge University Press (1927)

Neal, B. G., *Plastic Methods of Structural Analysis*, Chapman and Hall, London (1965)

Proceedings of the International Conferences on Fracture, First (Sendai, 1965), Second (Brighton, 1969), Third (Munich, 1973)

Tetelman, A. S. and McEvily, A. J., *Fracture of Structural Materials*, Wiley, New York (1967)

Timoshenko, S. P., *Strength of Materials*, Van Nostrand, New York (1956)

Practical Fracture Mechanics for Structural Steel, UKAEA/Chapman and Hall (1969)

INDEX